Graduate Texts in Mathematics 67

Springer Science+Business Media, LLC

Graduate Texts in Mathematics

(continued after index)

Jean-Pierre Serre

Local Fields

Translated from the French by
Marvin Jay Greenberg

Springer

Jean-Pierre Serre
Collège de France
3 rue d'Ulm
75005 Paris, France

Marvin Jay Greenberg
University of California at Santa Cruz
Mathematics Department
Santa Cruz, CA 95064

Mathematics Subject Classifications (1991): 11R37, 11R34, 12G05, 20J06

With 1 Figure

Library of Congress Cataloging-in-Publication Data

Serre, Jean-Pierre.
Local fields.

(Graduate texts in mathematics; 67)
Translation of Corps Locaux.
Bibliography: p.
Includes index.
1. Class field theory. 2. Homology theory.
I. Title. II. Series.
QA247.S4613 512′.74 79-12643

L'edition originale a été publieé en France sous le titre *Corps locaux* par HERMANN, éditeurs des sciences et des arts, Paris.

Originally published by Springer-Verlag New York Berlin Heidelberg in 1979
Softcover reprint of the hardcover 1st edition 1979

9 8 7 6 5 4 3

ISBN 978-1-4757-5675-3 ISBN 978-1-4757-5673-9 (eBook)
DOI 10.1007/978-1-4757-5673-9 SPIN 10761519

Contents

Part Two
RAMIFICATION

Part Three
GROUP COHOMOLOGY

Part Four
LOCAL CLASS FIELD THEORY

Introduction

The goal of this book is to present local class field theory from the cohomological point of view, following the method inaugurated by Hochschild and developed by Artin-Tate. This theory is about extensions—primarily abelian—of "local" (i.e., complete for a discrete valuation) fields with finite residue field. For example, such fields are obtained by completing an algebraic number field; that is one of the aspects of "localisation".

The chapters are grouped in "parts". There are three preliminary parts: the first two on the general theory of local fields, the third on group cohomology. Local class field theory, strictly speaking, does not appear until the fourth part.

Here is a more precise outline of the contents of these four parts:

The first contains basic definitions and results on discrete valuation rings, Dedekind domains (which are their "globalisation") and the completion process. The prerequisite for this part is a knowledge of elementary notions of algebra and topology, which may be found for instance in Bourbaki.

The second part is concerned with ramification phenomena (different, discriminant, ramification groups, Artin representation). Just as in the first part, no assumptions are made here about the residue fields. It is in this setting that the "norm" map is studied; I have expressed the results in terms of "additive polynomials" and of "multiplicative polynomials", since using the language of algebraic geometry would have led me too far astray.

The third part (group cohomology) is more of a summary—and an incomplete one at that—than a systematic presentation, which would have filled an entire volume by itself. In the two first chapters, I do not give complete proofs, but refer the reader to the work of Cartan-Eilenberg [13] as well as to Grothendieck's "Tôhoku" [26]. The next two chapters (theorem of Tate-Nakayama, Galois cohomology) are developed specifically for arithmetic applications, and there the proofs are essentially complete. The last chapter (class formations) is drawn with little change from the Artin-Tate seminar [8]—a seminar which I have also used in many other places.

The last part (local class field theory) is devoted to the case of a finite or, more generally, quasi-finite residue field; it combines the results of the three first parts. (The logical relations among the different chapters are made more precise in the *Leitfaden* below.) Besides standard results, this part includes a theorem of Dwork [21] as well as several computations of "local symbols".

This book would not have been written without the assistance of Michel Demazure, who drafted a first version with me in the form of lecture notes ("Homologie des groupes—Applications arithmétiques", Collège de France, 1958–1959). I thank him most heartily.

Leitfaden

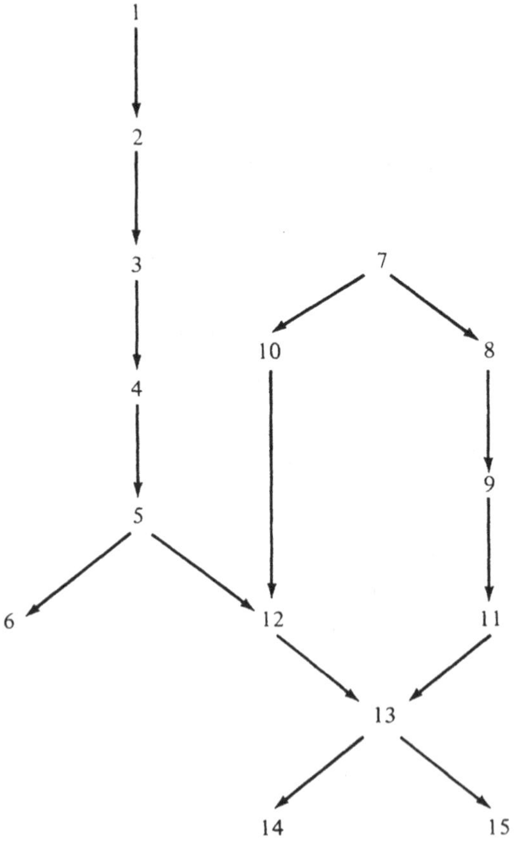

PART ONE

LOCAL FIELDS (BASIC FACTS)

CHAPTER I

Discrete Valuation Rings and Dedekind Domains

§1. Definition of Discrete Valuation Ring

A ring A is called a *discrete valuation ring* if it is a principal ideal domain (Bourbaki, *Alg.*, Chap. VII) that has a unique non-zero prime ideal $\mathfrak{m}(A)$. [Recall that an ideal $\mathfrak{p}$ of a commutative ring A is called *prime* if the quotient ring $A/\mathfrak{p}$ is an integral domain.]

The field $A/\mathfrak{m}(A)$ is called the *residue field* of A. The invertible elements of A are those elements that do not belong to $\mathfrak{m}(A)$; they form a multiplicative group and are often called the *units* of A (or of the field of fractions of A).

In a principal ideal domain, the non-zero prime ideals are the ideals of the form πA, where π is an irreducible element. The definition above comes down to saying that A has one and only one irreducible element, up to multiplication by an invertible element; such an element is called a *uniformizing element* of A (or *uniformizer*; Weil [123] calls it a "prime element").

The non-zero ideals of A are of the form $\mathfrak{m}(A) = \pi^n A$, where π is a uniformizing element. If $x \neq 0$ is any element of A, one can write $x = \pi^n u$, with $n \in \mathbf{N}$ and u invertible; the integer n is called the *valuation* (or the *order*) of x and is denoted $v(x)$; it does not depend on the choice of π.

Let K be the field of fractions of A, K^* the multiplicative group of non-zero elements of K. If $x = a/b$ is any element of K^*, one can again write x in the form $\pi^n u$, with $n \in \mathbf{Z}$ this time, and set $v(x) = n$. The following properties are easily verified:

a) *The map* $v\colon K^* \to \mathbf{Z}$ *is a surjective homomorphism.*
b) *One has* $v(x + y) \geq \operatorname{Inf}(v(x), v(y))$.

(We make the convention that $v(0) = +\infty$.)

The knowledge of the function v determines the ring A: it is the set of those $x \in K$ such that $v(x) \geq 0$; similarly, $\mathfrak{m}(A)$ is the set of those $x \in K$ such that $v(x) > 0$. One could therefore have begun with v. More precisely:

Proposition 1. *Let* K *be a field, and let* $v: K^* \to \mathbf{Z}$ *be a homomorphism having properties* a) *and* b) *above. Then the set* A *of* $x \in K$ *such that* $v(x) \geq 0$ *is a discrete valuation ring having* v *as its associated valuation.*

Indeed, let π be an element such that $v(\pi) = 1$. Every $x \in A$ can be written in the form $x = \pi^n u$, with $n = v(x)$, and $v(u) = 0$, i.e., u invertible. Every non-zero ideal of A is therefore of the form $\pi^n A$, with $n \geq 0$, which shows that A is indeed a discrete valuation ring. □

EXAMPLES OF DISCRETE VALUATION RINGS

1) Let p be a prime number, and let $\mathbf{Z}_{(p)}$ be the subset of the field $\mathbf{Q}$ of rationals consisting of the fractions r/s, where s is not divisible by p; this is a discrete valuation ring with residue field the field $\mathbf{F}_p$ of p elements. If v_p denotes the associated valuation, $v_p(x)$ is none other than the exponent of p in the decomposition of x into prime factors.

An analogous procedure applies to any principal ideal domain (and even to any Dedekind domain, cf. §3).

2) Let k be a field, and let $k((T))$ be the field of *formal power series in one variable* over k. For every non-zero formal series

$$f(T) = \sum_{n \geq n_0} a_n T^n, \qquad a_{n_0} \neq 0,$$

one defines the *order* $v(f)$ of f to be the integer n_0 (cf. Bourbaki, *Alg*., Chap. IV). One obtains thereby a discrete valuation of $k((T))$, whose valuation ring is $k[[T]]$, the set of formal series with non-negative exponents; its residue field is k.

3) Let V be a normal algebraic variety, of dimension n, and let W be an irreducible subvariety of V, of dimension $n - 1$. Let $A_{V/W}$ be the local ring of V along W (i.e., the set of rational functions f on V which are defined at least at one point of W). The normality hypothesis shows that $A_{V/W}$ is integrally closed; the dimension hypothesis shows that it is a one-dimensional local ring; therefore it is a discrete valuation ring (cf. §2, prop. 3); its residue field is the field of rational functions on W. If v_W denotes the associated valuation, and if f is a rational function on V, the integer $v_W(f)$ is called the "order" of f along W; it is the multiplicity of W in the divisor of zeros and poles of f.

4) Let S be a *Riemann surface* (i.e., a one-dimensional complex manifold), and let $P \in S$. The ring $\mathfrak{H}_P$ of functions holomorphic in a neighborhood (unspecified) of P is a discrete valuation ring, isomorphic to the subring of convergent power series in $\mathbf{C}[[T]]$; its residue field is $\mathbf{C}$.

§2. Characterisations of Discrete Valuation Rings

Proposition 2. *Let* A *be a commutative ring. In order that* A *be a discrete valuation ring, it is necessary and sufficient that it be a Noetherian local ring, and that its maximal ideal be generated by a non-nilpotent element.*

[Recall that a ring A is called *local* if it has a unique maximal ideal, *Noetherian* if every increasing sequence of ideals is stationary (or, equivalently, if every ideal of A is finitely generated).]

It is clear that a discrete valuation ring has the stated properties. Conversely, suppose that A has these properties, and let π be a generator of the maximal ideal $\mathfrak{m}(A)$ of A. Let $\mathfrak{u}$ be the ideal of the ring A formed by the elements x such that $x\pi^m = 0$ for m sufficiently large; since A is Noetherian, $\mathfrak{u}$ is finitely generated, hence there exists a fixed N such that $x\pi^N = 0$ for all $x \in \mathfrak{u}$. Let us prove now that the intersection of the powers $\mathfrak{m}(A)^n$ is zero (this is in fact valid in every Noetherian local ring, cf. Bourbaki, *Alg. comm.*, Chap. III, §3). Let $y \in \bigcap \mathfrak{m}(A)^n$; one can write $y = \pi^n x_n$ for all n, whence

$$\pi^n(x_n - \pi x_{n+1}) = 0 \quad \text{and} \quad x_n - \pi x_{n+1} \in \mathfrak{u}.$$

The sequence of ideals $\mathfrak{u} + Ax_n$ being increasing, it follows that $x_{n+1} \in \mathfrak{u} + Ax_n$ for n large, whence $x_{n+1} = z + tx_n$, $z \in \mathfrak{u}$, and as $x_n = \pi x_{n+1} + z'$, $z' \in \mathfrak{u}$, one gets $(1 - \pi t)x_{n+1} \in \mathfrak{u}$; but $1 - \pi t$ does not belong to $\mathfrak{m}(A)$, therefore is invertible (A being local); hence x_{n+1} belongs to $\mathfrak{u}$ for n large enough, and, taking $n + 1 \geq N$, one sees that $y = \pi^{n+1}x_{n+1}$ is zero, which proves

$$\bigcap \mathfrak{m}(A)^n = 0.$$

By hypothesis, none of the $\mathfrak{m}(A)^n$ is zero. If y is a non-zero element of A, y can therefore be written in the form $\pi^n u$, with u not in $\mathfrak{m}(A)$, i.e., u invertible.

This writing is clearly unique; it shows that A is an integral domain. Furthermore, if one sets $n = v(y)$, one checks easily that the function v extends to a discrete valuation of the field of fractions of A with A as its valuation ring. □

Remark. When one knows in advance that A is an integral domain (which is often the case), one has $\mathfrak{u} = 0$, $\pi x_n = x_{n+1}$, and the proof above becomes much simpler.

Proposition 3. *Let* A *be a Noetherian integral domain. In order that* A *be a discrete valuation ring, it is necessary and sufficient that it satisfy the two following conditions*:

(*i*) A *is integrally closed.*
(*ii*) A *has a unique non-zero prime ideal.*

[Recall that an element x of a ring containing A is called *integral* over A if it satisfies an equation "of integral dependence":

$$(*) \qquad x^n + a_1 x^{n-1} + \cdots + a_n = 0, \qquad a_i \in \mathrm{A}.$$

One says that A is *integrally closed* in a ring B containing it if every element of B integral over A belongs to A. One says that A is *integrally closed* if it is an integral domain integrally closed in its field of fractions. Cf. Bourbaki, *Alg. comm.*, Chap. V, §1.]

It is clear that a discrete valuation ring satisfies (*ii*). Let us show that it satisfies (*i*). Let K be the field of fractions of A, and let x be an element of K satisfying an equation of type (*), and suppose x were not in A. That means $v(x) = -m$, with $m > 0$. In equation (*), the first term has valuation $-nm$, while the valuation of the others is $\geq -(n-1)m$, which is $> -nm$; that is a contradiction, according to the following lemma:

Lemma 1. *Let* A *be a discrete valuation ring, and let x_i be elements of the field of fractions of* A *such that $v(x_i) > v(x_1)$ for $i \geq 2$. One then has*

$$x_1 + x_2 + \cdots + x_n \neq 0.$$

One can assume $x_1 = 1$ (dividing by x_1 if necessary), whence $v(x_i) \geq 1$ for $i \geq 2$, i.e., $x_i \in \mathfrak{m}(\mathrm{A})$; as $x_1 \notin \mathfrak{m}(\mathrm{A})$, it follows that $x_1 + \cdots + x_n \notin \mathfrak{m}(\mathrm{A})$, which proves the lemma.

[This proof also shows that $x_1 + \cdots + x_n$ has the same valuation as x_1.]

Let us now show that a Noetherian integral domain satisfying (*i*) and (*ii*) is a discrete valuation ring. Condition (*ii*) shows that A is a local ring whose maximal ideal $\mathfrak{m}$ is $\neq 0$. Let $\mathfrak{m}'$ be the set of $x \in \mathrm{K}$ such that $x\mathfrak{m} \subset \mathrm{A}$ (i.e., $xy \in \mathrm{A}$ for every $y \in \mathfrak{m}$); it is a sub-A-module of K containing A. If y is a nonzero element of $\mathfrak{m}$, it is clear that $\mathfrak{m}' \subset y^{-1}\mathrm{A}$, and as A is Noetherian, this shows that $\mathfrak{m}'$ is a finitely generated A-module (that is what one calls a "fractional ideal" of K with respect to A). Let $\mathfrak{m}.\mathfrak{m}'$ be the product of $\mathfrak{m}$ and $\mathfrak{m}'$, i.e., the set of all $\sum x_i y_i$, $x_i \in \mathfrak{m}$, $y_i \in \mathfrak{m}'$; by definition of $\mathfrak{m}'$, one has $\mathfrak{m}.\mathfrak{m}' \subset \mathrm{A}$; on the other hand, since $\mathrm{A} \subset \mathfrak{m}'$, one has $\mathfrak{m}.\mathfrak{m}' \supset \mathfrak{m}$; since $\mathfrak{m}.\mathfrak{m}'$ is an ideal, one has either $\mathfrak{m}.\mathfrak{m}' = \mathfrak{m}$ or $\mathfrak{m}.\mathfrak{m}' = \mathrm{A}$. We will successively show:

I. *If* $\mathfrak{m}.\mathfrak{m}' = \mathrm{A}$, *the ideal* $\mathfrak{m}$ *is principal.*
II. *If* $\mathfrak{m}.\mathfrak{m}' = \mathfrak{m}$, *and if* (*i*) *is satisfied, then* $\mathfrak{m}' = \mathrm{A}$.
III. *If* (*ii*) *is satisfied, then* $\mathfrak{m}' \neq \mathrm{A}$.

By combining II and III, one sees that $\mathfrak{m}.\mathfrak{m}' = \mathfrak{m}$ is impossible, whence, by I, $\mathfrak{m}$ must be principal, therefore A is a discrete valuation ring (prop. 2).

It remains to prove assertions I, II, III.

Proof of I. If $\mathfrak{m}.\mathfrak{m}' = \mathrm{A}$, one has a relation $\sum x_i y_i = 1$, with $x_i \in \mathfrak{m}$, $y_i \in \mathfrak{m}'$. The products $x_i y_i$ all belong to A; at least one of them—say xy—does not belong to $\mathfrak{m}$, therefore is an invertible element u. Replacing x by xu^{-1}, one obtains a relation $xy = 1$, with $x \in \mathfrak{m}$ and $y \in \mathfrak{m}'$. If $z \in \mathfrak{m}$, one has $z = x(yz)$, with $yz \in \mathrm{A}$ since $y \in \mathfrak{m}'$; therefore z is a multiple of x, which shows that $\mathfrak{m}$ is indeed a principal ideal, generated by x.

PROOF OF II. Suppose $\mathfrak{m}.\mathfrak{m}' = \mathfrak{m}$, and let $x \in \mathfrak{m}'$. Then $x\mathfrak{m} \subset \mathfrak{m}$, whence, by iteration, $x^n\mathfrak{m} \subset \mathfrak{m}$ for all n, i.e., $x^n \in \mathfrak{m}'$. Let $\mathfrak{a}_n$ be the sub-A-module of K generated by the powers $\{1, x, \ldots, x^n\}$ of x; one has $\mathfrak{a}_n \subset \mathfrak{a}_{n+1}$, and all the $\mathfrak{a}_n$ are contained in the finitely generated A-module $\mathfrak{m}'$. Since A is Noetherian, one gets $\mathfrak{a}_{n-1} = \mathfrak{a}_n$ for n large, i.e., $x^n \in \mathfrak{a}_{n-1}$. One can then write $x^n = b_0 + b_1 x + \cdots + b_{n-1}x^{n-1}$, $b_i \in \mathrm{A}$, which shows that x is integral over A. Condition (*i*) then implies $x \in \mathrm{A}$, hence $\mathfrak{m}' = \mathrm{A}$.

PROOF OF III. Let x be a non-zero element of $\mathfrak{m}$, and form the ring A_x of fractions of the type y/x^n, with $y \in \mathrm{A}$, and $n \geq 0$ arbitrary. Condition (*ii*) implies $\mathrm{A}_x = \mathrm{K}$: indeed, if not, A_x would not be a field, and would contain a non-zero maximal ideal $\mathfrak{p}$; as x is invertible in A_x, one would have $x \notin \mathfrak{p}$, which shows that $\mathfrak{p} \cap \mathrm{A} \neq \mathfrak{m}$. On the other hand, if y/x^n is a non-zero element of $\mathfrak{p}$, one has $y \in \mathfrak{p} \cap \mathrm{A}$, so that $\mathfrak{p} \cap \mathrm{A} \neq 0$. But since $\mathfrak{p}$ is prime, so is $\mathfrak{p} \cap \mathrm{A}$, which contradicts (*ii*).

Thus every element of K can be written in the form y/x^n; let us apply this to $1/z$, with $z \neq 0$ in A. We get $1/z = y/x^n$, whence $x^n = yz \in z\mathrm{A}$. Therefore every element of $\mathfrak{m}$ has a power belonging to the ideal $z\mathrm{A}$. Let $x_1, \ldots, x_k$ generate $\mathfrak{m}$, and let n be large enough so that $x_i^n \in z\mathrm{A}$ for all i; if one chooses $\mathrm{N} > k(n-1)$, all the monomials in the x_i of total degree N contain an x_i^n as factor, therefore belong to $z\mathrm{A}$; as the ideal $\mathfrak{m}^{\mathrm{N}}$ is generated by these monomials, one has $\mathfrak{m}^{\mathrm{N}} \subset z\mathrm{A}$. Apply this with $z \in \mathfrak{m}$: one concludes that there is a smallest integer $\mathrm{N} \geq 1$ such that $\mathfrak{m}^{\mathrm{N}} \subset z\mathrm{A}$; choose $y \in \mathfrak{m}^{\mathrm{N}-1}$, $y \notin z\mathrm{A}$ (putting $\mathfrak{m}^0 = \mathrm{A}$ by convention). One then has $\mathfrak{m}y \subset z\mathrm{A}$, whence $y/z \in \mathfrak{m}'$, and $y/z \notin \mathrm{A}$, which indeed proves that $\mathfrak{m}' \neq \mathrm{A}$. □

Remark. The construction of $\mathfrak{m}'$ does not use the hypotheses made on A and $\mathfrak{m}$; for every non-zero ideal $\mathfrak{a}$ of an integral domain A, one can define $\mathfrak{a}'$ as the set of $x \in \mathrm{K}$ such that $x\mathfrak{a} \subset \mathrm{A}$; if A is Noetherian, this is a fractional ideal. When $\mathfrak{a}\mathfrak{a}' = \mathrm{A}$, one says that $\mathfrak{a}$ is *invertible*. The proof of I shows that every invertible ideal of a local ring is principal.

§3. Dedekind Domains

Reminder. Let A be an integral domain, K its field of fractions, and let S be a subset of A that is multiplicatively stable and contains 1 (such a set will be called *multiplicative*); suppose also that 0 does not belong to S. The set of those elements of K of the form x/s, $x \in A$, $s \in S$ is a ring that will be denoted $S^{-1}A$. The map $\mathfrak{p}' \to \mathfrak{p}' \cap A$ is a bijection of the set of prime ideals of $S^{-1}A$ onto the set of those prime ideals of A that do not meet S.

This applies notably when $S = A - \mathfrak{p}$, where $\mathfrak{p}$ is a prime ideal of A. The ring $S^{-1}A$ is then denoted $A_\mathfrak{p}$; it is a local ring with maximal ideal $\mathfrak{p}A_\mathfrak{p}$ and residue field the field of fractions of $A/\mathfrak{p}$; the prime ideals of $A_\mathfrak{p}$ correspond to those prime ideals of A that are contained in $\mathfrak{p}$. One says that $A_\mathfrak{p}$ is the *localisation of* A *at* $\mathfrak{p}$, cf. Bourbaki, *Alg. comm.*, Chap. II, §2.

Proposition 4. *If* A *is a Noetherian integral domain, the following two properties are equivalent*:

(*i*) *For every prime ideal* $\mathfrak{p} \neq 0$ *of* A, $A_\mathfrak{p}$ *is a discrete valuation ring.*
(*ii*) A *is integrally closed and of dimension* ≤ 1.

[An integral domain A is said to be *of dimension* ≤ 1 if every non-zero prime ideal of A is maximal; equivalently, if $\mathfrak{p}$ and $\mathfrak{p}'$ are two prime ideals of A such that $\mathfrak{p} \subset \mathfrak{p}'$, then $\mathfrak{p} = 0$ or $\mathfrak{p} = \mathfrak{p}'$.]

(*i*) *implies* (*ii*): If $\mathfrak{p} \subset \mathfrak{p}'$, then $A_{\mathfrak{p}'}$ contains the prime ideal $\mathfrak{p}A_{\mathfrak{p}'}$, which implies $\mathfrak{p} = 0$ or $\mathfrak{p} = \mathfrak{p}'$ (cf. prop. 3, (*ii*)). On the other hand, if a is integral over A, it is *a fortiori* integral over each $A_\mathfrak{p}$, and by prop. 3, (*i*), it belongs to all the $A_\mathfrak{p}$. If one writes a in the form $a = b/c$, with $b, c \in A$ and $c \neq 0$, and if $\mathfrak{a}$ is the ideal of those $x \in A$ such that $xb \subset cA$, the ideal $\mathfrak{a}$ is not contained in any prime ideal $\mathfrak{p}$, whence $\mathfrak{a} = A$ and $a \in A$.

(*ii*) *implies* (*i*): It is clear that the $A_\mathfrak{p}$ satisfy condition (*ii*) of prop. 3, so that it suffices to prove they are integrally closed. Let x be integral over $A_\mathfrak{p}$. Multiplying by a common denominator of the coefficients of the equation of integral dependence of x over $A_\mathfrak{p}$, one can write the latter in the form:

$$sx^n + a_1x^{n-1} + \cdots + a_n = 0, \quad \text{with } a_i \in A,\ s \in A - \mathfrak{p}.$$

Multiplying by s^{n-1}, one obtains an equation of integral dependence for sx over A, which implies $sx \in A$, whence $x \in A_\mathfrak{p}$. □

Remark. The proof above actually establishes the following result:

Let A *be a subring of a field* K, S *a multiplicative subset of* A *not containing* 0. *In order that an element of* K *be integral over* $S^{-1}A$, *it is necessary and sufficient that it be of the form* a'/s, *where* a' *is integral over* A *and* s *belongs to* S. (Passage to rings of fractions commutes with integral closure.)

Definition. *A Noetherian integral domain which has the two equivalent properties of prop.* 4 *is called a Dedekind domain.*

EXAMPLES. Every principal ideal domain is Dedekind. The ring of integers of an algebraic number field is Dedekind (apply prop. 9 below to the ring **Z**).

If V is an affine algebraic variety, defined over an algebraically closed field k, the coordinate ring $k[V]$ of V is a Dedekind domain if and only if V is non-singular, irreducible and of dimension ≤ 1.

Proposition 5. *In a Dedekind domain, every non-zero fractional ideal is invertible.*

[If K is the field of fractions of A, a *fractional ideal* $\mathfrak{a}$ of A is a sub-A-module of K finitely generated over A. One says $\mathfrak{a}$ is *invertible* if there exists $\mathfrak{a}' \subset K$ with $\mathfrak{a}.\mathfrak{a}' = A$.]

In a discrete valuation ring, a fractional ideal has the form $\pi^n A$, where $n \in \mathbf{Z}$, and is therefore invertible. The proposition follows from this by localisation, taking into account that:

$$(\mathfrak{a}.\mathfrak{b})_\mathfrak{p} = \mathfrak{a}_\mathfrak{p}\mathfrak{b}_\mathfrak{p}; \qquad (\mathfrak{a}+\mathfrak{b})_\mathfrak{p} = \mathfrak{a}_\mathfrak{p} + \mathfrak{b}_\mathfrak{p}; \qquad (\mathfrak{a}:\mathfrak{b})_\mathfrak{p} = (\mathfrak{a}_\mathfrak{p}:\mathfrak{b}_\mathfrak{p})$$

if $\mathfrak{b}$ is finitely generated. □

[$(\mathfrak{a}:\mathfrak{b})$ denotes the ideal of those $x \in K$ such that $x\mathfrak{b} \subset \mathfrak{a}$. If $\mathfrak{a}' = (A:\mathfrak{a})$, to say that $\mathfrak{a}$ is invertible amounts to saying that $\mathfrak{a}.\mathfrak{a}' = A$.]

Corollary. *The non-zero fractional ideals of a Dedekind domain form a group under multiplication.*

This group is called the *ideal group* of the ring.

Proposition 6. *If $x \in A$, $x \neq 0$, then only finitely many prime ideals contain x.*

Indeed, the ideals containing x satisfy the descending chain condition: if $Ax \subset \mathfrak{a} \subset \mathfrak{a}' \subset A$, one has $Ax^{-1} \supset \mathfrak{a}^{-1} \supset \mathfrak{a}'^{-1} \supset A$, and A is Noetherian.

It follows that if $x \in \mathfrak{p}_1, \mathfrak{p}_2, \ldots, \mathfrak{p}_k, \ldots$, the sequence

$$\mathfrak{p}_1 \supset \mathfrak{p}_1 \cap \mathfrak{p}_2 \supset \cdots \supset \mathfrak{p}_1 \cap \mathfrak{p}_2 \cap \cdots \cap \mathfrak{p}_k \supset \cdots$$

is stationary, which means that from some point onward, one has

$$\mathfrak{p}_i \supset \mathfrak{p}_1 \cap \mathfrak{p}_2 \cdots \cap \mathfrak{p}_k \supset \mathfrak{p}_1\mathfrak{p}_2 \cdots \mathfrak{p}_k$$

which, as the $\mathfrak{p}_j$ are prime, shows that $\mathfrak{p}_i$ is one of the $\mathfrak{p}_1, \ldots, \mathfrak{p}_k$. □

Corollary. *If one denotes by $v_\mathfrak{p}$ the valuation of K defined by $A_\mathfrak{p}$, then for every $x \in K^*$, the numbers $v_\mathfrak{p}(x)$ are almost all zero* (i.e., zero except for a finite number).

Now let $\mathfrak{a}$ be an arbitrary fractional ideal of A; it is contained in only finitely many prime ideals $\mathfrak{p}$. The image $\mathfrak{a}_\mathfrak{p}$ of $\mathfrak{a}$ in $A_\mathfrak{p}$ has the form $\mathfrak{a}_\mathfrak{p} = (\mathfrak{p}A_\mathfrak{p})^{v_\mathfrak{p}(\mathfrak{a})}$, where the $v_\mathfrak{p}(\mathfrak{a})$ are rational integers, almost all zero.

If one considers the ideal $\mathfrak{a}_1 = \prod_{\mathfrak{p}} \mathfrak{p}^{v_{\mathfrak{p}}(\mathfrak{a})}$ and the ideal $\mathfrak{a}_2$ of those x such that $v_{\mathfrak{p}}(x) \geq v_{\mathfrak{p}}(\mathfrak{a})$ for all $\mathfrak{p}$, the three ideals $\mathfrak{a}$, $\mathfrak{a}_1$, and $\mathfrak{a}_2$ are equal *locally* (i.e., have the same images in all the $A_{\mathfrak{p}}$). An elementary argument shows that they must then be equal, whence:

Proposition 7. *Every fractional ideal* $\mathfrak{a}$ *of* A *can be written uniquely in the form*:

$$\mathfrak{a} = \prod \mathfrak{p}^{v_{\mathfrak{p}}(\mathfrak{a})},$$

where the $v_{\mathfrak{p}}(\mathfrak{a})$ *are integers almost all zero.*

The following formulas are immediate:

$$v_{\mathfrak{p}}(\mathfrak{a}.\mathfrak{b}) = v_{\mathfrak{p}}(\mathfrak{a}) + v_{\mathfrak{p}}(\mathfrak{b})$$
$$v_{\mathfrak{p}}((\mathfrak{b}:\mathfrak{a})) = v_{\mathfrak{p}}(\mathfrak{b}.\mathfrak{a}^{-1}) = v_{\mathfrak{p}}(\mathfrak{b}) - v_{\mathfrak{p}}(\mathfrak{a})$$
$$v_{\mathfrak{p}}(\mathfrak{a} + \mathfrak{b}) = \mathrm{Inf}(v_{\mathfrak{p}}(\mathfrak{a}), v_{\mathfrak{p}}(\mathfrak{b}))$$
$$v_{\mathfrak{p}}(xA) = v_{\mathfrak{p}}(x).$$

Furthermore:

Approximation Lemma. *Let k be a positive integer. For every i, $1 \leq i \leq k$, let $\mathfrak{p}_i$ be distinct prime ideals of* A, *x_i elements of* K, *and n_i integers. Then there exists an $x \in$ K such that $v_{\mathfrak{p}_i}(x - x_i) \geq n_i$ for all i, and $v_{\mathfrak{q}}(x) \geq 0$ for $\mathfrak{q} \neq \mathfrak{p}_1, \ldots, \mathfrak{p}_k$.*

Suppose first that the x_i belong to A, and let us seek a solution x belonging to A. By linearity, one may assume that $x_2 = \cdots = x_k = 0$. Increasing the n_i if necessary, one may also assume $n_i \geq 0$. Put

$$\mathfrak{a} = \mathfrak{p}_1^{n_1} + \mathfrak{p}_2^{n_2} \cdots \mathfrak{p}_k^{n_k}.$$

One has $v_{\mathfrak{p}}(\mathfrak{a}) = 0$ for all $\mathfrak{p}$, whence $\mathfrak{a} = A$. It follows that

$$x_1 = x + y, \quad \text{with } y \in \mathfrak{p}_1^{n_1},\ x \in \mathfrak{p}_2^{n_2} \cdots \mathfrak{p}_k^{n_k}.$$

and the element x has the desired properties.

In the general case, one writes $x_i = a_i/s$, with $a_i \in A$, $s \in A$, $s \neq 0$, and $x = a/s$. The element a must fulfill the conditions:

$$v_{\mathfrak{p}_i}(a - a_i) \geq n_i + v_{\mathfrak{p}_i}(s), \qquad 1 \leq i \leq k,$$
$$v_{\mathfrak{q}}(a) \geq v_{\mathfrak{q}}(s) \quad \text{for } \mathfrak{q} \neq \mathfrak{p}_1, \ldots, \mathfrak{p}_k.$$

These conditions are of the type envisaged above (if one adds to the family $\{\mathfrak{p}_i\}$ the prime ideals $\mathfrak{q}$ for which $v_{\mathfrak{q}}(s) > 0$); the existence of a then follows from the previous case. □

Corollary. *A Dedekind domain with only finitely many prime ideals is principal.*

It sufficies to show that all its prime ideals are principal. Now if $\mathfrak{p}$ is one of them, there exists an $x \in A$ with $v_{\mathfrak{p}}(x) = 1$ and $v_{\mathfrak{q}}(x) = 0$ for $\mathfrak{q} \neq \mathfrak{p}$, i.e., with $xA = \mathfrak{p}$. □

§4. Extensions

Throughout this paragraph, K is a field and L a finite extension of K; its degree $[L:K]$ will be denoted by n.

We are also given a Noetherian integrally closed domain A, having K as field of fractions. We denote by B the *integral closure* of A in L (i.e., the set of elements of L that are integral over A). According to the remark that followed proposition 4, we have $K.B = L$. In particular, the field of fractions of B is L.

We make the following hypothesis:

(F) *The ring* B *is a finitely generated* A-*module.*

This hypothesis implies that B is a Noetherian integrally closed domain.

Proposition 8. *Hypothesis* (F) *is satisfied when* L/K *is a separable extension.*

Let $\mathrm{Tr}: L \to K$ be the *trace* map (Bourbaki, *Alg.*, Chap. V, §10, no. 6). One knows (*loc. cit.*, prop. 12) that $\mathrm{Tr}(xy)$ is a symmetric non-degenerate K-bilinear form on L. If $x \in B$, the conjugates of x with respect to K (in a suitable extension of L) are integral over A, and so is their sum $\mathrm{Tr}(x)$; as $\mathrm{Tr}(x) \in K$, it follows that $\mathrm{Tr}(x) \in A$.

Next let $\{e_i\}$ be a basis of L over K, with $e_i \in B$, and let V be the free A-module spanned by the e_i. For every sub-A-module M of L, let M^* be the set of those $x \in L$ such that $\mathrm{Tr}(xy) \in A$ for all $y \in M$. Obviously one has:

$$V \subset B \subset B^* \subset V^*.$$

Since V^* is the free module spanned by the basis dual to $\{e_i\}$ (with respect to the bilinear form $\mathrm{Tr}(xy)$), it follows from the Noetherian hypothesis on A that B is finitely generated as an A-module. □

Remarks. 1) The same proof shows that B^* is a finitely generated B-module, i.e., a fractional ideal of B. Its inverse is called the *different* of B over A, cf. Chap. III, §3.

2) One can show that hypothesis (F) is satisfied when A is an algebra of finite type over a field (cf. Bourbaki, *Alg. comm.*, Chap. V), or when A is a complete discrete valuation ring (cf. Chap. II, §2).

Proposition 9. *If* A *is Dedekind then* B *is Dedekind.*

One knows already, thanks to hypothesis (F), that B is Noetherian and integrally closed. According to proposition 4, it suffices to show that B is of dimension ≤ 1. Let $\mathfrak{P}_0 \subset \mathfrak{P}_1 \subset \mathfrak{P}_2$ be a chain of distinct prime ideals of B. The next lemma shows that the $\mathfrak{P}_i \cap A$ are distinct (contradicting the fact that A is of dimension ≤ 1):

Lemma 2. *Let* $A \subset B$ *be rings, with* B *integral over* A. *If* $\mathfrak{P} \subset \mathfrak{Q}$ *are prime ideals of* B *such that* $\mathfrak{P} \cap A = \mathfrak{Q} \cap A$, *then* $\mathfrak{P} = \mathfrak{Q}$.

Passing to the quotient by $\mathfrak{P}$, one may assume $\mathfrak{P} = 0$. If $\mathfrak{Q} \neq \mathfrak{P}$, there is a non-zero $x \in \mathfrak{Q}$. Let

$$x^n + a_{n-1}x^{n-1} + \cdots + a_0 = 0, \qquad a_i \in A,$$

be its minimal equation over A. One has $a_0 \neq 0$, and a_0 belongs to the ideal of B generated by x, therefore to $\mathfrak{Q} \cap A = \mathfrak{P} \cap A$, which is absurd. □

Remark. One can show that prop. 9 remains valid even when hypothesis (F) fails (cf. Bourbaki, *Alg. comm.*, Chap. VII).

Let us keep the hypotheses of prop. 9. If $\mathfrak{P}$ is a non-zero prime ideal of B, and if $\mathfrak{p} = \mathfrak{P} \cap A$, we will say that $\mathfrak{P}$ *divides* $\mathfrak{p}$ (or that $\mathfrak{P}$ is "above" $\mathfrak{p}$), and we will write $\mathfrak{P}|\mathfrak{p}$. This relation is also equivalent to saying that $\mathfrak{P}$ *contains* the ideal $\mathfrak{p}B$ generated by $\mathfrak{p}$. Denote by $e_{\mathfrak{P}}$ the exponent of $\mathfrak{P}$ in the decomposition of $\mathfrak{p}B$ into prime ideals. Thus:

$$e_{\mathfrak{P}} = v_{\mathfrak{P}}(\mathfrak{p}B), \qquad \mathfrak{p}B = \prod_{\mathfrak{P}|\mathfrak{p}} \mathfrak{P}^{e_{\mathfrak{P}}}.$$

The integer $e_{\mathfrak{P}}$ is called the *ramification index* of $\mathfrak{P}$ in the extension L/K.

On the other hand, if $\mathfrak{P}$ divides $\mathfrak{p}$, the field $B/\mathfrak{P}$ is an extension of the field $A/\mathfrak{p}$. As B is finitely generated over A, $B/\mathfrak{P}$ is an extension of $A/\mathfrak{p}$ of finite degree. The degree of this extension is called the *residue degree* of $\mathfrak{P}$ in the extension L/K, and is denoted $f_{\mathfrak{P}}$. Thus:

$$f_{\mathfrak{P}} = [B/\mathfrak{P} : A/\mathfrak{p}].$$

[When we want to specify K, we write $e_{\mathfrak{P}/\mathfrak{p}}$ and $f_{\mathfrak{P}/\mathfrak{p}}$, instead of $e_{\mathfrak{P}}$ and $f_{\mathfrak{P}}$.]

When there is only one prime ideal $\mathfrak{P}$ which divides $\mathfrak{p}$ and $f_{\mathfrak{P}} = 1$, one says that L/K is *totally ramified* at $\mathfrak{p}$.

When $e_{\mathfrak{P}} = 1$ and $B/\mathfrak{P}$ is separable over $A/\mathfrak{p}$, one says that L/K is *unramified* at $\mathfrak{P}$. If L/K is unramified for all the prime ideals $\mathfrak{P}$ dividing $\mathfrak{p}$, one says that L/K is unramified above $\mathfrak{p}$ (or "at $\mathfrak{p}$"); cf. Chap. III, §5.

Proposition 10. *Let* $\mathfrak{p}$ *be a non-zero prime ideal of* A, *the ring* $B/\mathfrak{p}B$ *is an* $A/\mathfrak{p}$-*algebra of degree* $n = [L:K]$, *isomorphic to the product* $\prod_{\mathfrak{P}|\mathfrak{p}} B/\mathfrak{P}^{e_{\mathfrak{P}}}$. *We have the formula*:

$$n = \sum_{\mathfrak{P}|\mathfrak{p}} e_{\mathfrak{P}} f_{\mathfrak{P}}.$$

Let $S = A - \mathfrak{p}$, $A' = S^{-1}A$, and $B' = S^{-1}B$. The ring $A' = A_{\mathfrak{p}}$ is a discrete valuation ring, and B' is its integral closure in L (cf. the remark after prop. 4). One has $A'/\mathfrak{p}A' = A/\mathfrak{p}$, and one sees easily that $B'/\mathfrak{p}B' = B/\mathfrak{p}B$. As A' is principal, hypothesis (F) shows that B' is a free module of rank $n = [L:K]$ and $B'/\mathfrak{p}B'$ is free of rank n over $A'/\mathfrak{p}A'$. Thus $B/\mathfrak{p}B$ is an algebra of degree n.

Since $\mathfrak{p}B = \bigcap \mathfrak{P}^{e_{\mathfrak{P}}}$, the canonical map

$$B/\mathfrak{p}B \to \prod_{\mathfrak{P}|\mathfrak{p}} B/\mathfrak{P}^{e_{\mathfrak{P}}}$$

is injective; the approximation lemma shows that it is surjective; hence it is an isomorphism. By comparing degrees, one sees that n is the sum of the degrees

$$n_{\mathfrak{P}} = [B/\mathfrak{P}^{e_{\mathfrak{P}}} : A/\mathfrak{p}].$$

One has $n_{\mathfrak{P}} = \sum_{i=0}^{i=e_{\mathfrak{P}}-1} [\mathfrak{P}^i/\mathfrak{P}^{i+1} : A/\mathfrak{p}] = e_{\mathfrak{P}} . [B/\mathfrak{P} : A/\mathfrak{p}] = e_{\mathfrak{P}} f_{\mathfrak{P}}$, which proves the proposition. □

Corollary. *The number of prime ideals $\mathfrak{P}$ of B which divide a prime ideal $\mathfrak{p}$ of A is at least 1 and at most n. If A has only finitely many ideals, then so has B (which is therefore principal).*

Remark. When hypothesis (F) is not satisfied, the sum of the $e_{\mathfrak{P}} f_{\mathfrak{P}}$ is still equal to the degree of $B/\mathfrak{p}B$, but this degree can be $< n$.

Let $\mathfrak{P}$ be a non-zero prime ideal of B, and let $\mathfrak{p} = A \cap \mathfrak{P}$. Clearly $v_{\mathfrak{P}}(x) = e_{\mathfrak{P}} v_{\mathfrak{p}}(x)$ if $x \in K$. One says (by abuse of language) that the valuation $v_{\mathfrak{P}}$ *prolongs* (or "extends") the valuation $v_{\mathfrak{p}}$ with index $e_{\mathfrak{P}}$. Conversely:

Proposition 11. *Let w be a discrete valuation of L which prolongs $v_{\mathfrak{p}}$ with index e. Then there is a prime divisor $\mathfrak{P}$ of $\mathfrak{p}$ with $w = v_{\mathfrak{P}}$ and $e = e_{\mathfrak{P}}$.*

Let W be the ring of w, and let $\mathfrak{Q}$ be its maximal ideal. This ring is integrally closed with field of fractions L, and contains A; hence it contains B. Let $\mathfrak{P} = \mathfrak{Q} \cap B$. Obviously $\mathfrak{P} \cap A = \mathfrak{p}$, so that $\mathfrak{P}$ divides $\mathfrak{p}$. The ring W thus contains $B_{\mathfrak{P}}$. But one checks immediately that every discrete valuation ring is a *maximal* subring of its field of fractions. Hence $W = B_{\mathfrak{P}}$, so that $w = v_{\mathfrak{P}}$ and $e = e_{\mathfrak{P}}$. □

§5. The Norm and Inclusion Homomorphisms

We keep the hypotheses of the preceding paragraph. We denote by I_A and I_B the ideal groups of A and of B. We will define two homomorphisms

$$i: I_A \to I_B,$$
$$N: I_A \to I_B.$$

As I_A (resp. I_B) is the free abelian group generated by the non-zero prime ideals $\mathfrak{p}$ of A (resp. $\mathfrak{P}$ of B), it suffices to define $i(\mathfrak{p})$ and $N(\mathfrak{P})$. Put:

$$i(\mathfrak{p}) = \mathfrak{p}B = \prod_{\mathfrak{P}|\mathfrak{p}} \mathfrak{P}^{e_{\mathfrak{P}}}$$

$$N(\mathfrak{P}) = \mathfrak{p}^{f_{\mathfrak{P}}} \quad \text{if } \mathfrak{P}|\mathfrak{p}.$$

By prop. 10, one has $N(i(\mathfrak{a})) = \mathfrak{a}^n$ for every $\mathfrak{a} \in I_A$. The homomorphism i assigns to an ideal $\mathfrak{a}$ of A the ideal $\mathfrak{a}B$ of B generated by $\mathfrak{a}$.

These two homomorphisms may be interpreted in a more suggestive manner by means of suitable "Grothendieck groups":

Let $\mathscr{C}_A$ be the category of A-modules of finite length. If $M \in \mathscr{C}_A$ and if M is of length m, M has a composition series:

$$0 = M_0 \subset M_1 \subset \cdots \subset M_m = M,$$

each M_i/M_{i-1} being isomorphic to a simple A-module, i.e., to a quotient $A/\mathfrak{p}_i$, where $\mathfrak{p}_i$ is a non-zero prime ideal of A (ignoring the trivial case $A = K$). By the Jordan-Hölder theorem, the sequence of $A/\mathfrak{p}_i$ depends only on M (up to order), and one can put:

$$\chi_A(M) = \prod \mathfrak{p}_i.$$

EXAMPLE. When $M = \mathfrak{b}/\mathfrak{a}$, where $\mathfrak{a}$ and $\mathfrak{b}$ are non-zero fractional ideals with $\mathfrak{a} \subset \mathfrak{b}$, one has $\chi_A(M) = \mathfrak{a} \,.\, \mathfrak{b}^{-1}$. In particular, $\chi_A(A/\mathfrak{a}) = \mathfrak{a}$ if $\mathfrak{a} \subset A$.

The map $\chi_A : \mathscr{C}_A \to I_A$ is "multiplicative": if one has an exact sequence:

$$0 \to M' \to M \to M'' \to 0$$

of A-modules of finite length, one has $\chi_A(M) = \chi_A(M')\chi_A(M'')$. Conversely, every multiplicative map $f : \mathscr{C}_A \to G$, where G is a commutative group, can be put uniquely into the form $g \circ \chi_A$, where g is a homomorphism of I_A into G (it suffices to define $g(\mathfrak{p})$ to be $f(A/\mathfrak{p})$). In other words, χ_A identifies the "*Grothendieck group*" of $\mathscr{C}_A$ with the group I_A.

Similarly define $\mathscr{C}_B$ and $\chi_B : \mathscr{C}_B \to I_B$. Clearly every B-module of finite length is of finite length as an A-module. One thus defines an exact functor $\mathscr{C}_B \to \mathscr{C}_A$, hence a homomorphism of I_B into I_A. This homomorphism is none other than the *norm*. In other words:

Proposition 12. *If* M *is a* B-*module of finite length, then* $\chi_A(M) = N(\chi_B(M))$.

By linearity, it suffices to consider the case $M = B/\mathfrak{P}$, which case follows from the definition of the norm. □

On the other hand, every A-module M of finite length defines by tensor product with B a module M_B of finite length. The functor $\mathscr{C}_A \to \mathscr{C}_B$ thus defined is still exact (by localisation, one reduces to the case where A is principal, and B is then a free A-module). Hence one obtains again a homomorphism $I_A \to I_B$ which coincides with the inclusion:

Proposition 13. *If* M *is an* A*-module of finite length, then* $\chi_B(M_B) = i(\chi_A(M))$.

By linearity, it suffices to consider the case $M = A/\mathfrak{p}$, whence $M_B = B/\mathfrak{p}B$, and the proposition is clear. □

The next proposition shows that the restriction of N to principal ideals coincides with the usual norm map (defined in Bourbaki, *Alg.*, Chap. V):

Proposition 14. *If* $x \in L$, *then* $N(xB) = N_{L/K}(x)A$.

One may assume x integral over A, and, by localising, that A is principal. The ring B is then a free A-module of rank n. Let u_x be multiplication by x in B. One has $N_{L/K}(x) = \det(u_x)$ and $N(xB) = \chi_A(B/xB) = \chi_A(\mathrm{Coker}\ u_x)$. One is thus reduced to:

Lemma 3. *Let* A *be a principal ideal domain and* $u: A^n \to A^n$ *a linear map with* $\det(u) \neq 0$. *Then* $\det(u)A = \chi_A(\mathrm{Coker}\ u)$.

The ideal $\det(u)A$ does not change when one multiplies u by an invertible linear map; hence one may reduce by the theory of elementary divisors to the case where u is diagonal (Bourbaki, *Alg.*, Chap. VII, §4, no. 5, prop. 4). The proof is then carried out by induction on n, the case $n = 1$ being the property already remarked: $\chi_A(A/\mathfrak{a}) = \mathfrak{a}$. □

§6. Example: Simple Extensions

In this paragraph, we place ourselves once again in the *local case*. Thus let A be a local ring with residue field k. Let n be a positive integer, and let $f \in A[X]$ be a monic polynomial of degree n. Let B_f be the quotient ring of A[X] by the principal ideal (f) generated by f. It is an A-algebra that is free and of finite type over A, with $\{1, X, \ldots, X^{n-1}\}$ as basis. We first determine its maximal ideals. Toward that end, denote by $\mathfrak{m}$ the maximal ideal of A, and put $\bar{B}_f = B_f/\mathfrak{m}B_f = A[X]/(\mathfrak{m}, f)$. If one denotes by $\bar{f}$ the image of $f \in k[X]$ by reduction mod $\mathfrak{m}$, one then has

$$\bar{B}_f = k[X]/(\bar{f}).$$

Let $\bar{f} = \prod_{i \in I} \varphi_i^{e_i}$ be the decomposition of the polynomial $\bar{f}$ into irreducible factors in $k[X]$, and, for each i, choose a polynomial $g_i \in A[X]$ with $\bar{g}_i = \varphi_i$. With this notation, we have:

Lemma 4. *Let* $\mathfrak{m}_i = (\mathfrak{m}, g_i)$ *be the ideal of* B_f *generated by* $\mathfrak{m}$ *and the canonical image of* g_i *in* B_f*; the ideals* $\mathfrak{m}_i$, $i \in I$, *are maximal and distinct, and every maximal ideal of* B_f *is equal to one of them. The quotient* $B_f/\mathfrak{m}_i$ *is isomorphic to the field* $k_i = k[X]/(\varphi_i)$.

By definition, $\mathfrak{m}_i$ is the inverse image in B_f of the ideal $\bar{\mathfrak{m}}_i$ of $\bar{B}_f$ generated by φ_i; as $\bar{B}_f/(\varphi_i) = k_i = k[X]/(\varphi_i)$, it is clear that $\mathfrak{m}_i$ is maximal and that $B_f/\mathfrak{m}_i = k_i$. In order to show that every maximal ideal $\mathfrak{n}$ of B_f is equal to one of the $\mathfrak{m}_i$, it suffices to prove that $\mathfrak{n}$ contains $\mathfrak{m}$ (for $\mathfrak{n}$ would then be the inverse image of one of the maximal ideals (φ_i) of $\bar{B}_f$). If not, one would have $\mathfrak{n} + \mathfrak{m}B_f = B_f$, and as B_f is a finitely generated A-module, Nakayama's lemma (Bourbaki, *Alg.*, Chap. VIII, §6, no. 3) would show that $\mathfrak{n} = B_f$, which is absurd. □

Suppose now that A is a *discrete valuation ring*; we give two special cases in which B_f is also a discrete valuation ring.

(i) *Unramified case*

Proposition 15. *If* A *is a discrete valuation ring, and if* $\bar{f}$ *is irreducible, then* B_f *is a discrete valuation ring with maximal ideal* $\mathfrak{m}B_f$ *and residue field* $k[X]/(\bar{f})$.

By lemma 4, B_f is local with maximal ideal $\mathfrak{m}B_f$ and residue field $k[X]/(\bar{f})$. Moreover, if π generates $\mathfrak{m}$, the image of π in B_f generates $\mathfrak{m}B_f$ and is not nilpotent. By prop. 2, B_f is a discrete valuation ring. □

Corollary 1. *If* K *is the field of fractions of* A, *the polynomial* f *is irreducible in* K[X]. *If* L *denotes the field* $K[X]/(f)$, *then the ring* B_f *is the integral closure of* A *in* L.

One has $K[X]/(f) = B_f \otimes_A K$. As B_f is an integral domain, so is $B_f \otimes_A K$, hence $K[X]/(f)$ is a field. As B_f is integrally closed and has L as its field of fractions, it is the integral closure of A in L. □

Corollary 2. *If* $\bar{f}$ *is a separable polynomial, the extension* L/K *is unramified.*

Obvious.

Proposition 15 admits the following converse:

Proposition 16. *Let* A *be a discrete valuation ring,* K *its field of fractions, and let* L *be an extension of* K *of finite degree* n. *Let* B *be the integral closure of* A *in* L. *Suppose that* B *is a discrete valuation ring and that the residue field* $\bar{L}$ *of* B *is a simple extension of degree* n *of the residue field* $k = \bar{K}$ *of* A. *Let* x *be any* *element of* B *whose image* $\bar{x}$ *in* $\bar{\mathrm{L}}$ *generates* $\bar{\mathrm{L}}$ *over* k, *and let* f *be the characteristic polynomial of* x *over* K. *Then the homomorphism of* A[X] *into* B *that maps* X *onto* x *defines by passage to the quotient an isomorphism of* B_f *onto* B.

The coefficients of f are integral over A and belong to K; as A is integrally closed, they belong to A. Furthermore, the equation $f(x) = 0$ shows that the map $\mathrm{A}[\mathrm{X}] \to \mathrm{B}$ factors into $\mathrm{A}[\mathrm{X}] \to \mathrm{B}_f \to \mathrm{B}$. Since $\bar{f}(\bar{x}) = 0$ and $\bar{x}$ is of degree n over k, one concludes that $\bar{f}$ is the minimal polynomial of $\bar{x}$ over k, hence is irreducible. The conditions of cor. 1 above thus hold, and the proposition follows from it. □

(ii) *Totally ramified case*

Proposition 17. *Suppose* A *is a discrete valuation ring and that* f *has the following form*:

$$f = \mathrm{X}^n + a_1\mathrm{X}^{n-1} + \cdots + a_n, \qquad a_i \in \mathfrak{m},\ a_n \notin \mathfrak{m}^2.$$

Then B_f *is a discrete valuation ring, with maximal ideal generated by the image* x *of* X *and with residue field* k.

[A polynomial having the form above is called an "Eisenstein polynomial."]

One has $\bar{f} = \mathrm{X}^n$. Lemma 4 then shows that B_f is local with maximal ideal generated by $(\mathfrak{m}, x)$. Furthermore, the element $\pi = a_n$ uniformizes A. Since:

$$-\pi = x^n + a_1x^{n-1} + \cdots + a_{n-1}x,$$

one sees that π belongs to the ideal (x), and it follows that $(\mathfrak{m}, x) = (x)$. As π is not nilpotent, neither is x, and prop. 2 shows that B_f is indeed a discrete valuation ring. □

As before, one deduces:

Corollary. *The polynomial* f *is irreducible in* K[X], *and if* $\mathrm{L} = \mathrm{K}[\mathrm{X}]/(f)$, *then* B_f *is the integral closure of* A *in* L.

Here again there is a converse:

Proposition 18. *Let* A *be a discrete valuation ring,* K *its field of fractions, and let* L *be a finite extension of* K *of degree* n. *Let* B *be the integral closure of* A *in* L. *Suppose that* B *is a discrete valuation ring, and that the associated valuation prolongs that of* A *with ramification index* n. *Let* x *be a uniformizing element of* B, *and let* f *be the characteristic polynomial of* x *over* K. *Then* f *is an Eisenstein polynomial, and the homomorphism of* A[X] *into* B *that maps* X *onto* x *defines by passage to the quotient an isomorphism of* B_f *onto* B.

One sees as in case (i) that the coefficients of f belong to A. Write f in the form:

$$f = a_0 X^n + \cdots + a_n, \qquad a_i \in A, a_0 = 1.$$

Since $f(x) = 0$, one has:

$$a_0 x^n + \cdots + a_n = 0.$$

Let w be the discrete valuation associated to B. One has $w(x) = 1$, and $w(a) \equiv 0 \bmod n$ for all $a \in A$. Let $r = \inf(w(a_i x^{n-i}))$, $0 \leq i \leq n$. By lemma 1 of §1, there are two integers i and j, with $0 \leq i < j \leq n$, such that

$$r = w(a_i x^{n-i}) = w(a_j x^{n-j}).$$

From this one deduces that $j - i = w(a_j/a_i) \equiv 0 \bmod. n$, which is only possible if $i = 0$ and $j = n$, so that $r = n$, $w(a_n) = n$ and $w(a_i) \geq n - i$ for all $i \geq 1$; thus f is an Eisenstein polynomial, and the proposition follows from the corollary to proposition 17. □

Exercise

With the notation of lemma 4, show that if $e_i = 1$, the local ring $(B_f)_{\mathfrak{m}_i}$ is a discrete valuation ring.

§7. Galois Extensions

We return now to the hypotheses and notation of paragraphs 4 and 5, and we further assume that L/K is a *Galois* extension. Its Galois group will be denoted G(L/K).

Proposition 19. *The group* G(L/K) *acts transitively on the set of prime ideals* $\mathfrak{P}$ *of* B *dividing a given prime ideal* $\mathfrak{p}$ *of* A.

Let $\mathfrak{P}|\mathfrak{p}$, and suppose there were a prime ideal $\mathfrak{P}'$ of B over $\mathfrak{p}$ distinct from all the $s(\mathfrak{P})$, $s \in G(L/K)$. By the approximation lemma, there exists $a \in \mathfrak{P}'$, $a \notin s(\mathfrak{P})$ for all s. If $x = N_{L/K}(a)$, one has $x \in A$, and $x = \prod s(a)$, whence $x \notin \mathfrak{P}$, $x \in \mathfrak{P}'$, which contradicts $\mathfrak{P} \cap A = \mathfrak{P}' \cap A$. □

Corollary. *Let* $\mathfrak{p}$ *be a non-zero prime ideal of* A. *The integers* $e_{\mathfrak{P}}$ *and* $f_{\mathfrak{P}}$ *(for* $\mathfrak{P}$ *dividing* $\mathfrak{p}$*) depend only on* $\mathfrak{p}$. *If one denotes them by* $e_{\mathfrak{p}}, f_{\mathfrak{p}}$, *and if* $g_{\mathfrak{p}}$ *denotes the number of prime ideals* $\mathfrak{P}$ *dividing* $\mathfrak{p}$, *then*

$$n = e_{\mathfrak{p}} f_{\mathfrak{p}} g_{\mathfrak{p}}.$$

This follows from proposition 10.

The subgroup of $G(L/K)$ consisting of those s such that $s(\mathfrak{P}) = \mathfrak{P}$ is called the *decomposition group* of $\mathfrak{P}$ in L/K; we denote it by $D_{\mathfrak{P}}(L/K)$, or sometimes simply by D. If $\mathfrak{P}'$ is another prime ideal of B over the same ideal $\mathfrak{p}$ of A, prop. 19 shows that $D_{\mathfrak{P}'}(L/K)$ is conjugate to $D_{\mathfrak{P}}(L/K)$. The index of D in $G(L/K)$ is equal to the number $g_{\mathfrak{p}}$ of prime ideals of B dividing $\mathfrak{p}$.

We now fix the ideal $\mathfrak{P}$, and we write G, D, e, f, g instead of $G(L/K)$, $D_{\mathfrak{P}}(L/K)$, $e_{\mathfrak{p}}$, $f_{\mathfrak{p}}$, $g_{\mathfrak{p}}$. By Galois theory, the group D corresponds to an extension K_D of K contained in L; this extension is only Galois when D is normal in G. We have:

$$[K_D:K] = g, \qquad [L:K_D] = ef, \qquad G(L/K_D) = D.$$

If E is a field between K and L, let $B_E = E \cap B$ be the integral closure of A in E, $\mathfrak{P}_E = \mathfrak{P} \cap B_E$, and let $\bar{E}$ be the residue field $B_E/\mathfrak{P}_E$. This applies in particular to K and L, defining the fields $\bar{K}$ and $\bar{L}$. If $s \in D$, s defines by passage to the quotient a $\bar{K}$-automorphism $\bar{s}$ of $\bar{L}$. We thus obtain a homomorphism

$$\varepsilon: D \to G(\bar{L}/\bar{K})$$

whose kernel is called the *inertia group* of $\mathfrak{P}$, and is denoted $T_{\mathfrak{P}}(L/K)$, or simply T. Corresponding to it is a Galois extension K_T/K_D, with Galois group D/T; one has $G(L/K_T) = T$.

Proposition 20. *The residue extension* $\bar{L}/\bar{K}$ *is normal and the homomorphism*

$$\varepsilon: D \to G(\bar{L}/\bar{K})$$

defines an isomorphism of D/T *onto* $G(\bar{L}/\bar{K})$.

We first show that $\bar{L}/\bar{K}$ is normal. Let $\bar{a} \in \bar{L}$, and let $a \in B$ represent $\bar{a}$. Let $P(X) = \prod(X - s(a))$, where s runs through G; this is a monic polynomial with coefficients in A, which has a as a root. The reduced polynomial $\bar{P}(X)$ has the $s(a)$ as its roots; that suffices to prove that $\bar{L}/\bar{K}$ is normal (cf. Bourbaki, *Alg.*, Chap V, §6, cor. 3 to prop. 9). Consider next the map ε. Choose $\bar{a}$ to be a generator of the largest separable extension $\bar{L}_s$ of $\bar{K}$ within $\bar{L}$; the approximation lemma of §3 shows that there exists a representative a of $\bar{a}$ which belongs to all the prime ideals $s(\mathfrak{P})$, $s \notin D$. We again form the polynomial $P(X) = \prod(X - s(a))$. The non-zero roots of $\bar{P}(X)$ all have the form $\overline{s(a)}$, with $s \in D$; it follows that every conjugate of $\bar{a}$ is equal to one of the $\overline{s(a)}$, with $s \in D$, which proves the surjectivity of ε. □

We continue to denote by $\bar{L}_s$ the largest separable extension of $\bar{K}$ in $\bar{L}$. We have just shown that it is a Galois extension of $\bar{K}$ with Galois group D/T. Put:

$$f_0 = [\bar{L}_s:\bar{K}] = [\bar{L}:\bar{K}]_s \qquad p^s = [\bar{L}:\bar{L}_s] = [\bar{L}:\bar{K}]_i,$$

so that

$$f = f_0 p^s.$$

Proposition 21. *With notation as above, let w, w_T, w_D, v be the discrete valuations defined by the ideals $\mathfrak{P}, \mathfrak{P}_T, \mathfrak{P}_D, \mathfrak{p}$. Then:*

a) $[L:K_T] = ep^s$, $[K_T:K_D] = f_0$, $[K_D:K] = g$.
b) w prolongs w_T with index e; w_T and w_D prolong v with index 1.
c) $\bar{K}_T = \bar{L}_s$, $\bar{K}_D = \bar{K}$. *In particular,* $[\bar{L}:\bar{K}_T] = p^s$, $[\bar{K}_T:\bar{K}_D] = f_0$, $[\bar{K}_D:\bar{K}] = 1$.

We know that the order of D is ef, and we've just seen that the order of D/T is f_0; the order of T is thus ep^s, which proves *a*).

On the other hand, we can apply prop. 20 to the group T: it tells us that $\bar{L}$ is purely inseparable over $\bar{K}_T$. In particular, every $x \in \bar{L}_s$ is purely inseparable over $\bar{K}_T$; as x is separable over $\bar{K}$ which is contained in $\bar{K}_T$, we must have $x \in \bar{K}_T$. Thus $\bar{K}_T$ contain $\bar{L}_s$, and we have $[\bar{L}:\bar{K}_T] \leq p^s$, i.e., $f(L/K_T) \leqslant p^s$; but it is clear that $e(L/K_T) \leq e$. As $[L:K_T] = ep^s$, we must have $\bar{L} = \bar{K}_T$ and $e(L/K_T) = e$, which proves *b*) and the first formula of *c*). The second one is a consequence of prop. 20, applied to the group D/T operating on B_T. □

Corollary. *If $\bar{L}/\bar{K}$ is separable then it is a Galois extension with Galois group* D/T, *and we have* $\bar{K}_T = \bar{L}$, $[L:K_T] = e$, $[K_T = K_D] = f$, $[K_D:K] = g$.

Indeed, $p^s = 1$ in that case.

Remark. The residue extension $\bar{L}/\bar{K}$ is separable in each of the following cases (which cover most of the applications):

1) $\bar{K}$ is perfect.
2) The order of the inertia group T is prime to the characteristic p of the residue field $\bar{K}$ (indeed, we have seen that the order of this group is divisible by p^s).

With the same hypotheses as in prop. 21, let E be a subfield of L containing K; the groups D(L/E) and T(L/E) are well-defined; similarly, when E/K is Galois, the groups D(E/K) and T(E/K) are well-defined.

Proposition 22

a) $D(L/E) = D(L/K) \cap G(L/E)$ *and* $T(L/E) = T(L/K) \cap G(L/E)$.
b) If E/K is Galois, the diagram below is commutative, and its rows and columns are exact:

$$\begin{array}{ccccccccc} & & 1 & & 1 & & 1 & & \\ & & \downarrow & & \downarrow & & \downarrow & & \\ 1 & \to & T(L/E) & \to & T(L/K) & \to & T(E/K) & \to & 1 \\ & & \downarrow & & \downarrow & & \downarrow & & \\ 1 & \to & D(L/E) & \to & D(L/K) & \to & D(E/K) & \to & 1 \\ & & \downarrow & & \downarrow & & \downarrow & & \\ 1 & \to & G(\bar{L}/\bar{E}) & \to & G(\bar{L}/\bar{K}) & \to & G(\bar{E}/\bar{K}) & \to & 1 \\ & & \downarrow & & \downarrow & & \downarrow & & \\ & & 1 & & 1 & & 1 & & \end{array}$$

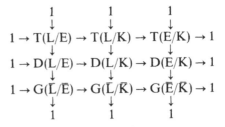

Assertion *a*) is immediate, as well as the commutativity of diagram *b*). Exactness of the columns follows from prop. 20, and exactness of the third row from Galois theory applied to the residue fields $\bar{L}$, $\bar{E}$, $\bar{K}$. If $s \in D(E/K)$, there exists $t \in G(L/K)$ which induces s on E; the ideals $\mathfrak{P}$ and $t(\mathfrak{P})$ have the same restriction to E; by prop. 19, there exists $t' \in G(L/E)$ such that $t't(\mathfrak{P}) = \mathfrak{P}$; the element $t't$ belongs to D(L/K) and induces s on L, which shows that $D(L/K) \to D(E/K)$ is surjective. The second row of the diagram is thus exact, and a little diagram-chasing shows that consequently the first row is also. □

Remark. When one wants to study the decomposition or inertia groups above a given prime ideal $\mathfrak{p}$ of A, one may, if one wishes, replace A by the discrete valuation ring $A_{\mathfrak{p}}$; this reduction to the local case can be pushed further: one may even replace $A_{\mathfrak{p}}$ by its *completion* (cf. Chap. II).

§8. Frobenius Substitution

Let L/K be a Galois extension, A a Dedekind domain with field of fractions K, and let B be the integral closure of A in L. Let $\mathfrak{P}$ be a prime ideal of B, $\mathfrak{P} \neq 0$, and let $\mathfrak{p} = \mathfrak{P} \cap A$. Assume that L/K is *unramified* at $\mathfrak{P}$ and that $A/\mathfrak{p}$ is a *finite field* with q elements. The inertia group $T_{\mathfrak{P}}(L/K)$ is then reduced to $\{1\}$, and the decomposition group $D_{\mathfrak{P}}(L/K)$ can be identified with the Galois group of the residue extension $\bar{L}/\bar{K}$. Since $\bar{K} = \mathbf{F}_q$, the latter group is cyclic and generated by the map $x \to x^q$ (cf. Bourbaki, *Alg*., Chap. V). Let $s_{\mathfrak{P}}$ be the element of $D_{\mathfrak{P}}(L/K)$ corresponding to this generator; it is characterised by the following property:

$$s_{\mathfrak{P}}(b) \equiv b^q \ \text{mod.}\ \mathfrak{P} \quad \text{for all } b \in B.$$

The element $s_{\mathfrak{P}}$ is called the *Frobenius substitution* of $\mathfrak{P}$ (or attached to $\mathfrak{P}$). Its definition shows that it generates the decomposition group of $\mathfrak{P}$; its order is equal to $f_{\mathfrak{P}}$. It is often denoted $(\mathfrak{P}, L/K)$. Here are two samples of functorial properties that it enjoys (a third will be seen in Chap. VII, §8):

Proposition 23. *Let* E *be a subfield of* L *containing* K, *and let* $\mathfrak{P}_E = \mathfrak{P} \cap E$. *Then*:

a) $(\mathfrak{P}, L/E) = (\mathfrak{P}, L/K)^f$, *with* $f = [\bar{E} : \bar{K}]$.
b) *If* E *is Galois over* K, *the image of* $(\mathfrak{P}, L/K)$ *in* G(E/K) *is* $(\mathfrak{P}_E, E/K)$.

Immediate. □

Returning to the extension L/K, if $t \in G(L/K)$, one has (by transport of structure) the formula:

$$(t(\mathfrak{P}), L/K) = t(\mathfrak{P}, L/K)t^{-1}.$$

In particular, if G(L/K) is *abelian*, $(\mathfrak{P}, L/K)$ depends only on $\mathfrak{p} = \mathfrak{P} \cap A$; it is the *Artin symbol* of $\mathfrak{p}$ and is denoted $(\mathfrak{p}, L/K)$. One defines by linearity the Artin symbol for any ideal $\mathfrak{a}$ of A that does not contain a ramified prime, and one denotes it again by $(\mathfrak{a}, L/K)$ [the notations

$$\left(\frac{L/K}{\mathfrak{a}}\right), \quad \text{or simply} \quad \left(\frac{L}{\mathfrak{a}}\right),$$

are also found in the literature].

We state without proof:

Artin Reciprocity Law (cf. [3], [75], [94], [123]). *Let* L *be a finite abelian extension of a number field* K, A *the ring of integers of* K, *and* $\mathfrak{p}_i$ *the prime ideals of* A *that ramify in* L/K. *Then there exist positive integers* n_i *such that the conditions*

(*i*) $v_{\mathfrak{p}_i}(x - 1) \geq n_i$ *for all i*,
(*ii*) *x is positive in every real embedding of* K *that is not induced by a real embedding of* L,

imply $(xA, L/K) = 1$.

Furthermore, every automorphism $s \in G(L/K)$ *is of the form* $(\mathfrak{a}, L/K)$ *for a suitable ideal* $\mathfrak{a}$ (in fact, one even has $s = (\mathfrak{p}, L/K)$ for infinitely many prime ideals $\mathfrak{p}$ of A).

Example. Let n be a positive integer, $K = \mathbf{Q}$, and let $L = \mathbf{Q}(\zeta_n)$ be the field of nth roots of unity. The Galois group G(L/K) is a subgroup $G'(n)$ of the group $G(n)$ of invertible elements of $\mathbf{Z}/n\mathbf{Z}$ (cf. Bourbaki, *Alg.*, Chap. V); if $x \in G'(n)$, the automorphism σ_x associated to x transforms a root of unity ζ_n into its xth power. If $(p, n) = 1$, one sees easily (e.g., by using the results of Chap. IV, §4) that p is unramified, and that the Artin symbol $(p, L/K)$ is equal to σ_p. It follows by linearity that *the Artin symbol of a positive integer m prime to n is equal to* σ_m. Consequently, $G'(n) = G(n)$, that is to say

$$[L:K] = \varphi(n)$$

(irreducibility of the cyclotomic polynomial). Moreover, if $m > 0$, and if $m \equiv 1 \bmod n$, one gets $(m, L/K) = 1$, which verifies the Artin reciprocity law for this case. [The fact that $s = (p, L/K)$ for infinitely many primes p is equivalent to Dirichlet's theorem on the infinity of prime numbers belonging to an arithmetic progression.]

Once the Artin symbol has been determined in $\mathbf{Q}(\zeta_n)/\mathbf{Q}$, prop. 23 gives it for every subfield E of $\mathbf{Q}(\zeta_n)$. Such a field is abelian over $\mathbf{Q}$. Conversely, every finite abelian extension of $\mathbf{Q}$ can be obtained in this way (theorem of Kronecker-Weber). In particular, every quadratic field $\mathbf{Q}(\sqrt{d})$ can be embedded in a suitable field $\mathbf{Q}(\zeta_n)$; this result can also be checked by various elementary methods (Gauss sums, for example). Thus one has a procedure for determining the Artin symbol $(p, \mathbf{Q}(\sqrt{d})/\mathbf{Q})$; by comparing the result with that given by a direct computation, one obtains the *quadratic reciprocity law*. For more details, see Hasse [34], §27, or Weyl [68], Chap. III, §11.

CHAPTER II

Completion

§1. Absolute Values and the Topology Defined by a Discrete Valuation

Let K be a field on which a discrete valuation v is defined, having valuation ring A. If a is any real number between 0 and 1, we put

$$\|x\| = a^{v(x)} \quad \text{for } x \neq 0,$$
$$\|0\| = 0.$$

We then have the formulas

$$\|x \,.\, y\| = \|x\| \cdot \|y\|$$
$$\|x + y\| \leq \sup(\|x\|, \|y\|)$$
$$\|x\| = 0 \quad \text{if and only if } x = 0.$$

Thus we see that $\|x\|$ is an *absolute value* on K (in the sense of Bourbaki, *Top. gen.*, Chap. IX, §3); it is in fact an *ultrametric* absolute value. Conversely, it is easy to show that every ultrametric absolute value of a field K has the form $a^{v(x)}$, where v is a *real* valuation of K, i.e., a valuation whose ordered group of values is an additive subgroup of **R**. As for the non-ultrametric absolute values, it can be shown (Ostrowski's theorem) that they have the form:

$$\|x\| = |f(x)|^c, \quad \text{with } 0 < c \leq 1,$$

where $f: \mathrm{K} \to \mathbf{C}$ is an isomorphism of K onto a subfield of the field of complex numbers.

Returning now to the case where v is discrete, let $\hat{\mathrm{K}}$ be the *completion* of K for the topology defined by its absolute value (the topology does not depend on the choice of the number a). It is known (Bourbaki, *loc. cit.*) that $\hat{K}$ is a valued field whose absolute value extends that of K. If one writes the absolute value in the form

$$\|x\| = a^{\hat{v}(x)}, \qquad x \in \hat{K},$$

the function $\hat{v}(x)$ is integer-valued, and one checks immediately that it is a discrete valuation on $\hat{K}$, whose valuation ring is the closure $\hat{A}$ of A in $\hat{K}$. If π is a uniformizing element of A, the ideals $\pi^n A$ form a base for the neighborhoods of zero in K, hence also in A, which shows that the topology on A coincides with its natural topology as a *local ring*; thus one has

$$\hat{A} = \varprojlim A/\pi^n A \quad \text{(projective limit)}$$

The element π is a uniformizer for $\hat{A}$, and one has $\hat{A}/\pi^n\hat{A} = A/\pi^n A$. In particular, the residue fields of A and $\hat{A}$ coincide.

Proposition 1. *In order that* K *be locally compact, it is necessary and sufficient that its residue field* $\bar{K} = A/\pi A$ *be a finite field and* K *be complete.*

If K is locally compact, it is complete. And as the $\pi^n A$ form a fundamental system of closed neighborhoods of 0, at least one of them is compact, so multiplying by π^{-n} shows that A is compact. The quotient $\bar{K} = A/\pi A$, being both compact and discrete, must be finite.

Conversely, if $\bar{K}$ is finite, the $A/\pi^n A$ are finite; hence $\hat{A}$, being the projective limit of finite rings, is compact; if in addition K is complete, one has $A = \hat{A}$, so that K is indeed locally compact. □

EXAMPLES. 1) The field $\mathbf{Q}_p$, completion of $\mathbf{Q}$ for the topology defined by the p-adic valuation, is a locally compact field with residue field $\mathbf{F}_p$.

2) If F is a finite field, the field F((T)) of formal power series is locally compact.

When K satisfies the conditions of prop. 1, there is a canonical way to choose the number a: one takes $a = q^{-1}$, where q is the number of elements in the residue field $\bar{K}$. The corresponding absolute value is said to be *normalised.* The next proposition characterises it "analytically":

Proposition 2. *Let* K *be a field satisfying the conditions of prop.* 1, *and let* μ *be a Haar measure on the locally compact additive group underlying* K. *Then for every measurable subset* E *of* K *and every* $x \in K$ *one has*

$$\mu(xE) = \|x\|\mu(E),$$

where $\|x\|$ *denotes the normalised absolute value of* x.

One may assume $x \neq 0$; the homothety $y \mapsto xy$ is then an automorphism of the additive group of K, hence transforms the Haar measure μ into one of its multiples $\chi(x).\mu$, and one must verify that the multiplier $\chi(x)$ is equal to $\|x\|$. Since $\chi(x)$ and $\|x\|$ are multiplicative, one can assume $x \in A$. Taking $E = A$, one sees that E is the union of $(A:xA)$ cosets module xE, whence $\mu(E) = (A:xA).\mu(xE)$, and $\chi(x) = 1/(A:xA)$. Since $(A:xA)$ is equal to $q^{v(x)}$, one gets

$$\chi(x) = q^{-v(x)} = \|x\|. \quad \square$$

Remark. One can carry out the same normalisation for a locally compact valued field K whose absolute value is not ultrametric; by Ostrowski's theorem (cited above), one has $K = \mathbf{R}$ or $K = \mathbf{C}$; in the first case one recovers the usual absolute value, whereas in the second case one gets its square (which is not an absolute value in the strict sense, because it doesn't satisfy the triangle inequality). These normalisations are necessary for the *product formula*: let K be a number field and P the set of its normalised absolute values (ultrametric or not); then

$$\prod_{\mathfrak{p} \in P} \|x\|_{\mathfrak{p}} = 1 \quad \text{for all } x \in K^*$$

(this infinite product is meaningful, for almost all its terms are equal to 1). To prove this formula, one checks it first for $K = \mathbf{Q}$ by a direct computation; then one uses the following result (equivalent, in the ultrametric case, to the formula $\sum e_i f_i = n$):

$$\|N_{K/Q}(x)\|_p = \prod_{\mathfrak{p} | p} \|x\|_{\mathfrak{p}}, \qquad x \in K^*.$$

An analogous formula is valid for algebraic function fields in one variable.

§2. Extensions of a Complete Field

Proposition 3. *Let* K *be a field on which a discrete valuation* v *is defined, having valuation ring* A. *Assume* K *to be complete in the topology defined by* v. *Let* L/K *be a finite extension of* K, *and let* B *be the integral closure of* A *in* L (cf. Chap. I, §4). *Then* B *is a discrete valuation ring and is a free* A*-module of rank* $n = [L:K]$; *also,* L *is complete in the topology defined by* B.

We begin with the case L/K *separable.* Condition (F) of Chap. I, §4 is then automatically satisfied; as A is principal, it follows that B is a free A-module of rank n. Let $\mathfrak{P}_i$ be the prime ideals of B, with w_i the corresponding valuations. Each w_i defines (as in the preceding §) a *norm* on L, which makes L a Hausdorff *topological vector space* over K; as K is complete, it follows (cf. Bourbaki, *Esp. Vect. Top.*, Chap. I, §2, th. 2) that the topology $\mathcal{T}_i$ defined by w_i is actually the product topology on L (identified with K^n), hence does not depend on the index i. But w_i is determined by $\mathcal{T}_i$: the ring of w_i is the set of those x such that x^{-n} does not converge to zero for $\mathcal{T}_i$. Thus there is only one w_i, which shows that B is a discrete valuation ring. As K is complete, so is K^n, hence L. Once this case has been treated, a straightforward "dèvissage" argument reduces one to the case where L/K is *purely inseparable*. In that case, there is a power q of the exponent characteristic such that $x^q \in K$ for all $x \in L$. Put $v'(x) = v(x^q)$; the map $v': L^* \to \mathbf{Z}$ is a homomorphism. If m denotes the positive generator of the subgroup $v'(L^*)$, the function $w = (1/m)v$ is a discrete valuation of L. It is immediate that its valuation ring is B; the same argument as above shows that the topology defined by w coincides with that of K^n, making L into a *complete* field. It remains to prove that B is an A-module of finite type. Let π be a uniformizer of A, and let $\bar{B} = B/\pi B$. Let b_i be elements of B whose images $\bar{b}_i$ in $\bar{B}$ are linearly independent over $\bar{K} = A/\pi A$. We claim that the b_i are linearly independent over A: for if one had a relation $\sum a_i b_i = 0$ that was non-trivial, one could assume that at least one of the a_i was not divisible by π, and reducing mod. πB, one would obtain a non-trivial relation among the $\bar{b}_i$. In particular, the number of b_i is $\leq n$. Suppose now that the $\bar{b}_i$ form a *basis* of $\bar{B}$ and let E be the sub-A-module of B spanned by the b_i. Every $b \in B$ can then be written in the form $b = b_0 + \pi b_1$, with $b_0 \in E$ and $b_1 \in B$; applying this to b_1 and iterating this procedure, one gets b into the form:

$$b = b_0 + \pi b_1 + \pi^2 b_2 + \cdots, \qquad b_i \in E,$$

and since A is complete, this shows that $b \in E$. □

Corollary 1. *If* e (resp. f) *denotes the ramification index* (resp. *the residue degree*) *of* L *over* K, *then* $ef = n$.

That follows from prop. 10 of Chapter I, which is applicable because we have shown that B is an A-module of finite type. □

Corollary 2. *There is a unique valuation* w *of* L *that prolongs* v.

This is just a reformulation of part of the proposition. □

Corollary 3. *Two elements of* L *that are conjugate over* K *have the same valuation.*

Enlarging L if necessary, we can assume L/K to be normal. If $s \in G(L/K)$, $w \circ s$ prolongs v, hence coincides with w (cor. 2); the corollary then results from the fact that the conjugates of $x \in L$ are none other than the $s(x)$, $s \in G(L/K)$. □

Corollary 4. *For every* $x \in L$, $w(x) = (1/f)v(N_{L/K}(x))$.

Here again one reduces to the case L/K normal, where the assertion results from cor. 3. [One could just as well directly apply prop. 14 of Chap. I.]

In terms of *absolute values*, cor. 4 means that the topology of L can be defined by the norm

$$\|x\|_L = \|N_{L/K}(x)\|_K.$$

Note that if K is locally compact, and if $\| \ \|_K$ is normalised, so is $\| \ \|_L$.

Remark. It is possible to take the formula above as the *definition* of $\| \ \|_L$; one must then prove directly that it is an ultrametric absolute value, which can be done by means of "Hensel's lemma" (cf. van der Waerden [65], §77); one could also make use of the existence of at least one valuation prolonging v, which is a general fact (cf. Bourbaki, *Alg. comm.*, Chap. VI). These methods have the advantage of applying to arbitrary "valuations of rank 1", not necessarily discrete.

Exercises

1. (Krasner's lemma) Let E/K be a finite Galois extension of a complete field K. Prolong the valuation of K to E. Let $x \in E$ and let $\{x_1, \ldots, x_n\}$ be the set of conjugates of x over K, with $x = x_1$. Let $y \in E$ be such that $\|y - x\| < \|y - x_i\|$ for $i \geq 2$. Show that x belongs to the field K(y). (Note that if x_i is conjugate to x over K(y), then $\|y - x\| = \|y - x_i\|$, according to cor. 3.)

2. Let K be a complete field, and let $f(X) \in K[X]$ be a separable irreducible polynomial of degree n. Let L/K be the extension of degree n defined by f. Show that for every polynomial $h(X)$ of degree n that is close enough to f, $h(X)$ is irreducible and the extension L_h/K defined by h is isomorphic to L. (Apply exer. 1 to the roots x_i of f and to a root y of h.)

3. With the hypotheses of prop. 3, show directly that B is an A-module of finite type by using exer. 8 of Bourbaki, *Alg.*, Chap. VII, §3.

4. Let K be a field complete under a discrete valuation v, and let Ω be an algebraic closure of K.
   a) Let S be the set of subextensions E of Ω with the property that for every finite subextension E′ of E, $e(E'/K) = 1$. Show that S has maximal elements. If K_0 is maximal, show that v prolongs to a discrete valuation of K_0, and that the residue field of K_0 is the algebraic closure of that of K (use prop. 15 of Chap. I).
   b) Let L/K be a totally ramified extension within Ω, and let K_0/K be a maximal extension as in a). Show that L and K_0 are linearly disjoint over K. If L/K is Galois with group G, deduce that the extension L_0/K_0, where $L_0 = K_0L$, is Galois with group G.

§3. Extension and Completion

Theorem 1. *Let* L/K *be an extension of finite degree n, v a discrete valuation of* K *with ring* A, *and* B *the integral closure of* A *in* L. *Suppose that the* A-*module* B *is finitely generated. Let* w_i *be the different prolongations of v to* L, *and let* e_i, f_i *be the corresponding numbers* (cf. Chap. I, §4). *Let* $\hat{\mathrm{K}}$ *and* $\hat{\mathrm{L}}_i$ *be the completions of* K *and* L *for* v *and the* w_i.

(i) *The field* $\hat{\mathrm{L}}_i$ *is an extension of* K *of degree* $n_i = e_i f_i$.
(ii) *The valuation* $\hat{w}_i$ *is the unique valuation of* $\hat{\mathrm{L}}_i$ *prolonging* $\hat{v}$, *and*

$$e_i = e(\hat{\mathrm{L}}_i/\hat{\mathrm{K}}) \quad \textit{and} \quad f_i = f(\hat{\mathrm{L}}_i/\hat{\mathrm{K}}).$$

(iii) *The canonical homomorphism* $\varphi : \mathrm{L} \otimes_{\mathrm{K}} \hat{\mathrm{K}} \to \prod_i \hat{\mathrm{L}}_i$ *is an isomorphism.*

Statement (ii) is evident, taking §2 into account, and it implies statement (i). On the other hand, the product topology makes $\prod \hat{\mathrm{L}}_i$ into a Hausdorff topological vector space of dimension n over $\hat{\mathrm{K}}$; by the approximation lemma (Chap. I, §3), $\varphi(\mathrm{L})$ is dense in $\prod \hat{\mathrm{L}}_i$, hence also $\varphi(\mathrm{L} \otimes_{\mathrm{K}} \hat{\mathrm{K}})$. It follows (cf. Bourbaki, *Esp. Vect. Top.*, Chap. I, §2, cor. 1 to th. 2) that φ is surjective, hence bijective, since $\mathrm{L} \otimes_{\mathrm{K}} \hat{\mathrm{K}}$ and $\prod \hat{\mathrm{L}}_i$ are both n-dimensional vector spaces over $\hat{\mathrm{K}}$. □

Corollary 1. *The fields* $\hat{\mathrm{L}}_i$ *are the composites of the extensions* $\hat{\mathrm{K}}$ *and* L *of* K.

One knows that those composites are the quotient fields of the tensor product $\mathrm{L} \otimes_{\mathrm{K}} \hat{\mathrm{K}}$ (cf. Bourbaki, *Alg.*, Chap. VIII, §8).

Corollary 2. *If* $x \in \mathrm{L}$, *the characteristic polynomial* F *of* x *in* L/K *is equal to the product of the characteristic polynomials* F_i *of* x *in the* $\hat{\mathrm{L}}_i/\hat{\mathrm{K}}$. *In particular, if* Tr *and* N (resp. Tr_i *and* N_i) *denote the trace and norm in* L/K (resp. *in* $\hat{\mathrm{L}}_i/\hat{\mathrm{K}}$), *then*

$$\mathrm{Tr}(x) = \sum \mathrm{Tr}_i(x), \qquad \mathrm{N}(x) = \prod \mathrm{N}_i(x).$$

The polynomial F is also the characteristic polynomial of x in the $\hat{\mathrm{K}}$-algebra $\mathrm{L} \otimes_{\mathrm{K}} \hat{\mathrm{K}}$. The formula $\mathrm{F} = \prod \mathrm{F}_i$ follows from the isomorphism (iii), and the trace and norm formulas are an immediate consequence (cf. Bourbaki, *Alg.*, Chap. VIII, §12, no. 2).

Corollary 3. *If* L/K *is separable* (in which case the finiteness hypothesis made on B is automatically satisfied), *the* $\hat{\mathrm{L}}_i/\hat{\mathrm{K}}$ *are also.*

For we have $\hat{\mathrm{L}}_i = \mathrm{L}\hat{\mathrm{K}}$.

Corollary 4. *If* L/K *is Galois with group* G, *and if* D_i *denotes the decomposition group of* w_i *in* G (cf. Chap. I, §7), *the extension* $\hat{\mathrm{L}}_i/\hat{\mathrm{K}}$ *is Galois with Galois group* D_i.

Every element of D_i extends by continuity to a $\hat{\mathrm{K}}$-automorphism of $\hat{\mathrm{L}}_i$, and the corollary results from the fact that D_i has order $[\hat{\mathrm{L}}_i : \hat{\mathrm{K}}]$.

(The isomorphism $\varphi : \mathrm{L} \otimes_{\mathrm{K}} \hat{\mathrm{K}} \to \prod \hat{\mathrm{L}}_i$ merely expresses the decomposition of $\mathrm{L} \otimes_{\mathrm{K}} \hat{\mathrm{K}}$ considered as a "Galois algebra" in the sense of Hasse, in this case.)

Let us now go on to the valuation rings themselves:

Proposition 4. *With the hypotheses and notation of theorem* 1, *let* B_i *be the ring of the valuation* w_i. *The canonical homomorphism*

$$\varphi: B \otimes_A \hat{A} \to \prod_i \hat{B}_i$$

is then an isomorphism.

Both sides are free $\hat{A}$-modules of rank n. To show that φ is bijective, it suffices therefore to see that it is when one reduces modulo the maximal ideal $\hat{\mathfrak{m}}$ of $\hat{A}$. One gets $B/\mathfrak{m}B$ for the left side, and $\prod B/\mathfrak{m}_i^{e_i}B$ for the right ($\mathfrak{m}$ and $\mathfrak{m}_i$ denoting the ideals of v and the w_i), whence the result follows at once. □

Remark. The ring $B \otimes_A \hat{A}$ is none other than the *completion* $\hat{B}$ of B for the natural topology on the semi-local ring B. Its decomposition into direct factors $\hat{B}_i$ is a special case of a general property of semi-local rings (cf. Bourbaki, *Alg. comm.*, Chap. III, §2, no. 12).

Exercises

1. Let K be a field on which is defined a discrete valuation v having ring A. Suppose that every finite purely inseparable extension L/K satisfies the finiteness condition (F) of Chap. I, §4, relative to A. Show that $\hat{K}$ is then a separable extension of K. (Use th. 1 of Bourbaki, *Alg.*, Chap. VIII, §7).

2. Keeping the hypotheses and notation of theorem 1, except that the hypothesis "B is of finite type over A" is replaced by its negation, show that (i) and (ii) remain valid, that φ is surjective, and that its kernel is a non-zero nilpotent ideal of $L \otimes_K \hat{K}$.

§4. Structure of Complete Discrete Valuation Rings I: Equal Characteristic Case

Let A be a complete discrete valuation ring, with field of fractions K and residue field $\bar{K}$. Let S be a system of representatives of $\bar{K}$ in A, π a uniformizer of A.

Proposition 5. *Every element* $a \in A$ *can be written uniquely as a convergent series*

$$(*) \qquad a = \sum_{n=0}^{\infty} s_n \pi^n, \quad \textit{with } s_n \in S.$$

Similarly, every element $x \in K$ can be written as

$$x = \sum_{n \gg -\infty} s_n \pi^n, \quad \text{with } s_n \in S,$$

the series requiring only finitely many terms with negative exponents.

The second assertion results from the first by multiplying by a suitable negative power of π. Thus let $a \in A$; by definition of S, there is an $s_0 \in S$ such that $a - s_0 \equiv 0 \bmod. \pi$; if one writes $a = s_0 + \pi a_1$ and applies the same procedure to a_1, one obtains an $s_1 \in S$ such that

$$a = s_0 + s_1 \pi + a_2 \pi^2,$$

and so on. The series $\sum s_n \pi^n$ converges to a and one sees easily that it is unique. Conversely, every series of the form $\sum s_n \pi^n$ is convergent, since its general term converges to zero and A is complete. □

EXAMPLE. If $A = \mathbf{Z}_p$, one may take S to be the set of non-negative integers less than p; one may also—and this is preferable—take S to consist of 0 and the $(p-1)$st roots of unity; cf. prop. 8.

Prop. 5 shows that addition and multiplication in A are determined by the decomposition of $s + s'$ and ss' into the form $(*)$. In particular, *if* S *is a subfield of* K (necessarily isomorphic to $\bar{K}$), the ring A may be identified with the ring $\bar{K}[[T]]$ of formal series with coefficients in $\bar{K}$. Evidently this is only possible if K and $\bar{K}$ have the same characteristic. Conversely:

Theorem 2. *Let* A *be a complete discrete valuation ring with residue field* $\bar{K}$. *Suppose that* A *and* $\bar{K}$ *have the same characteristic and that* $\bar{K}$ *is perfect. Then* A *is isomorphic to* $\bar{K}[[T]]$.

[In fact, this result remains valid even when $\bar{K}$ is imperfect; see the end of this §.]

It all comes down to showing that A contains a system of representatives which is a field. We will distinguish two cases, depending on the characteristic:

(i) *The characteristic of* $\bar{K}$ is 0.

The existence of a field of representatives is then true for local rings that are far more general than discrete valuation rings. More precisely:

Proposition 6. *Let* A *be a local ring that is Hausdorff and complete for the topology defined by a decreasing sequence* $\mathfrak{a}_1 \supset \mathfrak{a}_2 \supset \cdots$ *of ideals such that* $\mathfrak{a}_n . \mathfrak{a}_m \subset \mathfrak{a}_{n+m}$. *Suppose that* $\bar{K} = A/\mathfrak{a}_1$ *is a field of characteristic zero. Then* A *contains a system of representatives of* $\bar{K}$ *which is a field.*

[Note that the first hypothesis on A is satisfied if A is a Noetherian local ring, complete in its natural topology as a local ring.]

As $\mathbf{Z} \to A \to \bar{K}$ is injective, the homomorphism $\mathbf{Z} \to A$ extends to $\mathbf{Q}$, and we see that A contains $\mathbf{Q}$. By Zorn's lemma, there exists a maximal subfield S of A; if $\bar{S}$ denotes its image in $\bar{K}$, we will show that $\bar{S} = \bar{K}$.

We first show that $\bar{K}$ is *algebraic* over $\bar{S}$; if not, there would exist an $a \in A$ whose image $\bar{a}$ in $\bar{K}$ is transcendental over $\bar{S}$; the subring $S[a]$ of A maps into $\bar{S}[\bar{a}]$, hence is isomorphic to $S[X]$, and $S[a] \cap \mathfrak{a}_1 = 0$; one concludes that A contains the field S(a) of rational functions in a, contradicting the maximality of S.

Thus any $\lambda \in \bar{K}$ has a minimal polynomial $\bar{f}(X)$ over $\bar{S}$; since the characteristic is 0, λ is a simple root of $\bar{f}$. Let $f \in S[X]$ be the polynomial corresponding to $\bar{f}$ under the isomorphism $\bar{S} \to S$. By prop. 7 below, there is an $x \in A$ such that $\bar{x} = \lambda$ and $f(x) = 0$, and one can lift $\bar{S}[\lambda]$ into A by sending λ to x; by the maximality of S, we must have $\lambda \in \bar{S}$, which shows that $\bar{K} = \bar{S}$. □

It remains to prove the following proposition, which is a special case of "Hensel's lemma" (Bourbaki, *Alg. comm.*, Chap III):

Proposition 7. *Let* A *be a local ring that is Hausdorff and complete for the topology defined by a decreasing sequence* $\mathfrak{a}_1 \supset \mathfrak{a}_2 \supset \cdots$ *of ideals such that* $\mathfrak{a}_n \cdot \mathfrak{a}_m \subset \mathfrak{a}_{n+m}$. *Suppose that* $\mathfrak{a}_1$ *is the maximal ideal of* A, *and let* $\bar{K} = A/\mathfrak{a}_1$. *Let* $f(X)$ *be a polynomial with coefficients in* A *such that the reduced polynomial* $\bar{f} \in \bar{K}[X]$ *has a simple root* λ *in* $\bar{K}$. *Then* f *has a unique root* x *in* A *such that* $\bar{x} = \lambda$.

If x is such a root, one has $f(X) = (X - x)g(X)$, with $\bar{g}(\lambda) \neq 0$; if x' is also such a root, substituting x' for X yields $0 = (x' - x)g(x')$. As $g(x')$ has $\bar{g}(\lambda)$ as its reduction mod. $\mathfrak{a}_1$, $g(x')$ is invertible, hence $x = x'$, which proves uniqueness of the solution.

To prove existence, we use Newton's approximation method. Let $x_1 \in A$ be such that $\bar{x}_1 = \lambda$; one has $f(x_1) \equiv 0 \bmod. \mathfrak{a}_1$.

Suppose we have found $x_n \in A$ such that $\bar{x}_n = \lambda$, $f(x_n) \equiv 0 \bmod. \mathfrak{a}_n$; let us show that one can find $x_{n+1} \in A$, $x_{n+1} \equiv x_n \bmod. \mathfrak{a}_n$ and $f(x_{n+1}) \equiv 0 \bmod. \mathfrak{a}_{n+1}$. That will prove the lemma by setting $x = \lim x_n$. To find x_{n+1}, write $x_{n+1} = x_n + h$, with $h \in \mathfrak{a}_n$, and apply Taylor's formula:

$$f(x_{n+1}) = f(x_n) + h \,.\, f'(x_n) + h^2 \,.\, y, \quad \text{with } y \in A.$$

One has $h^2 \,.\, y \in \mathfrak{a}_n \,.\, \mathfrak{a}_n \subset \mathfrak{a}_{n+1}$, and it all comes down to finding $h \in \mathfrak{a}_n$ such that

$$f(x_n) + h \,.\, f'(x_n) \equiv 0 \bmod. \mathfrak{a}_{n+1}.$$

But since λ is a simple root of $\bar{f}$, one has $\bar{f}'(\lambda) \neq 0$, and $f'(x_n)$ is *invertible* in A; as $f(x_n) \in \mathfrak{a}_n$, the equation above can be solved. □

Theorem 2 is therefore proved in characteristic zero.

(ii) *The fields* K *and* $\bar{K}$ *have characteristic* $p \neq 0$.

Here again, we are going to obtain a result valid for much more general rings.

We will say that a ring A of characteristic p is *perfect* if the endomorphism $x \to x^p$ of A is an *automorphism* (i.e., is surjective). Every element $x \in A$ then has a unique pth root, denoted $x^{p^{-1}}$. When A is a field, this is the usual definition of a perfect field.

Proposition 8. *Let* A *be a ring that is Hausdorff and complete for the topology defined by a decreasing sequence* $\mathfrak{a}_1 \supset \mathfrak{a}_2 \supset \cdots$ *of ideals such that* $\mathfrak{a}_n . \mathfrak{a}_m \subset \mathfrak{a}_{n+m}$. *Assume that the residue ring* $\bar{\mathrm{K}} = \mathrm{A}/\mathfrak{a}_1$ *is a perfect ring of characteristic* p. *Then*:

(i) *There exists one and only one system of representatives* $f : \bar{\mathrm{K}} \to \mathrm{A}$ *which commutes with* pth *powers*: $f(\lambda^p) = f(\lambda)^p$.
(ii) *In order that* $a \in \mathrm{A}$ *belong to* $\mathrm{S} = f(\bar{\mathrm{K}})$, *it is necessary and sufficient that* a *be a* p^nth *power for all* $n \geq 0$.
(iii) *This system of representatives is multiplicative, i.e., one has* $f(\lambda\mu) = f(\lambda) . f(\mu)$ *for all* $\lambda, \mu \in \bar{\mathrm{K}}$.
(iv) *If* A *has characteristic* p, *this system of representatives is additive, i.e.,* $f(\lambda + \mu) = f(\lambda) + f(\mu)$.

Let $\lambda \in \bar{\mathrm{K}}$; for all $n \geq 0$, denote by L_n the inverse image of $\lambda^{p^{-n}}$ in A, and by U_n the set of all x^{p^n}, $x \in \mathrm{L}_n$; the U_n are contained in the residue class L_0 of λ, and they form a decreasing sequence. We will show that they form a Cauchy filter base in A. Indeed, if $a = x^{p^n}$ and $b = y^{p^n}$, one shows by induction on n that $a \equiv b \bmod. \mathfrak{a}_{n+1}$, making use of the following lemma:

Lemma 1. *If* $a \equiv b \bmod. \mathfrak{a}_n$, *then* $a^p \equiv b^p \bmod. \mathfrak{a}_{n+1}$.

This lemma results from the binomial formula, taking into account that $p \in a_1$, whence $p\mathfrak{a}_n \subset \mathfrak{a}_{n+1}$. □

Since the U_n form a Cauchy filter base and A is complete, one can set $f(\lambda) = \lim \mathrm{U}_n$. This defines a system of representatives. If $\lambda = \mu^p$, the pth power operation in A maps $\mathrm{U}_n(\mu)$ into $\mathrm{U}_{n+1}(\lambda)$, so passing to the limit shows that it maps $f(\mu)$ on $f(\lambda)$, and f does commute with the pth power. Conversely, if f' is a system of representatives having this property, $f'(\lambda)$ is a p^nth power for all n, hence $f'(\lambda) \in \mathrm{U}_n(\lambda)$ for all n; as the U_n form a Cauchy filter base, this implies the uniqueness of f' as well as the fact that the intersection of the U_n is non-empty and equal to $f(\lambda)$. This establishes (i) and (ii).

As for (iii), note that if x and y are p^nth powers for all n, so is xy; the same reasoning holds for (iv), taking into account that $(x + y)^{p^n} = x^{p^n} + y^{p^n}$ if A has characteristic p. □

The system of representatives of prop. 8 is called *the multiplicative system of representatives*, because of property (iii).

The application of prop. 8 to theorem 2 is immediate: if $\bar{K}$ is a perfect field, and if A has characteristic p, properties (iii) and (iv) show that $S = f(\bar{K})$ is a field. One sees also that it is *unique*. [When $\bar{K}$ is imperfect, one can still show that there exists a field S of representatives, but this field is no longer unique in general: one can lift arbitrarily the elements of a "p-base" of $\bar{K}$. For more details on these questions and those treated in the following §, see Cohen [18] and Roquette [52].]

EXERCISE

Let k be a perfect field of characteristic p. Show that every finite purely inseparable extension of $k((T))$ is isomorphic to an extension of the form $k((T^{q^{-1}}))$, where q is a power of p.

§5. Structure of Complete Discrete Valuation Rings II: Unequal Characteristic Case

Let A be a complete discrete valuation ring, with field of fractions K and residue field $\bar{K}$. Suppose that the characteristics of A and $\bar{K}$ are different, i.e., that A has characteristic zero and $\bar{K}$ has characteristic $p \neq 0$. One can then identify $\mathbf{Z}$ with a subring of A, and $p \in \mathbf{Z}$ with an element of A. Since p goes to zero in $\bar{K}$, one has $v(p) \geq 1$, where v is the discrete valuation attached to A. The integer $e = v(p)$ is called *the absolute ramification index* of A. Observe that the injection $\mathbf{Z} \to A$ extends by continuity to an injection of the ring $\mathbf{Z}_p$ of p-adic integers into A; when the residue field $\bar{K}$ is a finite field with $q = p^f$ elements, prop. 5 shows that A is a free $\mathbf{Z}_p$-module of rank $n = ef$, and K is an extension of degree n of the p-adic field $\mathbf{Q}_p$; the integer e can then be interpreted as the ramification index of the extension $K/\mathbf{Q}_p$, which justifies the terminology.

Returning to the general case, we will say that A is *absolutely unramified* if $e = 1$, i.e., if p is a local uniformizer of A. It is for such rings that one has a structure theorem:

Theorem 3. *For every perfect field k of characteristic p, there exists a complete discrete valuation ring and only one (up to unique isomorphism) which is absolutely unramified and has k as its residue field.*

In what follows, this ring will be denoted W(k). It is "unique" in the following sense: if A_1 and A_2 satisfy the conditions of the theorem, there is a *unique* isomorphism $g: A_1 \to A_2$ which makes commutative the diagram:

$$\begin{array}{ccc} A_1 & \xrightarrow{g} & A_2 \\ & \searrow \quad \swarrow & \\ & k & \end{array}$$

In the ramified case, one has:

Theorem 4. *Let* A *be a complete discrete valuation ring of characteristic unequal to that of its residue field k. Let e be its absolute ramification index. Then there exists a unique homomorphism of* W(k) *into* A *which makes commutative the diagram*:

$$\begin{array}{ccc} W(k) & \rightarrow & A \\ & \searrow \; k \; \swarrow & \end{array}$$

This homomorphism is injective, and A *is a free* W(k)*-module of rank equal to e.*

[By applying prop. 18 of Chap. I, one sees that A is obtained by adjoining to W(k) an element π satisfying an "Eisenstein equation":

$$\pi^e + a_1\pi^{e-1} + \cdots + a_e = 0, \qquad a_i \in W(k),$$

with the a_i being divisible by p and a_e not being divisible by p^2. Conversely, according to prop. 17 of Chap. I, such an equation does define a totally ramified extension of W(k) of degree e.]

We are going to prove ths. 3 and 4 by a method due to Lazard ([42], [43]). Here again we will obtain results valid for rings more general than discrete valuation rings: the rings provided with a filtration $\mathfrak{a}_1 \supset \mathfrak{a}_2 \supset \cdots$ satisfying the hypotheses of prop. 8; such a ring will be called a *p-ring*. We will call a p-ring A *strict* (Lazard says "p-adic", but this terminology could lead to confusion) if the filtration $\mathfrak{a}_n$ provided is its p-adic filtration $\mathfrak{a}_n = p^n A$ and if p is not a zero-divisor in A. A p-ring always has a system of multiplicative representatives $f: A/\mathfrak{a}_1 \rightarrow A$ (cf. prop. 8), and for every sequence $\alpha_0, \ldots, \alpha_n, \ldots,$ of elements of $A/\mathfrak{a}_1$, the series

$$(**) \qquad \sum_{i=0}^{\infty} f(\alpha_i) \cdot p^i$$

converges to an element $a \in A$. When A is strict, one sees by arguing as in prop. 5, that every element $a \in A$ can be uniquely expressed in the form of a series of type (**); the α_i which occur in this series will be called the *coordinates* of a.

EXAMPLE OF A STRICT p-RING. Let X_α be a family of indeterminates, and let S be the ring of $p^{-\infty}$-polynomials in the X_α with integer coefficients, i.e., the union of the rings $\mathbf{Z}[X_\alpha^{p^{-n}}]$ for all n. If one provides S with the p-adic filtration $\{p^n S\}_{n \geq 0}$ and completes it, one obtains a strict p-ring that will be denoted $\hat{S} = \hat{\mathbf{Z}}[X_\alpha^{p^{-\infty}}]$. The residue ring $\hat{S}/p\hat{S}$ is the ring $\mathbf{F}_p[X_\alpha^{p^{-\infty}}]$; it is perfect of characteristic p. Note that the X_α are multiplicative representatives in $\hat{S}$ since they admit p^nth roots for all n.

Let us apply this to the case in which the indeterminates are $X_0, \ldots, X_n, \ldots,$ and $Y_0, \ldots, Y_n, \ldots$; in the ring $\hat{\mathbf{Z}}[X_i^{p^{-\infty}}, Y_i^{p^{-\infty}}]$ thus obtained, consider the two elements

$$x = \sum_{i=0}^{\infty} X_i p^i \quad \text{and} \quad y = \sum_{i=0}^{\infty} Y_i p^i.$$

If * denotes one of the operations $+, \times, -$, the composite $x * y$ is an element of the ring, therefore can be written in a unique way in the form:

$$x * y = \sum_{i=0}^{\infty} f(Q_i^*) . p^i, \quad \text{with } Q_i^* \in \mathbf{F}_p[X_i^{p^{-\infty}}, Y_i^{p^{-\infty}}].$$

The Q_i^* are $p^{-\infty}$-polynomials with coefficients in the prime field $\mathbf{F}_p$; one can speak of the *value* of such a polynomial when elements of a perfect ring k of characteristic p are substituted for the indeterminates. We will see that these functions allow us to determine the structure of a strict p-ring. More precisely:

Proposition 9. *Let* A *be a p-ring with residue ring k and let* $f: k \to A$ *be the system of multiplicative representatives in* A. *Let* $\{\alpha_i\}$ *and* $\{\beta_i\}$ *be two sequences of elements of k. Then*

$$\sum_{i=0}^{\infty} f(\alpha_i) . p^i * \sum_{i=0}^{\infty} f(\beta_i) . p^i = \sum_{i=0}^{\infty} f(\gamma_i) \ p^i$$

with $\gamma_i = Q_i^*(\alpha_0, \alpha_1, \ldots ; \beta_0, \beta_1, \ldots)$.

One sees immediately that there is a homomorphism θ of $\mathbf{Z}[X_i^{p^{-\infty}}, Y_i^{p^{-\infty}}]$ into A which maps X_i to $f(\alpha_i)$ and Y_i to $f(\beta_i)$. This homomorphism extends by continuity to the completion $\hat{S} = \hat{\mathbf{Z}}[X_i^{p^{-\infty}}, Y_i^{p^{-\infty}}]$, and maps $x = \sum_{i=0}^{\infty} X_i p^i$ onto $a = \sum_{i=0}^{\infty} f(\alpha_i) . p^i$, and similarly for y. If one passes to the residue rings, θ defines a homomorphism $\bar{\theta}: \mathbf{F}_p[X_i^{p^{-\infty}}, Y_i^{p^{-\infty}}] \to k$ which maps the X_i onto the α_i and the Y_j onto the β_j. Also, θ commutes with the multiplicative representatives (that is a general property of p-rings, which results from the characterization of the multiplicative representatives as p^nth powers for all n and from the fact that θ is a homomorphism). Then:

$$\begin{aligned}\sum f(\alpha_i) . p^i * \sum f(\beta_i) . p^i &= \theta(x) * \theta(y) = \theta(x * y) \\ &= \sum \theta(f(Q_i^*)) . p^i \\ &= \sum f(\bar{\theta}(Q_i^*)) . p^i\end{aligned}$$

which proves the proposition, since $\bar{\theta}(Q_i^*)$ is none other than γ_i. □

Proposition 10. *Let* A *and* A′ *be two p-rings with residue rings k and k′, and suppose that* A *is strict. For every homomorphism* $\phi: k \to k'$, *there exists a unique homomorphism* $g: A \to A'$ *making commutative the diagram*:

$$\begin{array}{ccc} A & \xrightarrow{g} & A' \\ \downarrow & & \downarrow \\ k & \xrightarrow{\varphi} & k'. \end{array}$$

We have already remarked that every homomorphism of A into A′ commutes with multiplicative representatives. If $a \in A$ is an element with coordinates $\{\alpha_i\}$, we must have:

$$g(a) = \sum_{i=0}^{\infty} g(f_A(\alpha_i)) \cdot p^i = \sum_{i=0}^{\infty} f_{A'}(\varphi(\alpha_i)) \cdot p^i,$$

which proves the uniqueness of g. As for its existence, we take the preceding formula as its definition, and prop. 9 shows that it is indeed a ring homomorphism. □

Corollary. *Two strict p-rings having the same residue ring are canonically isomorphic.*

Lemma 2. *Let $\varphi: k \to k'$ be a surjective homomorphism, the rings k and k' being perfect of characteristic p. If there exists a strict p-ring A with residue ring k, then there also exists a strict p-ring A′ with residue ring k′.*

We define A′ as a quotient of A. If a and b are two elements of A with coordinates α_i, β_i in k, we write $a \equiv b$ if $\varphi(\alpha_i) = \varphi(\beta_i)$ for all i. If $a \equiv a'$ and $b \equiv b'$, prop. 9 shows that $a * b \equiv a' * b'$, and the quotient A′ of A by the equivalence relation just defined is a ring. If $x \in A'$ comes from an element $a \in A$ with coordinates α_i, the $\xi_i = \varphi(\alpha_i)$ depend only on x and are the *coordinates* of x; every sequence $(\xi_0, \xi_1, \ldots)$ of elements of k' forms the coordinates of a uniquely determined element $x' \in A'$. Multiplication by p in A′ transforms the element with coordinates $(\xi_0, \xi_1, \ldots)$ into the one with coordinates $(0, \xi_0, \xi_1, \ldots)$. It follows that p is not a zero-divisor in A′ and that $\bigcap p^n A' = 0$; the p-adic topology of A′ is thus Hausdorff; as A′ is a quotient of A, A′ is complete. Finally, one checks that the map which sends x' to its first coordinate ξ_0 induces an isomorphism of A'/pA' onto k'; this proves that A′ is a strict p-ring with residue ring k'. □

Theorem 5. *For every perfect ring k of characteristic p, there exists a unique strict p-ring* W(*k*) *with residue ring k.*

Uniqueness has already been proved. As for existence: if k has the form $\mathbf{F}_p[X_\alpha^{p^{-\infty}}]$, for an arbitrary family of indeterminates X_α, one takes $W(k) = \hat{\mathbf{Z}}[X_\alpha^{p^{-\infty}}]$ as above. The general case can be deduced from that by applying lemma 2 and remarking that every perfect ring is a quotient of a ring of type $\mathbf{F}_p[X_\alpha^{p^{-\infty}}]$. □

Proposition 10 shows that W(k) is a *functor* of k. More precisely, one has an isomorphism $\mathrm{Hom}(k, k') = \mathrm{Hom}(W(k), W(k'))$.

Proof of Theorems 3 and 4. Theorem 3 is a special case of theorem 5, once one remarks that a complete discrete valuation ring, absolutely unramified, with perfect residue field k, is nothing other than a strict p-ring with residue ring k. As for theorem 4, the existence and uniqueness of the homomorphism $g: W(k) \to A$ results from prop. 10 by remarking that A is a p-ring. The homomorphism g is injective because A has characteristic zero; finally, if π is a uniformizer of A, an argument similar to the one in the proof of prop. 5 shows that every element of A can be put uniquely into the form:

$$a = \sum_{i=0}^{\infty} \sum_{j=0}^{j=e-1} f(\alpha_{ij}) . \pi^j p^i, \qquad \alpha_{ij} \in k,$$

whence the fact that $\{1, \pi, \ldots, \pi^{e-1}\}$ is a basis of A considered as a $W(k)$-module (this also follows from prop. 18 of Chap. I). □

Remark. The functions Q_i^* that define the operations of $W(k)$ involve the p^nth roots of the X_n and Y_n. If one wishes to have polynomials in the usual sense, it is necessary to re-define the coordinates α_i of a by:

$$a = \sum_{i=0}^{\infty} f(\alpha_i)^{p^{-i}} p^i.$$

One is then led to introduce the "Witt vectors" which we study in the next §.

§6. Witt Vectors

Let p be a prime number, $(X_0, \ldots, X_n, \ldots)$ a sequence of indeterminates, and consider the following polynomials (called "Witt polynomials"):

$$W_0 = X_0,$$
$$W_1 = X_0^p + pX_1,$$
$$\vdots$$
$$W_n = \sum_{i=0}^{i=n} p^i X_i^{p^{n-i}} = X_0^{p^n} + pX_1^{p^{n-1}} + \cdots + p^n X_n.$$
$$\vdots$$

If $\mathbf{Z}'$ denotes the ring $\mathbf{Z}[p^{-1}]$, it is clear that the X_i can be expressed as polynomials with respect to the W_i with coefficients in $\mathbf{Z}'$:

$$X_0 = W_0, \qquad X_1 = p^{-1}W_1 - W_0^p, \ldots, \text{etc.}$$

Let $(Y_0, \ldots, Y_n, \ldots)$ be another sequence of indeterminates.

Theorem 6. *For every $\Phi \in \mathbf{Z}[X, Y]$, there exists a unique sequence $(\varphi_0, \ldots, \varphi_n, \ldots,)$ of elements of $\mathbf{Z}[X_0, \ldots, X_n, \ldots; Y_0, \ldots, Y_n, \ldots]$ such that:*

$$W_n(\varphi_0, \ldots, \varphi_n, \ldots) = \Phi(W_n(X_0, \ldots), W_n(Y_0, \ldots)), \qquad n = 0, 1, \ldots.$$

The existence and uniqueness of the φ_i are obvious *in the ring of polynomials with coefficients in* $\mathbf{Z}'$; thus it all comes down to showing that the coefficients of the φ_i have no denominators, i.e., are elements of $\mathbf{Z}$. This can be proved directly (cf. Witt [73]). But it is more convenient to follow Lazard (*loc. cit.*) in deducing it from the results of the preceding §:

We work once more in the ring $\hat{S} = \hat{\mathbf{Z}}[X_i^{p^{-\infty}}, Y_i^{p^{-\infty}}]$, and set:

$$x' = \sum_{i=0}^{\infty} X_i^{p^{-i}} p^i$$

$$y' = \sum_{i=0}^{\infty} Y_i^{p^{-i}} p^i.$$

Since $\Phi(x', y')$ is an element of $\hat{S}$, one can write it uniquely in the form:

$$\Phi(x', y') = \sum_{i=0}^{\infty} f(\bar{\psi}_i)^{p^{-i}} p^i, \qquad \bar{\psi}_i \in \mathbf{F}_p[X_i^{p^{-\infty}}, Y_i^{p^{-\infty}}].$$

Denote by ψ_i a representative of $\bar{\psi}_i$ in the ring $\hat{S}$. We will prove that *the* φ_i *have integer coefficients and that they are congruent* mod. p *to the* ψ_i.

First of all, the following congruence is obvious:

$$\Phi\left(\sum_{i \leq n} X_i^{p^{-i}} p^i, \sum_{i \leq n} Y_i^{p^{-i}} p^i\right) \equiv \sum_{i \leq n} f(\bar{\psi}_i(X, Y))^{p^{-i}} p^i \bmod. p^{n+1}.$$

Replacing X_i and Y_i by $X_i^{p^n}$ and $Y_i^{p^n}$ (which defines an automorphism of $\hat{S}$), we get:

$$\Phi(W_n(X), (W_n(Y)) \equiv \sum_{i \leq n} f(\bar{\psi}_i(X^{p^n}, Y^{p^n}))^{p^{-i}} p^i \bmod. p^{n+1}.$$

But one has $\bar{\psi}_i(X^{p^n}, Y^{p^n}) = \bar{\psi}_i(X, Y)^{p^n}$ since the coefficients of $\bar{\psi}_i$ belong to the field $\mathbf{F}_p$. As f commutes with the pth power, one sees that the above congruence can be written:

$$W_n(\varphi_0, \ldots, \varphi_n) \equiv \sum_{i \leq n} f(\bar{\psi}_i)^{p^{n-i}} p^i \bmod. p^{n+1}.$$

But $f(\bar{\psi}_i) \equiv \psi_i \bmod. p$, whence $f(\bar{\psi}_i)^{p^{n-i}} \equiv \psi_i^{p^{n-i}} \bmod. p^{n-i+1}$ (cf. §3, lemma 1). Thus:

$$W_n(\varphi_0, \ldots, \varphi_n) \equiv W_n(\psi_0, \ldots, \psi_n) \bmod. p^{n+1}.$$

Reasoning by induction on n, one may assume that φ_i has integer coefficients for $i < n$, and congruent mod. p to ψ_i (or, changing ψ_i if need be, one may even assume that $\varphi_i = \psi_i$ for $i < n$). The preceding congruence then yields

$$p^n \varphi_n \equiv p^n \psi_n \bmod. p^{n+1}.$$

Thus φ_n has integer coefficients and $\varphi_n \equiv \psi_n \bmod. p$. □

We now denote by $S_0, \ldots, S_n, \ldots,$ (resp. $P_0, \ldots, P_n, \ldots$) the polynomials $\varphi_0, \ldots, \varphi_n, \ldots$ associated by th. 6 with the polynomial

$$\Phi(X, Y) = X + Y \text{ (resp. } \Phi(X, Y) = X . Y).$$

If A is an arbitrary commutative ring, and if $\mathfrak{a} = (a_0, \ldots, a_n, \ldots)$, $\mathfrak{b} = (b_0, \ldots, b_n, \ldots)$ are elements of $A^{\mathbf{N}}$ ("Witt vectors with coefficients in A"), set:

$$\mathfrak{a} + \mathfrak{b} = (S_0(a, b), \ldots, S_n(a, b), \ldots)$$
$$\mathfrak{a} . \mathfrak{b} = (P_0(a, b), \ldots, P_n(a, b), \ldots).$$

Theorem 7. *The laws of composition defined above make* $A^{\mathbf{N}}$ *into a commutative unitary ring (called the ring of Witt vectors with coefficients in* A *and denoted* W(A)).

Note first that if one assigns to a Witt vector $\mathfrak{a} = (a_0, \ldots, a_n, \ldots)$ the element of the product ring $A^{\mathbf{N}}$ having the $W_n(\mathfrak{a})$ as coordinates, one gets a homomorphism

$$W_* : W(A) \to A^{\mathbf{N}},$$

by the very definition of the polynomials S and P.

The homomorphism W_* is an isomorphism if p is invertible in A, and in this case one sees that W(A) is indeed a commutative ring with unit element $1 = (1, 0, \ldots, 0, \ldots)$. But if the theorem is proved for one ring A, it is also valid for every subring and every quotient ring. As it is true for every polynomial ring $\mathbf{Z}'[T_\alpha]$, it holds for $\mathbf{Z}[T_\alpha]$, hence for all rings. □

EXAMPLES. We have

$$S_0(a, b) = a_0 + b_0, \qquad S_1(a, b) = a_1 + b_1 + \frac{a_0^p + b_0^p - (a_0 + b_0)^p}{p},$$

$$P_0(a, b) = a_0 . b_0, \qquad P_1(a, b) = b_0^p a_1 + b_1 a_0^p + p a_1 b_1.$$

Instead of considering vectors of infinite length, we can restrict ourselves to the consideration of vectors $(a_0, \ldots, a_{n-1})$ with n components. As the polynomials φ_i only involve variables of index $\leq i$, one concludes that these vectors form a ring $W_n(A)$, quotient of W(A), that one calls *the ring of Witt vectors of length n*. One has $W_1(A) = A$. The ring W(A) is the *projective limit* of the rings $W_n(A)$ as $n \to +\infty$.

The Maps V and r

If $\mathfrak{a} = (a_0, \ldots, a_n, \ldots)$ is a Witt vector, one defines the vector $V\mathfrak{a}$ by:

$$V\mathfrak{a} = (0, a_0, \ldots, a_{n-1}, \ldots) \quad (\text{"shift"}).$$

The map $V : W(A) \to W(A)$ is *additive*: for it suffices to verify this when p is invertible in A, and in that case the homomorphism

$$W_* : W(A) \to A^{\mathbf{N}}$$

transforms V into the map which sends $(w_0, w_1, \ldots)$ to $(0, pw_0, \ldots)$.

By passage to the quotient, one deduces from V an additive map of $W_n(A)$ into $W_{n+1}(A)$. There are exact sequences

$$0 \to W_k(A) \xrightarrow{V^r} W_{k+r}(A) \to W_r(A) \to 0.$$

If $x \in A$, set

$$r(x) = (x, 0, \ldots, 0, \ldots).$$

This defines a map $r: A \to W(A)$. When p is invertible in A, W_* transforms r into the mapping that sends x to $(x, x^p, \ldots, x^{p^n}, \ldots)$. One deduces by the same reasoning as above the formulas:

$$r(xy) = r(x).r(y), \qquad x, y \in A$$

$$(a_0, a_1, \ldots) = \sum_{n=0}^{\infty} V^n(r(a_n)), \qquad a_i \in A$$

$$r(x).(a_0, \ldots) = (xa_0, x^p a_1, \ldots, x^{p^n} a_n, \ldots), \qquad x, a_i \in A.$$

Theorem 8. *If k is a perfect ring of characteristic p, $W(k)$ is a strict p-ring with residue ring k.*

Let H be the strict p-ring with residue ring k, and let $f: k \to H$ be the multiplicative system of representatives of H. Associate to a Witt vector $\mathfrak{a} = (a_0, \ldots, a_n, \ldots)$ the element $\theta(\mathfrak{a}) \in H$ defined by

$$\theta(\mathfrak{a}) = \sum_{i=0}^{\infty} f(a_i)^{p^{-i}} p^i.$$

The formulas

$$\theta(\mathfrak{a} + \mathfrak{b}) = \theta(\mathfrak{a}) + \theta(\mathfrak{b}), \qquad \theta(\mathfrak{a}.\mathfrak{b}) = \theta(\mathfrak{a}).\theta(\mathfrak{b})$$

are valid when $H = \hat{S}$, $\mathfrak{a} = (X_0, \ldots)$, $\mathfrak{b} = (Y_0, \ldots)$, as was seen in the course of the proof of th. 6. It follows easily that they are valid without any restriction on $\mathfrak{a}$ and $\mathfrak{b}$, i.e., that θ is a ring homomorphism. As θ is bijective, it is an *isomorphism* of W(A) onto H. □

Corollary. $W(\mathbf{F}_p) = \mathbf{Z}_p$ *and* $W_n(\mathbf{F}_p) = \mathbf{Z}/p^n\mathbf{Z}$.

Indeed, the ring $\mathbf{Z}_p$ of p-adic integers is a strict p-ring with residue ring the field $\mathbf{F}_p$. □

The Map F

Suppose that k is a ring of characteristic p (not necessarily perfect). The map $x \quad x^p$ is a homomorphism of k into k. Therefore it defines a map $F: W(k) \to W(k)$ given by the formula

$$F(a_0, a_1, \ldots) = (a_0^p, a_1^p, \ldots),$$

and this is a ring homomorphism.

Furthermore, one has the *identity* $\mathrm{VF} = p = \mathrm{FV}$: for it suffices to check this when k is perfect; in that case, applying the isomorphism θ above, one finds:

$$\theta(\mathrm{FV}\mathfrak{a}) = \sum_{i=0}^{\infty} f(a_i)^{p^{-i}} p^{i+1} = p\theta(\mathfrak{a}) = \theta(p\mathfrak{a}),$$

which gives the identity.

Remark. In Grothendieck's language of *schemes*, the preceding constructions define, for each n, a *ring scheme* W_n, affine and of finite type over Spec($\mathbf{Z}$). For any ring A, the ring $\mathrm{W}_n(\mathrm{A})$ is just the set of *points of* W_n *with values in* A.

PART TWO

RAMIFICATION

CHAPTER III

Discriminant and Different

Throughout this chapter, A denotes a Dedekind domain, K its field of fractions.

§1. Lattices

Let V be a finite dimensional vector space over K. A *lattice* of V (with respect to A) is a sub-A-module X of V that is finitely generated and spans V. If A is principal, this means that X is a free A-module of rank $[V:K]$; one can often reduce to this case by localisation, i.e., by replacing A with $A_{\mathfrak{p}}$ and X with $A_{\mathfrak{p}}X = X_{\mathfrak{p}}$.

Let X_1 and X_2 be two lattices of V; if $X_2 \subset X_1$, then X_1/X_2 is a module of finite length, and its invariant $\chi(X_1/X_2)$, which is a non-zero ideal of A, was defined in Chap. I, §5. We wish to extend this definition to get an invariant for any pair of lattices:

Lemma 1. *If* X_1 *and* X_2 *are lattices of* V, *then the fractional ideal* $\chi(X_1/X_3).\chi(X_2/X_3)^{-1}$, *defined for every lattice* $X_3 \subset X_1 \cap X_2$, *depends only on* X_1 *and* X_2.

Indeed, if one sets $X_4 = X_1 \cap X_2$, the exact sequence

$$0 \to X_4/X_3 \to X_1/X_3 \to X_1/X_4 \to 0$$

shows that $\chi(X_1/X_3) = \chi(X_4/X_3).\chi(X_1/X_4)$, and similarly

$$\chi(X_2/X_3) = \chi(X_4/X_3).\chi(X_2/X_4).$$

Hence

$$\chi(X_1/X_3).\chi(X_2/X_3)^{-1} = \chi(X_1/X_4).\chi(X_2/X_4)^{-1}. \quad \square$$

We may therefore associate to X_1 and X_2 the non-zero fractional ideal

$$\chi(X_1, X_2) = \chi(X_1/X_3).\chi(X_2/X_3)^{-1} \quad \text{for } X_3 \subset X_1 \cap X_2.$$

Proposition 1. *The following formulas are valid:*

(a) $\chi(X_1, X_2).\chi(X_2, X_3).\chi(X_3, X_1) = 1$.
(b) $\chi(X_1, X_2).\chi(X_2, X_1) = 1$.
(c) $\chi(X_1, X_2) = \chi(X_1/X_2)$ if $X_1 \supset X_2$.

(We denote by 1 the unit element of the group of non-zero fractional ideals of A, i.e., the ideal A.)

Formula (a) is proved by choosing a lattice X contained in $X_1 \cap X_2 \cap X_3$, and by writing $\chi(X_i, X_j)$ in the form $\chi(X_i/X).\chi(X_j/X)^{-1}$. Formulas (b) and (c) are trivial. $\square$

Proposition 2. *If u is a* K*-automorphism of* V *and* X *a lattice of* V, *then* $\chi(X, uX) = (\det(u))$ (principal ideal generated by $\det(u)$).

(The symbol uX denotes the image of X under u.)

By localising and multiplying u by a constant, we are reduced to the case where $X = A^n$ and $uX \subset X$; the proposition follows in that case from lemma 3 of Chap. I, §5. $\square$

This result suggests the following direct definition of the ideal $\chi(X, X')$:

Let $n = [V:K]$, and let $W = \bigwedge^n V$; it is a one dimensional vector space over K. To each lattice X of V let us associate $X_W = \bigwedge^n X$, which may be identified with a lattice of W; as $[W:K] = 1$, if D and D′ are two lattices of W, there is a unique non-zero fractional ideal $\mathfrak{a}$ of K such that $D' = \mathfrak{a}D$ (namely, $\chi(D, D')$). Applying this to $D = X_W$, $D' = X'_W$, we obtain an ideal *which is none other than* $\chi(X, X')$: this follows from localisation and applying proposition 2.

§2. Discriminant of a Lattice with Respect to a Bilinear Form

We now suppose that the vector space V is provided with a *non-degenerate bilinear form* $T(x, y)$.

Let $n = [V:K]$. It is known that T extends to a non-degenerate bilinear form (again denoted by T) on the exterior algebra of V, and, in particular, on $W = \bigwedge^n V$. This form induces an isomorphism

$$T: W \otimes_K W \to K.$$

Let X be a lattice of V, and let X_W be its nth exterior power, identified with a lattice of W. The image of $X_W \otimes_A X_W$ under T is a *non-zero fractional ideal* of K, which is called the *discriminant of* X *with respect to* T; we denote it by $\mathfrak{d}_{X,T}$ or simply $\mathfrak{d}_X$ when that does not lead to confusion.

Remark. The above definition shows that $\mathfrak{d}_X$ is isomorphic as an A-module to $X_W \otimes_A X_W$; its ideal class (modulo the principal ideals) is thus a *square*.

Proposition 3. *If* X *is a free* A-*module with basis* $S = \{e_1, \ldots, e_n\}$, *then* $\mathfrak{d}_{X,T}$ *is the principal ideal generated by the discriminant* $D_T(S)$ (in the sense of Bourbaki, *Alg.*, Chap. IX, §2).

[Recall that $D_T(S) = \det(T(e_i, e_j))$.]

Indeed, it is known that in this case X_W is a free A-module with basis $\{e\}$, where $e = e_1 \wedge \cdots \wedge e_n$, and that $T(e, e) = D_T(S)$, cf. Bourbaki, *loc. cit.* The image of $X_W \otimes_A X_W$ in K is thus generated by $D_T(S)$. □

Remark. We could have taken the formula $\mathfrak{d}_{X,T} = (\det(T(e_i, e_j)))$ as the *definition* of $\mathfrak{d}_{X,T}$, at least in the local case.

Proposition 4. *Let* X *be a lattice of* V, *and let* X_T^* *be the set of all* $y \in V$ *such that* $T(x, y) \in A$ *for all* $x \in V$. *Then* X_T^* *is a lattice of* V *and*

$$\mathfrak{d}_{X,T} = \chi(X_T^*, X).$$

Localising reduces us to the case where X is free with basis $\{e_i\}$; then X_T^* is free with the basis $\{e_i^*\}$ defined by the relations

$$T(e_i, e_j^*) = \delta_{ij} \quad \text{(cf. Bourbaki, } \textit{loc. cit.}\text{, p. 22).}$$

If one writes $e_i = \sum x_{ij} e_j^*$, prop. 2 shows that $\chi(X_T^*, X) = (\det(x_{ij}))$. As $T(e_i, e_j) = x_{ji}$, the desired formula is obtained. □

The next proposition shows how $\mathfrak{d}_{X,T}$ varies with X.

Proposition 5. *If* X *and* X′ *are lattices of* V, *then*

$$\mathfrak{d}_{X',T} = \mathfrak{d}_{X,T} \cdot \chi(X, X')^2.$$

Let $\mathfrak{a} = \chi(X, X')$. We saw in §1 that $X'_W = \mathfrak{a} . X_W$ in W; the image of $X'_W \otimes X'_W$ under the isomorphism $T: W \otimes W \to K$ is therefore equal to the product of $\mathfrak{a}^2$ by the image of $X_W \otimes X_W$. □

Corollary. *If* $X' \subset X$, *then* $\mathfrak{d}_{X',T} = \mathfrak{d}_{X,T}\mathfrak{a}^2$, *where* $\mathfrak{a}$ *is an ideal of* A; $\mathfrak{a} = 1$ *if and only if* $X' = X$.

Take $\mathfrak{a} = \chi(X/X')$; it is clear that $\mathfrak{a} = 1$ if and only if $X' = X$.

§3. Discriminant and Different of a Separable Extension

Let L be a finite separable extension of the field K. It is known that the homomorphism

$$\mathrm{Tr}: \mathrm{L} \to \mathrm{K}$$

is surjective and that the bilinear form $\mathrm{Tr}(xy)$ is non-degenerate on L. Thus the definitions and results of the preceding § are applicable to this form; in particular, the *discriminant* of a lattice of L (with respect to A) is defined; if this lattice is a free A-module with basis $\{e_i\}$, its discriminant is the ideal generated by $\det(\mathrm{Tr}(e_i e_j))$, and it is known (Bourbaki, *Alg.*, Chap. V, §10, prop. 12) that

$$\det(\mathrm{Tr}(e_i e_j)) = (\det(\sigma(e_i)))^2,$$

where σ runs through the set of K-monomorphisms of L into an algebraic closure of K.

In particular, this applies to the integral closure B of A in L, for prop. 8 of Chapter I shows that B is a lattice of L. The corresponding discriminant will be denoted $\mathfrak{d}_{\mathrm{B/A}}$, or sometimes $\mathfrak{d}_{\mathrm{L/K}}$ (when no confusion about A is possible).

Let B^* be the set of all $y \in \mathrm{L}$ such that $\mathrm{Tr}(xy) \in \mathrm{A}$ for all $x \in \mathrm{B}$; B^* is the lattice denoted $\mathrm{B}^*_{\mathrm{T}}$ in the preceding §. It is called the *codifferent* (or "inverse different") of B over A. It is a sub-B-module of L; one sees at once that it is the largest sub-B-module E of L such that $\mathrm{Tr}(\mathrm{E}) \subset \mathrm{A}$. In particular, as $\mathrm{Tr}(\mathrm{B}) \subset \mathrm{A}$, one has $\mathrm{B} \subset \mathrm{B}^*$. The codifferent is thus a fractional ideal of L with respect to B; its inverse is called the *different* of B over A (or of the extension L/K), and is denoted $\mathfrak{D}_{\mathrm{B/A}}$ or $\mathfrak{D}_{\mathrm{L/K}}$; it is a non-zero ideal of B. The different is related to the discriminant by the next proposition.

Proposition 6. $\mathfrak{d}_{\mathrm{B/A}} = \chi_{\mathrm{A}}(\mathrm{B}^*/\mathrm{B}) = \mathrm{N}_{\mathrm{L/K}}(\mathfrak{D}_{\mathrm{B/A}})$.

The equality $\mathfrak{d}_{\mathrm{B/A}} = \chi_{\mathrm{A}}(\mathrm{B}^*/\mathrm{B})$ follows from prop. 4. On the other hand, $\chi_{\mathrm{B}}(\mathrm{B}^*/\mathrm{B}) = \mathfrak{D}_{\mathrm{B/A}}$, and we know that $\chi_{\mathrm{A}} = \mathrm{N}_{\mathrm{L/K}} \circ \chi_{\mathrm{B}}$, cf. Chap. I, §5, prop. 12. □

Corollary. *The discriminant* $\mathfrak{d}_{\mathrm{B/A}}$ *is contained in* A.

Remark. The preceding proposition shows that the different determines the discriminant; the converse is not true in general (except, however, when there is only one prime ideal of B over each prime ideal of A, which is the case when one completes).

Proposition 7. *Let* $\mathfrak{a}$ (resp. $\mathfrak{b}$) *be a fractional ideal of* K (resp. L) *relative to* A (resp. B). *The following two properties are equivalent:*

(i) $\mathrm{Tr}(\mathfrak{b}) \subset \mathfrak{a}$.
(ii) $\mathfrak{b} \subset \mathfrak{a} . \mathfrak{D}^{-1}_{\mathrm{B/A}}$.

The case $\mathfrak{a} = 0$ is trivial. When $\mathfrak{a} \neq 0$, the proposition follows from the equivalences:

$$\begin{aligned}\mathrm{Tr}(\mathfrak{b}) \subset \mathfrak{a} &\Leftrightarrow \mathfrak{a}^{-1}\mathrm{Tr}(\mathfrak{b}) \subset A \Leftrightarrow \mathrm{Tr}(\mathfrak{a}^{-1}\mathfrak{b}) \subset A \\ &\Leftrightarrow \mathfrak{a}^{-1}\mathfrak{b} \subset \mathfrak{D}_{B/A}^{-1} \Leftrightarrow \mathfrak{b} \subset \mathfrak{a} . \mathfrak{D}_{B/A}^{-1}. \quad \square\end{aligned}$$

It is clear that the property stated in prop. 7 *characterizes* the different.

§4. Elementary Properties of the Different and Discriminant

We keep the notation of the preceding section: L denotes a finite separable extension of K, and B the integral closure of A in L.

(i) Transitivity

Proposition 8. *Let* M/L *be a separable extension of finite degree* n, C *the integral closure of* A *in* M. *Then*

$$\mathfrak{D}_{C/A} = \mathfrak{D}_{C/B} . \mathfrak{D}_{B/A} \quad \textit{and} \quad \mathfrak{d}_{C/A} = (\mathfrak{d}_{B/A})^n . N_{L/K}(\mathfrak{d}_{C/B}).$$

Put $\theta = \mathrm{Tr}_{M/K}$, $\theta' = \mathrm{Tr}_{L/K}$, $\theta'' = \mathrm{Tr}_{M/L}$; then $\theta = \theta' \circ \theta''$. Let $\mathfrak{c}$ be a fractional ideal of M with respect to C. We have:

$$\begin{aligned}\mathfrak{c} \subset \mathfrak{D}_{C/B}^{-1} &\Leftrightarrow \theta''(\mathfrak{c}) \subset B \Leftrightarrow \mathfrak{D}_{B/A}^{-1}.\theta''(\mathfrak{c}) \subset \mathfrak{D}_{B/A}^{-1} \Leftrightarrow \theta'(\mathfrak{D}_{B/A}^{-1}.\theta''(\mathfrak{c})) \subset A \\ &\Leftrightarrow \theta(\mathfrak{D}_{B/A}^{-1}.\mathfrak{c}) \subset A \Leftrightarrow \mathfrak{D}_{B/A}^{-1}.\mathfrak{c} \subset \mathfrak{D}_{C/A}^{-1} \Leftrightarrow \mathfrak{c} \subset \mathfrak{D}_{B/A}.\mathfrak{D}_{C/A}^{-1}.\end{aligned}$$

Comparing the first and last terms, one sees that $\mathfrak{D}_{C/B}^{-1} = \mathfrak{D}_{B/A} . \mathfrak{D}_{C/A}^{-1}$, which is the formula for the different. By applying $N_{M/K}$ to both sides, one obtains the formula for the discriminant. $\square$

(ii) Localisation

Proposition 9. *If* S *is a multiplicative subset of* A, *then*

$$S^{-1}\mathfrak{D}_{B/A} = \mathfrak{D}_{S^{-1}B/S^{-1}A} \quad \textit{and} \quad S^{-1}\mathfrak{d}_{B/A} = \mathfrak{d}_{S^{-1}B/S^{-1}A}.$$

In view of the formula $(S^{-1}b)^{-1} = S^{-1}b^{-1}$ (used several times in Chap. I), it suffices to show that $S^{-1}B^* = (S^{-1}B)^*$. If $x = s^{-1}y$, with $s \in S$, $y \in B^*$, then

$$\mathrm{Tr}(x) = s^{-1}\,\mathrm{Tr}(y) \in S^{-1}A;$$

as $S^{-1}B^*$ contains $S^{-1}B$ and is a fractional ideal, this shows that $S^{-1}B^* \subset (S^{-1}B)^*$. Conversely, let b_i be generators of B as an A-module, and let

$x \in (S^{-1}B)^*$; then $\mathrm{Tr}(xb_i) \in S^{-1}A$, whence $\mathrm{Tr}(xb_i) = s^{-1}a_i$, with $a_i \in A$ (there being only finitely many b_i), and $sx \in B^*$; this proves the opposite inclusion. □

(iii) Completion

Proposition 10. *Let* $\mathfrak{P}$ *be a prime ideal of* B, *and let* $\mathfrak{p} = \mathfrak{P} \cap A$. *Let* $\hat{\mathfrak{D}}_{\mathfrak{P}}$ *be the ideal generated by the different* $\mathfrak{D}_{B/A}$ *in the completion* $\hat{B}_{\mathfrak{P}}$; *then* $\hat{\mathfrak{D}}_{\mathfrak{P}}$ *is the different of the ring* $\hat{B}_{\mathfrak{P}}$ *with respect to the ring* $\hat{A}_{\mathfrak{p}}$.

(In other words, the exponent of $\mathfrak{P}$ in $\mathfrak{D}_{B/A}$ is equal to the exponent of $\mathfrak{P}\hat{B}_{\mathfrak{P}}$ in the different of $\hat{B}_{\mathfrak{P}}$ over $\hat{A}_{\mathfrak{p}}$: "the different is preserved by completion".)

By applying prop. 9 with $S = A - \mathfrak{p}$, we are reduced to the case where A is a *discrete valuation ring*; we denote by $\hat{A}$ (resp. $\hat{K}$) its completion (resp. that of the field K). Similarly, if $\{\mathfrak{P}_i\}_{i \in I}$ is the family of prime ideals of B over $\mathfrak{p}$, we denote by $\hat{B}_i$ (resp. $\hat{L}_i$) the completion of B (resp. of L) for the valuation defined by $\mathfrak{P}_i$.

We first work in the $\hat{K}$-algebra $L \otimes_K \hat{K} = \hat{L}$; let $\hat{B} = B \otimes_A \hat{A}$; it is an $\hat{A}$-lattice of $\hat{L}$. We have the non-degenerate bilinear form $\mathrm{Tr}(xy)$ on $\hat{L}$, induced by extension of scalars from the analogous form on L. One proves easily (e.g., by taking a basis) that the dual lattice $(\hat{B})^*$ of $\hat{B}$ is obtained by extension of scalars from the lattice $B^* = \mathfrak{D}_{B/A}^{-1}$. In other words:

$$(\hat{B})^* = (B^*)\hat{} = B^* \otimes_A \hat{A}.$$

On the other hand, we know (cf. Chap. II, §2) that $L \otimes_K \hat{K} = \prod_{i \in I} \hat{L}_i$ and that $B \otimes_A \hat{A} = \prod_{i \in I} \hat{B}_i$; if Tr_i denotes the trace in the extension $\hat{L}_i/\hat{K}$, the bilinear form $\mathrm{Tr}(xy)$ on $\hat{L}$ is the direct sum of the forms $\mathrm{Tr}_i(xy)$ on the $\hat{L}_i$. The obvious formula

$$\left(\prod_{i \in I} X_i\right)^* = \prod_{i \in I} (X_i)^*,$$

applied to $X_i = \hat{B}_i$, shows that $(\hat{B})^* = \prod_{i \in I}(\hat{B}_i^*)$. In other words, *the codifferent of* B *with respect to* A *generates in each of the* $\hat{L}_i$ *the codifferent of* $\hat{B}_i$ *with respect to* $\hat{A}$; by taking inverses, we get the same result for the different. □

Corollary. *Let* $\hat{\mathfrak{d}}$ *be the ideal of* $\hat{A}_{\mathfrak{p}}$ *generated by the discriminant* $\mathfrak{d}_{B/A}$, *and let* $\mathfrak{d}_{\mathfrak{P}}$ *be the discriminant of* $\hat{B}_{\mathfrak{P}}$ *with respect to* $\hat{A}_{\mathfrak{p}}$. *Then*

$$\hat{\mathfrak{d}} = \prod_{\mathfrak{P} \mid \mathfrak{p}} \mathfrak{d}_{\mathfrak{P}}.$$

This follows from prop. 10 by taking the norm.

EXERCISE

Let C be a subring of B, containing A, and having the same field of fractions as B.

a) Show that among all the ideals of B contained in C, there is a largest one, and that it is the annihilator of the C-module B/C; it is denoted $\mathfrak{f}_{C/B}$ (the "conductor" of B in C).
b) Show that $\mathfrak{f}_{C/B} = (B^*:C^*)$, i.e., that $\mathfrak{f}_{C/B}$ is the set of all $x \in L$ such that $xC^* \subset B^*$.
c) Suppose that C^*, considered as a fractional C-ideal, is invertible; let $\mathfrak{c}$ be its inverse (so that $\mathfrak{c}C^* = C$). Deduce from b) the formula

$$\mathfrak{f}_{C/B} = \mathfrak{c} \, . \, \mathfrak{D}_{B/A}^{-1}.$$

§5. Unramified Extensions

We keep the notation and hypotheses of §§3 and 4.

Theorem 1. *Let* $\mathfrak{P}$ *be a prime ideal of* B, *and let* $\mathfrak{p} = \mathfrak{P} \cap A$. *In order that the extension* L/K *be unramified at* $\mathfrak{P}$ (cf. Chap. I, §4), *it is necessary and sufficient that* $\mathfrak{P}$ *does not divide the different* $\mathfrak{D}_{B/A}$.

Propositions 9 and 10 permit us to reduce to the case where A is a complete discrete valuation ring, with residue field k; in this case, B is also a discrete valuation ring. To say that $\mathfrak{P}$ is unramified is then equivalent to saying that B/$\mathfrak{p}$B is a field (i.e., that $e_{\mathfrak{P}} = 1$), and that this field is a separable extension of k.

Let $\{x_i\}$ be a basis of B over A, and let $d = \det(\mathrm{Tr}(x_i x_j))$; we know from prop. 3 that $\mathfrak{d}_{B/A}$ is the principal ideal generated by d. For $\mathfrak{P}$ to be unramified, it is therefore necessary and sufficient that $\mathfrak{p}$ does not divide d, i.e., that the image $\bar{d}$ of d in k be non-zero. Now if we set $\bar{B} = B/\mathfrak{p}B$, the images $\bar{x}_i$ of the x_i in $\bar{B}$ form a basis of $\bar{B}$ over k and the discriminant of this basis is $\bar{d}$. According to a known result (Bourbaki, *Alg.*, Chap. IX, §2, prop. 5), the condition $\bar{d} \neq 0$ is equivalent to saying that $\bar{B}$ is a separable k-algebra, and since it is a local ring, that amounts to saying that it is a field, and that this field is a separable extension of k. □

Corollary 1. *Let* $\mathfrak{p}$ *be a prime ideal of* A. *In order that* L/K *be unramified at* $\mathfrak{p}$, *it is necessary and sufficient that* $\mathfrak{p}$ *does not divide the discriminant* $\mathfrak{d}_{B/A}$.

This follows from the fact that $\mathfrak{d}_{B/A} = N(\mathfrak{D}_{B/A})$.

Corollary 2. *Almost all the prime ideals of* B(*or of* A) *are unramified in the extension* L/K.

Obvious.

We now examine more closely the structure of unramified extensions, limiting ourselves to the case where A *is a complete discrete valuation ring*; we denote its residue field by k.

Theorem 2. *Let k'/k be a finite separable extension. Then there exists a finite unramified extension K'/K whose corresponding residue extension is isomorphic to k'/k; this extension is unique, up to unique isomorphism. It is Galois if and only if k'/k is.*

Since k'/k is finite and separable, it is generated by a primitive element ξ; let φ be its minimal polynomial over k, which has degree $n = [k':k]$. Let $f \in A[X]$ be a monic polynomial whose reduction $\bar{f}$ mod. $\mathfrak{p}$ is equal to φ. By prop. 15 of Chap. I, the ring $A' = A[X]/(f)$ is a discrete valuation ring that is unramified over A and has residue extension k'/k. Its field of fractions K' is the solution to the existence problem. The uniqueness assertions (which depend in an essential way on the fact that A is *complete*) result from the following more precise theorem.

Theorem 3. *Let K'/K be a finite unramified extension, with residue extension k'/k, and let K''/K be an arbitrary finite extension, with residue extension k''/k. The set of K-isomorphisms of K' into K'' is then in one-to-one correspondence (by reduction) with the set of k-isomorphisms of k' into k''.*

Let us denote by $\mathrm{Hom}_A^{al}(B, C)$ the set of A-algebra homomorphisms of B into C (for any ring A). If A' and A'' are the integral closures of the ground ring A in K' and K'' respectively, then

$$\mathrm{Hom}_K^{al}(K', K'') = \mathrm{Hom}_A^{al}(A', A''),$$

and we are reduced to showing that the canonical homomorphism

$$\theta: \mathrm{Hom}_A^{al}(A', A'') \to \mathrm{Hom}_k^{al}(k', k'')$$

is bijective.

Let $n = [K':K]$. By prop. 16 of Chapter I, there exists an $x \in A'$ such that the set $\{1, x, \ldots, x^{n-1}\}$ is a basis of A' over A; moreover, if f denotes the characteristic polynomial of x, the reduction $\bar{f}$ of f is the characteristic polynomial of the image $\bar{x}$ of x in k'. It follows that the elements of $\mathrm{Hom}_A^{al}(A', A'')$ (resp. of $\mathrm{Hom}_k^{al}(k', k'')$) correspond bijectively with the elements $a'' \in A''$ (resp. $\xi'' \in k''$) such that $f(a'') = 0$ (resp. $\bar{f}(\xi'') = 0$); as for the map θ, it corresponds to the reduction $a'' \mapsto \xi'' = \bar{a}''$. Thus it all comes down to showing that a root of $\bar{f}$ in k'' can be lifted uniquely to a root of f in A''; that is a consequence of prop. 7 of Chap. II (note that all the roots of $\bar{f}$ are simple, since this polynomial is irreducible and separable over k). □

Corollary 1. *Let k_s be the separable closure of k (i.e., the largest separable extension of k within a given algebraic closure of k), and let K_{nr} be the inductive* *(direct) limit of the unramified extensions of* K *that correspond to the finite subextensions of* k_s. *The field* K_{nr} *is Galois over* K *with residue field* k_s, *and* $G(K_{nr}/K) = G(k_s/k)$.

This is clear.

The extension K_{nr} is called the *maximal unramified extension* of K. It is unique, up to isomorphism. For example, if k is a finite field, $G(k_s/k)$ is isomorphic to the completion $\hat{\mathbf{Z}}$ of $\mathbf{Z}$ for the topology of subgroups of finite index, and the same holds for $G(K_{nr}/K)$.

Corollary 2. *Let* K''/K *be a finite extension, with residue extension* k''/k. *The subextensions* K'/K *of* K''/K *which are unramified over* K *are in one-to-one correspondence with the separable subextensions* k'/k *of* k''/k.

This is clear.

Corollary 3. *With the hypotheses of cor. 2, there exists a maximal unramified subextension* K'/K *of* K''/K. *Its residue extension* k'/k *is the largest separable subextension of* k''/k. *We have* $e(K''/K) = e(K''/K')$, $f(K''/K') = [k'':k]_i$, *and* $f(K'/K) = [k'':k]_s$.

This follows from cor. 2.

[Corollary 3 is applied most often in the case where k''/k is separable; then one obtains an intermediate field K' such that K''/K' is totally ramified ($f = 1$) and K'/K is unramified.]

Remarks. 1) When K and k have the same characteristic, K is isomorphic to $k((T))$—cf. Chap. II, §4, if k' is a finite separable extension of k, the corresponding unramified extension is $K' = k'((T)) = k' \otimes_k K$. When k is a perfect field of characteristic $p > 0$, one has similarly $K' = W(k') \otimes_{W(k)} K$, where $W(k)$ denotes the ring of Witt vectors with coefficients in k.

2) The results of this section extend to arbitrary complete local Noetherian rings; see Grothendieck's seminar ([29], §1).

§6. Computation of Different and Discriminant

We keep the notation and hypotheses of §§3 and 4.

Proposition 11. *Let* $n = [L:K]$ *and let* C *be a subring of* B *containing* A *and having an* A*-basis consisting of the powers* x^i, $0 \le i \le n-1$, *of a single element* x. *Let* f *be the characteristic polynomial of* x. *Then*

(i) f *has coefficients in* A.

(ii) C^* *is a free* A-*module, having as basis the* $x^i/f'(x)$, $0 \le i \le n-1$, *where* $f'(X)$ *denotes the derivative of* $f(X)$.

The coefficients of f are integral over A and belong to K; as A is integrally closed, they belong to A, which proves (i). (Note that the ring C is isomorphic to the ring $B_f = A[X]/(f)$, cf. Chap I, §6.)

The proof of (ii) will use the following result:

Lemma 2 (Euler). $\mathrm{Tr}(x^i/f'(x))=0$ *for* $i=0,\ldots,n-2$, *and* $\mathrm{Tr}(x^{n-1}/f'(x))=1$.

Let x_k, $k = 1, \ldots, n$, be the conjugates of x in a suitable extension of L. We must compute the sums $\sum_k (x_k)^i/f'(x_k)$. Now we have the identity

$$(*) \qquad \frac{1}{f(T)} = \sum_{k=1}^{k=n} \frac{1}{f'(x_k)(T - x_k)}.$$

[Decompose the rational function $1/f(T)$ into a sum of "simple elements" $a_k/(T - x_k)$, and determine the a_k by the usual procedure.]

If $1/f(T)$ is expanded in a power series of powers of $1/T$, the term of lowest degree is $1/T^n$, and comparison of this expansion with the right side of $(*)$ gives the desired formulas. □

Returning now to prop. 11, it suffices to show that the matrix $r_{ij} = \mathrm{Tr}(x^i . x^j/f'(x))$ is invertible over A. Now lemma 2 shows that $r_{ij} = 0$ if $i + j \le n - 2$, and $r_{ij} = 1$ if $i + j = n - 1$; for $i + j \ge n$, one has

$$r_{ij} = \mathrm{Tr}(x^n . x^{i+j-n}/f'(x)),$$

and since x^n is a linear combination of the x^i, $0 \le i < n$, with coefficients in A, an inductive argument shows that $r_{ij} \in A$. Computation of the determinant of a triangular matrix then gives $\det(r_{ij}) = (-1)^{n(n-1)/2}$. □

Keeping the hypotheses of prop. 11, consider the set $\mathfrak{r}$ of all $t \in C$ such that $tB \subset C$; $\mathfrak{r}$ is an ideal both of C and of B, called the *conductor* of B in C.

Corollary 1. *With the preceding notation and hypotheses,*

$$\mathfrak{r} = f'(x) . \mathfrak{D}_{B/A}^{-1}.$$

To simplify the writing, put $b = f'(x)$. We have the following equivalences for $t \in L$:

$$t \in \mathfrak{r} \Leftrightarrow tB \subset C \Leftrightarrow b^{-1}tB \subset C^* \Leftrightarrow \mathrm{Tr}(b^{-1}tB) \subset A \Leftrightarrow b^{-1}t \in \mathfrak{D}_{B/A}^{-1} \Leftrightarrow t \in b . \mathfrak{D}_{B/A}^{-1},$$

whence the corollary.

Corollary 2. *The different* $\mathfrak{D}_{B/A}$ *divides the principal ideal* $(f'(x))$. *For these two ideals to be equal, it is necessary and sufficient that* $B = C$ (*i.e., that* $B = A[x]$).

The first assertion results from the formula $(f'(x)) = \mathfrak{r} . \mathfrak{D}_{B/A}$. That same formula shows that $\mathfrak{D}_{B/A} = (f'(x))$ if and only if $\mathfrak{r} = 1$, i.e., B = C. □

Corollary 2 enables us to compute the different $\mathfrak{D}_{B/A}$ when B is of the form A[x]; we will next supply a condition for that to be the case:

Proposition 12. *Suppose that* B (*hence also* A) *is a discrete valuation ring; if* $\bar{L}$ *and* $\bar{K}$ *denote the residue fields of these two rings, suppose also that the extension* $\bar{L}/\bar{K}$ *is separable. Then* B *has a basis over* A *composed of the powers* $1, x, \ldots, x^{n-1}$ *of a single element* x.

(This proposition applies notably when A is a *complete* discrete valuation ring whose residue field is *perfect*.)

Let e be the ramification index of L/K, and let $f = [\bar{L}:\bar{K}]$, so that $n = ef$. Let π be a uniformizer of B, and let $x \in B$ represent a primitive element $\bar{x}$ for the extension $\bar{L}/\bar{K}$. We need two lemmas.

Lemma 3. *The products* $x^i\pi^j$, $0 \leq i < f, 0 \leq j < e$, *form a basis for the* A-*module* B.

The number of these products being ef, it suffices to show that they span B, and even that their classes span B/$\mathfrak{p}$B (where $\mathfrak{p}$ is the maximal ideal of A). We have $\mathfrak{p}B = \pi^e B$. Hence it is enough to show that if these elements generate B mod. π^m, with $m < e$, then they generate B mod. π^{m+1}; this is easy. □

Lemma 4. *The element* x *may be chosen so that there exists a monic polynomial* R(X), *of degree* f, *with coefficients in* A, *such that* R(x) *is a uniformizer of* B.

Let us first choose x such that $\bar{L} = \bar{K}(\bar{x})$. Let R be a monic polynomial with coefficients in A whose reduction $\bar{R}$ is the minimal polynomial of $\bar{x}$. If w denotes the valuation of B, then $w(R(x)) \geq 1$, since the image of R(x) in $\bar{L}$ is zero. If equality holds, the element x works; if not, $w(R(x)) \geq 2$. Let h be any element of valuation 1, and apply Taylor's formula:

$$R(x + h) = R(x) + h . R'(x) + h^2 . b, \quad \text{with } b \in B.$$

As $\bar{L}/\bar{K}$ is separable, we have $\bar{R}'(\bar{x}) \neq 0$, so that R'(x) is invertible and $h . R'(x)$ has valuation 1; since the other terms in R($x + h$) have valuation ≥ 2, we see that $w(R(x + h)) = 1$, and $x + h$ is the solution. □

We can now finish the proof of prop. 12. Choose an element x as in lemma 4, and set $\pi = R(x)$; lemma 3 shows that the products $x^i R(x)^j$, $0 \leq i < f$, $0 \leq j < e$, form a basis of B. Thus B = A[x], and since x satisfies a monic equation of degree n, it follows that the powers $1, x, \ldots, x^{n-1}$ form a basis.

□

Remark. Prop. 12 does not extend to the case where A is only assumed to be a discrete valuation ring: cf. exercise 3.

Proposition 13. *Let* $\mathfrak{P}$ *be a non-zero prime ideal of* B, *and let* $\mathfrak{p} = \mathfrak{P} \cap \mathrm{A}$. Let $\bar{\mathrm{L}}_{\mathfrak{P}}$ *and* $\bar{\mathrm{K}}_{\mathfrak{p}}$ *be the corresponding residue fields. Suppose that the residue extension is separable. Then the exponent of* $\mathfrak{P}$ *in the different* $\mathfrak{D}_{\mathrm{B}/\mathrm{A}}$ *is greater than or equal to* $e_{\mathfrak{P}} - 1$ ($e_{\mathfrak{P}}$ denoting the ramification index), *with equality taking place if and only if* $e_{\mathfrak{P}}$ *is prime to the characteristic of the residue field* $\bar{\mathrm{K}}_{\mathfrak{p}}$.

By localising and completing, we reduce to the case where A and B are complete discrete valuation rings; using cor. 3 to th. 3, we may further suppose that L/K is totally ramified. If π is a uniformizer of B, we know (cf. Chap. I, §6) that π satisfies an Eisenstein equation $f(\pi) = 0$, with

$$f = \mathrm{X}^e + a_1\mathrm{X}^{e-1} + \cdots + a_e, \qquad a_i \in \mathfrak{p},\ e = e_{\mathfrak{P}} = [\mathrm{L}:\mathrm{K}].$$

In addition, $\mathrm{B} = \mathrm{A}[\pi]$, which allows us to apply cor. 2 to prop. 11: the different $\mathfrak{D}_{\mathrm{B}/\mathrm{A}}$ is generated by $f'(\pi)$. Now we have

$$(*) \qquad f'(\pi) = e\pi^{e-1} + \sum_{i=1}^{i=e-1} (e-i)a_i\pi^{e-i-1}.$$

Let w be the valuation of B, so that $w(\pi) = 1$ and $w(a_i) \geq e$. Thus $f'(\pi) = e\pi^{e-1} + b$, with $w(b) \geq e$. Hence $w(f'(\pi)) \geq e - 1$, and we see that equality holds if and only if $w(e) = 0$, i.e., if e is prime to the residue characteristic. □

Remarks. 1) Prop. 13 gives a lower bound for the exponent of the different. It is useful also to have the following *upper* bound:

$$\text{exponent of } \mathfrak{P} \text{ in } \mathfrak{D}_{\mathrm{B}/\mathrm{A}} \leq e_{\mathfrak{P}} - 1 + w(e_{\mathfrak{P}}).$$

Namely, reducing to the totally ramified case as before, we have $w(a) \equiv 0 \bmod. e$ for all $a \in \mathrm{A}$, hence

$$w((e-i)a_i\pi^{e-i-1}) \equiv -i - 1 \bmod. e \quad \text{if } (e-i)a_i \neq 0.$$

This shows that all non-zero terms in expression (*) above for $f'(\pi)$ have *distinct* valuation: they are even distinct mod. e. Hence

$$w(f'(\pi)) = \operatorname*{Inf}_{0 \leq i < e} w((e-i)a_i\pi^{e-i-1}).$$

Taking $i = 0$ gives $w(f'(\pi)) \leq e - 1 + w(e)$. □

(This local result gives an *upper bound* for the discriminant of a number field, depending on the degree, and the primes at which it ramifies. This upper bound had been conjectured by Dedekind, at the end of his paper on the discriminant (*Werke* I, S.351), and proved by Hensel (*Crelle J.* 113 (1894), 61, and *Gött. Nach.* (1897), 247).)

2) When the residue characteristic divides $e_{\mathfrak{P}}$, the exponent of $\mathfrak{P}$ in the different no longer depends only on $e_{\mathfrak{P}}$; in order to compute it, the "ramification groups" are needed (see prop. 4, §1 of the next chapter).

EXERCISES

1. With the hypotheses of prop. 12, show that if $B = A[x]$, and if y is sufficiently near x, then $B = A[y]$.

2. In the general case, let $\mathfrak{P}$ be a prime ideal of B whose corresponding residue extension is separable. Show that there exists an $x \in B$, generating the extension L/K, such that the conductor $\mathfrak{r}$ of B in $A[x]$ is prime to $\mathfrak{P}$ (apply prop. 12 and exer. 1 above to the completions of B and of A).

3. Suppose that A is a discrete valuation ring and that B is "completely decomposed", i.e., that there are $n = [L:K]$ prime ideals of B above the prime ideal $\mathfrak{p}$ of A. Show that in order for there to exist an $x \in B$ such that $B = A[x]$, it is necessary and sufficient that $n \leq \mathrm{Card}(\bar{K})$, where $\bar{K}$ is the residue field of A.

§7. A Differential Characterisation of the Different

Let B be any commutative A-algebra. The map $(x, y) \mapsto xy$ defines a homomorphism

$$\theta: B \otimes_A B \to B.$$

Let I be its kernel, and put $\Omega_A(B) = I/I^2 = I \otimes_{B \otimes B} B$; this gives a B-module called the module of A-*differentials of the ring* B (this definition is due to E. Kähler). If dx denotes the image of $x \otimes 1 - 1 \otimes x$ in I/I^2, every element of $\Omega_A(B)$ can be written in the form $\sum y_i \, dx_i$, and we have

$$d(xy) = x\,dy + y\,dx$$
$$da = 0 \quad \text{if } a \in A.$$

The module $\Omega_A(B)$ is actually *universal* for the above properties (cf. [12], exposé 13, as well as [46]).

Proposition 14. *Let* A *be a Dedekind domain, with field of fractions* K; *let* L *be a finite separable extension of* K, *and let* B *be the integral closure of* A *in* L. *Suppose that for every prime ideal* $\mathfrak{P}$ *of* B, *the corresponding residue extension is separable. The* B*-module* $\Omega_A(B)$ *can then be generated by one element, and its annihilator is the different* $\mathfrak{D}_{B/A}$.

It is easy to show that if A′ is an arbitrary commutative A-algebra, and if $B' = B \otimes_A A'$, then $\Omega_{A'}(B') = \Omega_A(B) \otimes_A A'$. It follows that the module of differentials is "compatible" with localisation and completion, which reduces us to the case where A is a complete discrete valuation ring. By proposition 12, we then have $B = A[X]/(f)$, where f is a monic polynomial. If x denotes the image of X in B, one checks ([12], *loc. cit.*) that $\Omega_A(B)$ is generated by dx, and that the annihilator of dx is $f'(x)$. But by cor. 2 to prop. 11, $\mathfrak{D}_{B/A}$ is the principal ideal generated by $f'(x)$. □

Remark. It would be interesting to obtain a more direct proof of the preceding proposition, and likewise to study the "principal parts of order m" $P_m(B/A) = (B \otimes_A B)/I^{m+1}$.

EXERCISES

1. With the hypotheses of prop. 14, show that there is an element $x \in B$ such that dx generates $\Omega_A(B)$.
2. Translate prop. 14 and exer. 1 in terms of derivations.
3. Give an example of a separable extension L/K having a residue extension $\bar{L}/\bar{K}$ which is purely inseparable of height 1 and not simple. Show that for such an extension, $\Omega_A(B)$ cannot be generated by one element.

CHAPTER IV

Ramification Groups

Notation and Hypotheses

Throughout this chapter, K denotes a field *complete* under a discrete valuation v_K. The corresponding valuation ring is denoted A_K, its maximal ideal $\mathfrak{p}_K$, its residue field $\bar{K} = A_K/\mathfrak{p}_K$, and $U_K = A_K - \mathfrak{p}_K$ the multiplicative group of invertible elements of A_K.

If L is a finite separable extension of K, A_L denotes the integral closure of A_K in L; it is a complete discrete valuation ring (Chap. II, §2, prop. 3). We define v_L, $\mathfrak{p}_L$, U_L, $\bar{L}$ as above. *We will always assume that the residue extension $\bar{L}/\bar{K}$ is separable.* The ramification index of $\mathfrak{p}_L$ in L/K will be denoted $e_{L/K}$, and its residue degree $f_{L/K}$; we have $e_{L/K} \cdot f_{L/K} = [L:K]$, cf. Chap. I, §4.

§1. Definition of the Ramification Groups; First Properties

Let L/K be a *Galois extension* (satisfying the hypotheses above), and let $G = G(L/K)$ be its Galois group; G acts on the ring A_L. Let x be an element of A_L which generates A_L as an A_K-algebra (cf. Chap. III, §6, prop. 12).

Lemma 1. *Let $s \in G$, and let i be an integer ≥ -1. Then the three following conditions are equivalent:*

a) *s operates trivially on the quotient ring $A_L/\mathfrak{p}_L^{i+1}$.*
b) *$v_L(s(a) - a) \geq i + 1$ for all $a \in A_L$.*
c) *$v_L(s(x) - x) \geq i + 1$.*

The equivalence of a) and b) is trivial. On the other hand, the image x_i of x in $A_L/\mathfrak{p}_L^{i+1} = A_i$ generates A_i as an A_K-algebra. Hence $s(x_i) = x_i$ is the necessary and sufficient condition for s to operate trivially on A_i; this shows the equivalence of a) and c). □

Proposition 1. *For each integer $i \geq -1$, let G_i be the set of $s \in G$ satisfying conditions* a), b), c) *of lemma* 1. *Then the G_i form a decreasing sequence of normal subgroups of* G; $G_{-1} = G$, G_0 *is the inertia subgroup of* G (cf. Chap. I, §7), *and $G_i = \{1\}$ for i sufficiently large.*

Condition a) shows that the G_i are normal subgroups of G, and they obviously decrease with i. Condition c) shows that if

$$i \geq \sup\{v_L(s(x) - x)\} \quad \text{for } s \neq 1,$$

then G_i is trivial. The other assertions are clear. □

The group G_i is called *the* ith *ramification group of* G (or of L/K). The ramification groups define a filtration of G (in the sense of Bourbaki, *Alg. comm.*, Chap. III, §2); the quotient G/G_0 is the Galois group $G(\bar{L}/\bar{K})$ of the residue extension (cf. Chap. I, §7); the quotients G_i/G_{i+1}, $i \geq 0$, will be studied in the next section.

Still denoting by x an A_K-generator of A_L, we define a *function i_G on* G by the formula

$$i_G(s) = v_L(s(x) - x).$$

If $s \neq 1$, $i_G(s)$ is a non-negative integer; $i_G(1) = +\infty$. The function i_G has the following properties:

$$i_G(s) \geq i + 1 \Leftrightarrow s \in G_i$$
$$i_G(tst^{-1}) = i_G(s)$$
$$i_G(st) \geq \operatorname{Inf}(i_G(s), i_G(t)).$$

The first property merely expresses the definition of G_i; it shows that the definition of i_G does not depend on the choice of the generator x, and that *the knowledge of i_G is equivalent to that of the* G_i (in Boùrbaki's terminology, *loc. cit.*, i_G is equal to the *order function* of the filtration (G_i), increased by one unit). The two other properties follow from the first and from the fact that the G_i are normal subgroups of G.

Now let H be a *subgroup* of G, and let K′ be the subextension of L fixed by H. Then $G(L/K') = H$, and we will see that the ramification groups of G determine those of H:

Proposition 2. *For every $s \in H$, $i_H(s) = i_G(s)$, and $H_i = G_i \cap H$.*

This follows from condition a) of lemma 1. □

Corollary. *Let* K_r *be the largest unramified subextension of* L *over* K, *and let* H *be the corresponding subgroup of* G. *Then* H *is equal to the inertia group* G_0, *and the ramification groups of* G *of index* ≥ 0 *are equal to those of* H.

The fact that $H = G_0$ was proved in Chap. I, §7. Since $G_i \subset H_0$ for $i \geq 0$, prop. 2 shows that $G_i = H_i$. □

Remark. The extension L/K_r is *totally ramified*; the above corollary therefore reduces the study of the "higher" ramification groups (of index $i \geq 0$) to the totally ramified case; that is the method we use in §2.

Now suppose in addition that the subgroup H is *normal*, so that G/H may be identified with the Galois group of K′/K. We will see that the ramification groups of G determine those of G/H; the result can be stated most simply in terms of the function i:

Proposition 3. *For every* $\sigma \in G/H$,

$$i_{G/H}(\sigma) = \frac{1}{e'} \sum_{s \to \sigma} i_G(s),$$

where $e' = e_{L/K'}$.

(Proof after J. Tate). For $\sigma = 1$, both sides are equal to $+\infty$, so the equation holds. Suppose $\sigma \neq 1$. Let x (resp. y) be an A_K-generator of A_L (resp. $A_{K'}$). By definition, $e' . i_{G/H}(\sigma) = v_L(\sigma(y) - y)$, and $i_G(s) = v_L(s(x) - x)$. If we choose one $s \in G$ representing σ, the other representatives have the form st, $t \in H$. Hence it all comes down to showing that *the elements*

$$a = s(y) - y \quad \textit{and} \quad b = \prod_{t \in H} (st(x) - x)$$

generate the same ideal in A_L.

Let $f \in A_{K'}[X]$ be the minimal polynomial of x over the intermediate field K′. Then $f(X) = \prod_{t \in H} (X - t(x))$. Denote by $s(f)$ the polynomial obtained from f by transforming each of its coefficients by s. Clearly

$$s(f)(X) = \prod_{t \in H} (X - st(x)).$$

As $s(f) - f$ has all coefficients divisible by $s(y) - y$, one sees that $a = s(y) - y$ divides $s(f)(x) - f(x) = s(f)(x) = \pm b$.

It remains to show that b divides a. Toward that end, write y as a polynomial in x, with coefficients in $A_K : y = g(x)$. The polynomial $g(X) - y$ has x as root and has all its coefficients in $A_{K'}$; it is therefore divisible by the minimal polynomial f:

$$g(X) - y = f(X) . h(X), \quad \text{with } h \in A_{K'}[X].$$

Transform this equation by s and substitute x for X in the result; one gets

$$y - s(y) = s(f)(x) . s(h)(x),$$

which shows that $b = \pm s(f)(x)$ divides a. □

Corollary. *If* $H = G_j$ *for some integer* $j \geq 0$, *then* $(G/H)_i = G_i/H$ *for* $i \leq j$, *and* $(G/H)_i = \{1\}$ *for* $i \geq j$.

The G_i/H, $i \leq j$, form a decreasing filtration of G/H. If $\sigma \in G/H$, $\sigma \neq 1$, then there is a unique index $i < j$ such that $\sigma \in G_i/H$, $\sigma \notin G_{i+1}/H$. If $s \in G$ represents σ, it is clear that $s \in G_i$, $s \notin G_{i+1}$, whence $i_G(s) = i + 1$. Also, since $H \subset G_0$, the extension L/K′ is totally ramified and $e_{L/K'}$ is equal to the order of H. Prop. 3 then shows that $i_{G/H}(\sigma) = i + 1$, which proves that the filtration of the G_i/H coincides with that of the $(G/H)_i$ for $i \leq j$. On the other hand, as $(G/H)_j = G_j/H = \{1\}$, we get $(G/H)_i = \{1\}$ for $i \geq j$. □

Remark. For general normal subgroups H, prop. 3 still enables us to prove that the ramification groups of G/H are the images of those of G, *but it is necessary to modify the numbering.* We will return to this in §3.

We are now going to use the ramification groups to determine the *different* of a subextension of L/K.

Proposition 4. *If* $\mathfrak{D}_{L/K}$ *denotes the different of* L/K, *then*

$$v_L(\mathfrak{D}_{L/K}) = \sum_{s \neq 1} i_G(s) = \sum_{i=0}^{i=\infty} (\text{Card}(G_i) - 1).$$

[Here, and in the sequel, Card(S) denotes the number of elements in the finite set S. Observe that $\text{Card}(G_i) - 1 = 0$ for i sufficiently large, so that the infinite sum makes sense.]

Again let x be an A_K-generator of A_L, and let f be its minimal polynomial over K. According to cor. 2 to prop. 11 of Chap. III, $\mathfrak{D}_{L/K}$ is generated by $f'(x)$. But $f(X) = \prod_{s \in G}(X - s(x))$, whence

$$f'(x) = \prod_{s \neq 1} (x - s(x)),$$

and

$$v_L(\mathfrak{D}_{L/K}) = v_L(f'(x)) = \sum_{s \neq 1} i_G(s),$$

which proves the first equation.

To prove the second, put $r_i = \text{Card}(G_i) - 1$, and note that the function i_G takes the constant value i on $G_{i-1} - G_i$. Therefore

$$\sum_{s \neq 1} i_G(s) = \sum_{i=0}^{i=\infty} i(r_{i-1} - r_i) = (r_0 - r_1) + 2(r_1 - r_2) + 3(r_2 - r_3) + \cdots$$
$$= r_0 + r_1 + r_2 + \cdots,$$

□

Corollary. *If* K′ *is a subextension of* L, *corresponding to the subgroup* H *of* G, *then*

$$v_{K'}(\mathfrak{D}_{K'/K}) = \frac{1}{e'} \sum_{s \notin H} i_G(s), \quad \text{with } e' = e_{L/K'}.$$

Indeed, prop. 4 gives

$$v_L(\mathfrak{D}_{L/K}) = \sum_{s \neq 1} i_G(s), \qquad v_L(\mathfrak{D}_{L/K'}) = \sum_{\substack{s \neq 1 \\ s \in H}} i_G(s),$$

and the argument is concluded by applying transitivity of the different (Chap. III, §4, prop. 8). □

Remarks. 1) When one no longer assumes the residue extension $\bar{L}/\bar{K}$ to be separable, one is led to split in two the sequence G_n of ramification groups (cf. Zariski-Samuel [53], I, Chap. V, §10). I don't know whether it is possible to extend propositions 3 and 4 to this case.

2) The "globalisation" of the definitions and results of this Chapter is easy. To be precise, let A be a Dedekind domain with field of fractions E, let B be its integral closure in a Galois extension F of E, with Galois group G, and let $\mathfrak{P}$ be a prime ideal of B; let $\mathfrak{p} = \mathfrak{P} \cap A$. We know (Chap. II, §3) that the *decomposition group* $D_{\mathfrak{P}}$ of $\mathfrak{P}$ is the Galois group of the completed extension $\hat{L}_{\mathfrak{P}}/\hat{K}_{\mathfrak{P}}$; if the corresponding residue extension is separable, we can apply the definitions above to this extension, and define *the ramification groups* $G_i(\mathfrak{P})$ *of* G *relative to* $\mathfrak{P}$; they are normal subgroups of the decomposition group $D_{\mathfrak{P}}$. It is easy to see that

$$s \in G_i(\mathfrak{P}) \Leftrightarrow s(x) \equiv x \bmod. \mathfrak{P}^{i+1} \quad \text{for all } x \in B.$$

Prop. 2 shows that if $H \subset G$, then $H_i(P) = H \cap G_i(P)$. Similar translations of all the other results of this chapter can made; we leave them to the reader.

EXERCISE

Let L/K be a totally ramified Galois extension, and let L_0/K_0 be an extension deduced from it by the process of exer. 4 of Chap. II, §2. Show that the ramification groups of L/K coincide with those of $\hat{L}_0/\hat{K}_0$. (By this method, the study of ramification groups is reduced to the case where the residue field is algebraically closed.)

§2. The Quotients G_i/G_{i+1}, $i \geq 0$

Keeping the notation and hypotheses of section 1, we denote by K_r the largest unramified extension of K contained in L (cf. cor. to prop. 2). Denote by π a uniformizer of L.

Proposition 5. *Let i be a non-negative integer. In order that an element s of the inertia group G_0 belong to G_i, it is necessary and sufficient that*

$$s(\pi)/\pi \equiv 1 \text{ mod. } \mathfrak{p}_L^i.$$

Replacing G by G_0 and K by K_r reduces us to the case of a *totally ramified* extension. By prop. 18 of Chapter I, §6, the element π is then an A_K-generator of A_L. Therefore

$$\begin{aligned} i_G(s) &= v_L(s(\pi) - \pi) \\ &= 1 + v_L(s(\pi)/\pi - 1), \quad \text{since } v_L(\pi) = 1, \end{aligned}$$

whence the proposition follows. □

We now define a *filtration* of the group U_L of invertible elements of A_L by:

$$\begin{aligned} U_L^{(0)} &= U_L \\ U_L^{(i)} &= 1 + \mathfrak{p}_L^i \quad \text{for } i \geq 1. \end{aligned}$$

This gives a decreasing sequence of closed subgroups of U_L; these subgroups form a neighborhood base of 1 for the topology induced on U_L by L^*; since U_L is closed, hence complete, we have

$$U_L = \varprojlim U_L/U_L^{(i)}.$$

(Throughout the sequel, we will write U_L^i instead of $U_L^{(i)}$, except when there is a risk of confusion with the group of ith powers of elements of U_L.)

Proposition 6. (*a*) $U_L^0/U_L^1 = \bar{L}^*$ (*multiplicative group of the residue field* $\bar{L}$).
(*b*) *For $i \geq 1$, the group U_L^i/U_L^{i+1} is canonically isomorphic to the group $\mathfrak{p}_L^i/\mathfrak{p}_L^{i+1}$, which is itself isomorphic (non-canonically) to the additive group of the residue field* $\bar{L}$.

Assertion (a) is trivial. To prove (b), let correspond to each $x \in \mathfrak{P}_L^i$ the element $1 + x$ of U_L^i; this gives, by passage to the quotient, the canonical isomorphism. Moreover, as $\mathfrak{p}_L^i/\mathfrak{p}_L^{i+1}$ is a one-dimensional vector space over $\bar{L}$, (b) is immediate. □

Remark. The direct sum of the $\mathfrak{p}_L^i/\mathfrak{p}_L^{i+1}$ has a natural structure of graded $\bar{L}$-algebra (namely, the graded algebra associated to the $\mathfrak{p}_L$-adic filtration of A_L—cf. Bourbaki, *Alg. comm.*, Chap. III, §2). In particular, if we set $\Omega_L = \mathfrak{p}_L/\mathfrak{p}_L^2$, then there is a canonical map of the ith tensor power Ω_L^i of Ω_L into $\mathfrak{p}_L^i/\mathfrak{p}_L^{i+1}$; it is easy to see that this map is an isomorphism. Thus one can *canonically* identify U_L^i/U_L^{i+1} with Ω_L^i.

Let us now come back to the ramification groups. Prop. 5 can be translated by the equivalence

$$s \in G_i \Leftrightarrow s(\pi)/\pi \in U_L^i.$$

More precisely:

Proposition 7. *The map which, to $s \in G_i$, assigns $s(\pi)/\pi$, induces by passage to the quotient an isomorphism θ_i of the quotient group G_i/G_{i+1} onto a subgroup of the group U_L^i/U_L^{i+1}. This isomorphism is independent of the choice of uniformizer π.*

If π' is another uniformizer, then $\pi' = \pi u$, with $u \in U_L$, whence

$$s(\pi')/\pi' = s(\pi)/\pi \,.\, s(u)/u.$$

If $s \in G_i$, then $s(u) \equiv u \bmod. \mathfrak{p}_L^{i+1}$ whence $s(u)/u \equiv 1 \bmod. U_L^{i+1}$, which shows that θ_i does not depend on the choice of π. If $s, t \in G_i$, one can write

$$st(\pi)/\pi = s(\pi)/\pi \,.\, t(\pi)/\pi \,.\, s(u)/u, \quad \text{with } u = t(\pi)/\pi.$$

Since u is in U_L, one has $s(u)/u \equiv 1 \bmod. U_L^{i+1}$ as above, and one gets

$$st(\pi)/\pi \equiv s(\pi)/\pi \,.\, t(\pi)/\pi \bmod. U_L^{i+1},$$

which shows that θ_i is a homomorphism. The fact that it is injective is trivial. □

[Let us make explicit the definition of θ_i: If $s \in G_0$, then $s\pi = u\pi$, with $u \in U_L$, and $\theta_0(s) = \bar{u} \in \bar{L}^*$. If $s \in G_i$, $i \geq 1$, then $s\pi = \pi(1 + a)$, with $a \in \mathfrak{p}_L^i$, and $\theta_i(s)$ is equal to the class of a in $\mathfrak{p}_L^i/\mathfrak{p}_L^{i+1}$.]

Remark. It will be shown in the next chapter that *for some values of i*, the image of θ_i in U_L^i/U_L^{i+1} can be characterised as the kernel of a map obtained from the norm map N_{L/K_r} by passage to the quotient.

Corollary 1. *The group G_0/G_1 is cyclic, and is mapped isomorphically by θ_0 onto a subgroup of the group of roots of unity contained in $\bar{L}$. Its order is prime to the characteristic of the residue field $\bar{L}$.*

Indeed, $U_L/U_L^1 = \bar{L}^*$, which shows that $\theta_0(G_0/G_1)$ is a finite subgroup of the group of roots of unity in $\bar{L}$; the fact that it is cyclic and of order prime to the characteristic of $\bar{L}$ then follows.

Corollary 2. *If the characteristic of $\bar{L}$ is zero, then $G_1 = \{1\}$, and the group G_0 is cyclic.*

Indeed, if $i \geq 1$, U_L^i/U_L^{i+1} is isomorphic to $\bar{L}$, which has no non-trivial finite subgroups. Hence $G_i = G_{i+1}$, and since $G_i = \{1\}$ for large i, we do have $G_1 = \{1\}$, and $G_0 = G_0/G_1$ is cyclic by corollary 1.

Corollary 3. *If the characteristic of $\bar{L}$ is $p \neq 0$, the quotients G_i/G_{i+1}, $i \geq 1$, are abelian groups, and are direct products of cyclic groups of order p. The group G_1 is a p-group.*

(Recall that a *p-group* is a finite group whose order is a power of p.)

Indeed, for $i \geq 1$, U_L^i/U_L^{i+1} is isomorphic to the additive group of $\bar{L}$, and every subgroup of $\bar{L}$ is a vector space over the prime field of p elements, hence is a direct sum of cyclic groups of order p. As the order of G_1 is equal to the product of the orders of the G_i/G_{i+1} for $i \geq 1$, it is clear that G_1 is a p-group.

Corollary 4. *If the characteristic of* $\bar{L}$ *is* $p \neq 0$, *the inertia group* G_0 *has the following property:*

(R_p) *It is the semi-direct product of a cyclic group of order prime to* p *with a normal subgroup whose order is a power of* p.

By corollaries 1 and 3, G_1 is a p-group and G_0/G_1 is cyclic of order prime to p; thus it all comes down to showing the existence of a subgroup H of G_0 that projects isomorphically onto G_0/G_1. That existence is a general property of extensions of finite groups with relatively prime orders (cf. M. Hall [30], th. 15.2.2, for example).

Here is a direct proof: Let s be an element of G_0 whose image in G_0/G_1 generates this cyclic group. Let e_0 be the order of G_0/G_1, and let p^n be the order of G_1. As p is prime to e_0, there is an integer N such that $p^N \equiv 1$ mod. e_0; replacing N by one of its multiples if necessary, we may further assume $N \geq n$. Set

$$t = s^{p^N}.$$

Then $t^{e_0} = s^{e_0 p^N} = 1$, since $e_0 p^N$ is a multiple of the order of G_0. On the other hand, as $p^N \equiv 1 \bmod. e_0$, the image of t in G_0/G_1 is equal to that of s. It follows that the subgroup H of G_0 generated by t is cyclic of order e_0 and projects isomorphically onto G_0/G_1. □

Remark. Conversely, it can be shown that *every group satisfying* (R_p) *is the inertia group* for an extension of type considered here. It would be interesting to go further and to characterise the group G_0, *provided with the filtration of the* G_i, but that appears to be much more difficult.

Corollary 5. *The group* G_0 *is solvable. If* $\bar{K}$ *is a finite field, then* G *too is solvable.*

The first assertion follows from the solvability of p-groups, cyclic groups and extensions of solvable groups. The second follows from this and the fact that $G/G_0 = G(\bar{L}/\bar{K})$ is cyclic when $\bar{K}$ is finite.

Here is a simple application of cor. 2:

Proposition 8. *Let* k *be an algebraically closed field of characteristic zero, and let* $K = k((T))$. *Then the algebraic closure* K_a *of the field* K *is the union of the fields* $K_n = k((T^{1/n}))$, *for integral* $n \geq 1$.

Let $L \subset K_a$ be a Galois extension of K of finite degree, G its Galois group. Since $\bar{K} = k$ is algebraically closed, $G = G_0$ (Chap. I, §7, prop. 20); cor. 2 then shows that G is cyclic. Let L′ be another finite Galois extension of K, contained in K_a, whose degree is a multiple of that of L. As the composite extension L′L is cyclic, the group G(L′L/L′) is contained in the group G(L′L/L), which implies that L is contained in L′. Applying this result with $L' = K_n$, for appropriate n, shows that L is contained in the union of the K_n. □

Corollary. *The Galois group* $G(K_a/K)$ *is isomorphic to* $\hat{\mathbf{Z}}$.

Remark. Proposition 8 can be considered the formal analogue of "Puiseux's theorem". (See exer. 8.)

We will now give several properties of the *commutators* vis-à-vis the filtration $\{G_i\}$, cf. Speiser [61].

First of all, since all G_i are normal subgroups of G_0, the group G_0 acts on G_i/G_{i+1} by inner automorphisms. Let us determine these actions in terms of the isomorphisms θ_i of props. 6 and 7.

Proposition 9. *If* $s \in G_0$ *and* $\tau \in G_i/G_{i+1}$, $i \geq 1$, *then*

$$\theta_i(s\tau s^{-1}) = \theta_0(s)^i \theta_i(\tau).$$

[The formula makes sense because $\theta_i(\tau)$ belongs to $\mathfrak{p}_L^i/\mathfrak{p}_L^{i+1}$ which is a one-dimensional vector space over $\bar{L}$, and $\theta_0(s)$ is an element of the multiplicative group of $\bar{L}$.]

Let $t \in G_i$ be a representative of τ. Write $\theta_i(t)$ instead of $\theta_i(\tau)$ in what follows. Let $\pi' = s^{-1}(\pi)$. Then $t(\pi') = \pi'(1 + a)$, with $a \in \mathfrak{p}_L^i$, and $\theta_i(t)$ is the class $\bar{a}$ of $a \bmod. \mathfrak{p}_L^{i+1}$. Applying s to this equation gives

$$sts^{-1}(\pi) = st(\pi') = \pi(1 + s(a)),$$

which shows that $\theta_i(sts^{-1})$ is the class of $s(a) \bmod. \mathfrak{p}_L^{i+1}$. But one can write $a = b\pi^i$, so that if $s(\pi) = u\pi$, then $s(a) = s(b)u^i\pi^i$; as $s(b) \equiv b \bmod. \mathfrak{p}_L$, and $\bar{u} = \theta_0(s)$, one sees that the class of $s(a)$ is indeed equal to the product of $\theta_0(s)^i$ with the class of a. □

Corollary 1. *If* $s \in G_0$ *and* $t \in G_i$, $i \geq 1$, *then* $sts^{-1}t^{-1} \in G_{i+1}$ *if and only if* $s^i \in G_1$ *or* $t \in G_{i+1}$.

Indeed, $sts^{-1}t^{-1} \in G_{i+1}$ is equivalent to $sts^{-1} \equiv t \bmod. G_{i+1}$, which in turn is equivalent to

$$\theta_i(sts^{-1}) = \theta_i(t),$$

whence the result follows from the proposition.

Corollary 2. *Suppose* G *is abelian, and let* e_0 *be the order of* G_0/G_1. *Then for any integer i not divisible by* e_0, $G_i = G_{i+1}$.

Indeed, if $t \in G_i$, and if s is an element of G_0 which generates G_0/G_1, then by hypothesis $sts^{-1}t^{-1} = 1$, and since $s^i \notin G_1$, cor. 1 implies $t \in G_{i+1}$.

Consider next the commutator $sts^{-1}t^{-1}$ when $s \in G_i$, $t \in G_j$ and $i, j \geq 1$.

Proposition 10. *If* $s \in G_i$, $t \in G_j$, *and* $i, j \geq 1$, *then* $sts^{-1}t^{-1} \in G_{i+j+1}$.

We will also prove:

Proposition 11. *The integers* $i \geq 1$ *such that* $G_i \neq G_{i+1}$ *are all congruent to one another* mod. p (where p is the characteristic of the residue field $\bar{L}$.)

We begin by proving a weaker result:

Lemma 2. *With the hypotheses of prop.* 10, *we have* $sts^{-1}t^{-1} \in G_{i+j}$ *and*

$$\theta_{i+j}(sts^{-1}t^{-1}) = (j-i)\theta_i(s)\theta_j(t).$$

[This formula is meaningful because, as was already noted above, the direct sum of the $\mathfrak{p}_L^i/\mathfrak{p}_L^{i+1}$ has a natural structure of graded $\bar{L}$-algebra.]

We put $s(\pi) = \pi(1+a)$, $t(\pi) = \pi(1+b)$, $a \in \mathfrak{p}_L^i$, $b \in \mathfrak{p}_L^j$. Then

$$st(\pi) = \pi(1+a)(1+s(b)) = \pi(1+c),$$

with $c = a + s(b) + a\,.\,s(b)$.

Similarly, $ts(\pi) = \pi(1+d)$, with $d = b + t(a) + b\,.\,t(a)$.

Put $a = \alpha\pi^i$, $b = \beta\pi^j$, $\alpha, \beta \in A_L$. Then

$$s(b) = s(\beta)s(\pi)^j = s(\beta)\pi^j(1+a)^j.$$

Since $s \in G_i$, we have $s(\beta) \equiv \beta \bmod. \mathfrak{p}_L^{i+1}$, and since $a \in \mathfrak{p}_L^i$, we get

$$(1+a)^j \equiv 1 + ja \bmod. \mathfrak{p}_L^{i+1} \quad \text{(and even mod. } \mathfrak{p}_L^{2i}\text{)}.$$

Hence:

$$s(b) \equiv \beta\pi^j(1+ja) \bmod. \mathfrak{p}_L^{i+j+1}$$
$$\equiv b + j\,.\,ab \bmod. \mathfrak{p}_L^{i+j+1}$$

whence:

$$c \equiv a + b + (j+1)ab \bmod. \mathfrak{p}_L^{i+j+1},$$

and similarly:

$$d \equiv a + b + (i+1)ab \bmod. \mathfrak{p}_L^{i+j+1}.$$

Put $\pi' = ts(\pi)$. Then

$$sts^{-1}t^{-1}(\pi') = st(\pi) = \pi(1+c) = \pi'(1+c)(1+d)^{-1}$$
$$= \pi'(1+e),$$

with

$$e = (c - d)/(1 + d) \equiv (j - i)ab \mod. \mathfrak{p}_L^{i+j+1}.$$

We conclude first of all that $sts^{-1}t^{-1}$ belongs to G_{i+j}; secondly, as the class of a (resp. b, resp. e) is equal to $\theta_i(s)$ (resp. $\theta_j(t)$, resp. $\theta_{i+j}(sts^{-1}t^{-1})$), we do get

$$\theta_{i+j}(sts^{-1}t^{-1}) = (j - i)\theta_i(s)\theta_j(t). \quad \square$$

Let us now prove prop. 11. If $G_1 = \{1\}$, there is nothing to prove; if not, let j be the largest integer such that $G_j \neq \{1\}$, so that $G_{j+1} = \{1\}$. Let $i \geq 1$ be an integer for which $G_i \neq G_{i+1}$; we must show that $j \equiv i \mod. p$. Let s (resp. t) be an element of G_i (resp. G_j) not belonging to G_{i+1} (resp. G_{j+1}). By lemma 2, the commutator $sts^{-1}t^{-1}$ belongs to G_{i+j}; it is therefore equal to 1, and $\theta_{i+j}(sts^{-1}t^{-1}) = 0$. As $\theta_i(s)$ and $\theta_j(t)$ are non-zero, lemma 2 shows that $j - i \equiv 0 \mod. p$. $\square$

Let us return to prop. 10. If $s \in G_{i+1}$, or $t \in G_{j+1}$, lemma 2 shows that $sts^{-1}t^{-1} \in G_{i+j+1}$. Otherwise, prop. 11 tells us that $j \equiv i \mod. p$, and lemma 2 implies $\theta_{i+j}(sts^{-1}t^{-1}) = 0$, which means that $sts^{-1}t^{-1}$ belongs to G_{i+j+1}. $\square$

Remarks. 1) When G is *abelian*, one can prove more precise congruences than those given in prop. 11. We will come back to this in §3 as well as in Chapter VI.

2) The fact that $s \in G_i$, $t \in G_j$ implies $sts^{-1}t^{-1} \in G_{i+j}$ is a special case of results due to Lazard on filtered groups ([44], Chap. I, n° 3). In his terminology, prop. 10 means that *the Lie algebra* $\mathrm{gr}(G_1) = \sum_{i \geq 1} G_i/G_{i+1}$ *is abelian.*

Exercises

1. If G' is a quotient group of $G = G(L/K)$, show that G'_0 and G'_1 are the images in G' of G_0 and G_1. Deduce from this (by passage to the limit) a definition of G_0 and G_1 when G is the Galois group of an infinite extension.

2. Suppose that $\bar{K}$ is a perfect field. Let K_s be the separable closure of K, and let $G = G(K_s/K)$ be its Galois group. Define the subgroups G_0 and G_1 as in exer. 1.
   a) Let $\bar{K}_s$ be the separable (or algebraic—it is the same) closure of $\bar{K}$. Show that $G/G_0 = G(\bar{K}_s/\bar{K})$.
   b) For every integer $n \geq 1$, let μ_n be the group of nth roots of unity in $\bar{K}_s$. If m divides n, let $f_{mn}: \mu_n \to \mu_m$ be the homomorphism $x \mapsto x^{n/m}$, and let μ be the projective limit of the system (μ_n, f_{mn}). Show that G_0/G_1 is (canonically) isomorphic to μ. Deduce that it is (non-canonically) isomorphic to the product $\prod Z_q$ of the groups of q-adic integers, q running through the set of primes distinct from the characteristic of $\bar{K}$. Show that the isomorphism $G_0/G_1 = \mu$ is compatible with the operations of G/G_0 on G_0/G_1 and on μ.
   c) Deduce from the above the structure of the group G/G_1 when $\bar{K}$ is a finite field.

3. Suppose that the characteristic of $\bar{K}$ is $p \neq 0$. Put $e = v_L(p)$ (the absolute ramification index of L).

a) Let $s \in G_i$, $i \geq 1$, and let $s(\pi) = \pi(1 + a)$, with $a \in \mathfrak{p}_L^i$. Put $\varphi = s - 1$; this a K-linear map of L into itself. Show that $\varphi(x) \equiv jax \bmod. \mathfrak{p}_L^{i+j+1}$ if $x \in \mathfrak{p}_L^j$.

b) Put $\psi = s^p - 1$. Using the binomial formula, show that for all $x \in \mathfrak{p}_L^j$,

(i) $\psi(x) \equiv pjax \bmod. \mathfrak{p}_L^{i+j+e+1}$ if $i > e/(p-1)$,
(ii) $\psi(x) \equiv pjax + j(1 - j^{p-1})a^p x \bmod. \mathfrak{p}_L^{i+j+e+1}$ if $i = e/(p-1)$,
(iii) $\psi(x) \equiv j(1 - j^{p-1})a^p x \bmod. \mathfrak{p}_L^{j+pi+1}$ if $i < e/(p-1)$.

c) Suppose that $i > e/(p-1)$ and $s \notin G_{i+1}$. Deduce from b) that $s^p \in G_{i+e}$ and $s^p \notin G_{i+e+1}$. Show that this contradicts the fact that the order of s is a power of p, hence $G_i = \{1\}$ if $i > e/(p-1)$.

d) Using a similar argument, show that, when $i = e/(p-1)$, the group G_i is either trivial or cyclic of order p, the latter case being only possible if $i \equiv 0 \bmod. p$.

e) If $i < e/(p-1)$, and if $i \not\equiv 0 \bmod. p$, show that $s^p \in G_{pi+1}$. If $i \equiv 0 \bmod. p$, show that $s^p \in G_{pi}$, and that $\theta_{pi}(s^p) = \theta_i(s)^p$; deduce that for such a value of i, the group G_i/G_{i+1} is either trivial or cyclic of order p, the latter case being only possible if there exists an integer $h > 0$ such that $p^h i = e/(p-1)$.

f) Show that if the integers $i \geq 1$ such that $G_i \neq G_{i+1}$ are all divisible by p (cf. prop. 11), then these integers have the form

$$p^k i_0, \qquad 1 \leq k \leq h, \quad \text{where } p^h i_0 = e/(p-1),$$

and the group G_1 is cyclic of order p^h.

4. Suppose that K contains a primitive pth root of unity. Let L be the extension of K obtained by means of the equation $x^p = \pi$. Show that L is a cyclic totally ramified extension of K. If s is a generator of its Galois group, show that $s \in G_i$, $s \notin G_{i+1}$, where $i = e/(p-1)$, with e denoting the absolute ramification index of L (cf. exer. 3, d).

5. Let e_K be the absolute ramification index of K, and let n be a positive integer prime to p and (strictly) less than $pe_K/(p-1)$; let y be an element of valuation $-n$.
a) Show that the *Artin-Schreier equation*

$$x^p - x = y$$

is irreducible over K, and defines an extension L/K which is cyclic of degree p. (Show that if x is a root of this equation, then the other roots have the form $x + z_i$ $(0 \leq i < p)$, with $z_i \in A_L$ and $z_i \equiv i \bmod. \mathfrak{p}_L$.)
b) Let $G = G(L/K)$. Show that $G_n = G$ and $G_{n+1} = \{1\}$.
[Further information on exers. 3, 4, and 5 can be found in Ore [49] and MacKenzie-Whaples [45].]

6. Give a new proof of prop. 13 of Chap. III based on the expression for the different in terms of the ramification groups.

7. Let k be an algebraically closed field of characteristic zero, and let E be the field of formal series $\sum c_r T^r$, where the coefficients c_r belong to k, and where the exponents r are *rational numbers* tending toward $+\infty$. Show that E is algebraically closed. (Note that E is the completion of the algebraic closure of $k((T))$, according to prop. 8.)

8. Let k be a complete algebraically closed valued field of characteristic zero, and let $k\{\{T\}\}$ be the field of fractions of the ring of convergent series having coefficients in k. Show that the algebraic closure of $k\{\{T\}\}$ is the union of the fields $k\{\{T^{1/n}\}\}$.

§3. The Functions ϕ and ψ; Herbrand's Theorem

We keep the notation and hypotheses of the two preceding sections. If u is a real number ≥ -1, G_u denotes the ramification group G_i, where i is the smallest integer $\geq u$. Thus

$$s \in G_u \Leftrightarrow i_G(s) \geq u + 1.$$

Put

$$\varphi(u) = \int_0^u \frac{dt}{(G_0 : G_t)}.$$

When $-1 \leq t \leq 0$, our convention is that $(G_0 : G_t)$ is equal to $(G_{-1} : G_0)^{-1}$ for $t = -1$ and is equal to $1 = (G_0 : G_0)^{-1}$ for $-1 < t \leq 0$. The function $\varphi(u)$ is thus equal to u between -1 and 0.

[When we need to specify the extension L/K, we write $\varphi_{L/K}$ instead of φ.]

It is easy to make the function φ explicit: if $m \leq u \leq m + 1$, where m is a positive integer, then

$$\varphi(u) = \frac{1}{g_0}(g_1 + \cdots + g_m + (u - m)g_{m+1}), \quad \text{with } g_i = \text{Card}(G_i).$$

In particular,

$$\varphi(m) + 1 = \frac{1}{g_0} \sum_{i=0}^{i=m} g_i.$$

Proposition 12

a) *The function φ is continuous, piecewise linear, increasing, and concave.*
b) $\varphi(0) = 0$.
c) *If we denote by φ'_d and φ'_g the right and left derivatives of φ, then*

$$\varphi'_g(u) = \varphi'_d(u) = 1/(G_0 : G_u) \quad \textit{if } u \textit{ is not an integer},$$
$$\varphi'_g(u) = 1/(G_0 : G_u), \qquad \varphi'_d(u) = 1/(G_0 : G_{u+1}) \quad \textit{if } u \textit{ is an integer}.$$

The verification is immediate. Note that these properties characterise φ. The map φ is a homeomorphism of the half-line $[-1, +\infty[$ onto itself. Denote by ψ (or $\psi_{L/K}$) the inverse map.

Proposition 13

a) *The function ψ is continuous, piecewise linear, increasing and convex.*
b) $\psi(0) = 0$.
c) *If $v = \phi(u)$, then $\psi'_g(v) = 1/\varphi'_g(u)$, $\psi'_d(v) = 1/\varphi'_d(u)$. (In particular, ψ'_g and ψ'_d only take on integer values.)*
d) *If v is an integer, so is $u = \psi(v)$.*

Properties a), b), c) are immediate. To prove d), let m be an integer such that $m \leq u \leq m + 1$. Then

$$g_0 v = g_1 + \cdots + g_m + (u - m)g_{m+1}.$$

As G_{m+1} is contained in $G_0, \ldots, G_m$, its order g_{m+1} divides $g_0, \ldots, g_m$. It follows that $u - m$ is an integer, whence u is. □

We can now define the *upper numbering* of the ramification groups. Put:

$$G^v = G_{\psi(v)}$$

or, equivalently,

$$G^{\varphi(u)} = G_u.$$

We have $G^{-1} = G$, $G^0 = G_0$, and $G^v = \{1\}$ for v sufficiently large. The knowledge of the G^v is equivalent to that of the G_u: namely, it is easy to see that

$$\psi(v) = \int_0^v (G^0 : G^w)\, dw.$$

The determination of the ramification groups of a quotient group can now be simply stated as follows:

Proposition 14. *If* H *is a normal subgroup of* G, *then* $(G/H)^v = G^vH/H$ *for all* v.

(In other words, when the upper numbering is used, the ramification groups of G/H are the images of those of G; the upper numbering is adapted to quotients, just as the lower numbering is adapted to subgroups.)

Let $K' \subset L$ be the fixed field of H; then:

Proposition 15. *The functions* φ *and* ψ *satisfy the transitivity formulas*:

$$\varphi_{L/K} = \varphi_{K'/K} \circ \varphi_{L/K'} \quad \textit{and} \quad \psi_{L/K} = \psi_{L/K'} \circ \psi_{K'/K}.$$

We first prove several lemmas.

Lemma 3. $\varphi_{L/K}(u) = \dfrac{1}{g_0} \sum_{s \in G} \mathrm{Inf}(i_G(s), u + 1) - 1.$

Let $\theta(u)$ be the function defined by the right side of this equation; it is a continuous piecewise-linear function that vanishes at $u = 0$. If $m < u < m + 1$, where m is an integer, the derivative $\theta'(u)$ is equal to the number of $s \in G$ such that $i_G(s) \geq m + 2$, multiplied by $1/g_0$; thus $\theta'(u) = 1/(G_0 : G_{m+1})$, and as this is also the value of $\phi'(u)$, the functions θ and ϕ must coincide. □

Lemma 4. *Let* $\sigma \in G/H$, *and let* $j(\sigma)$ *be the upper bound of the integers* $i_G(s)$ *as* s *runs through the pre-image of* σ *in* G. *Then*

$$i_{G/H}(\sigma) - 1 = \varphi_{L/K'}(j(\sigma) - 1).$$

Let $s \in G$ have image σ and $i_G(s) = j(\sigma)$. Put $m = i_G(s)$. If $t \in H$ belongs to H_{m-1}, then $i_G(t) \geq m$ whence $i_G(st) \geq m$, so $i_G(st) = m$. If, on the other hand, t does not belong to H_{m-1}, then $i_G(t) < m$, and $i_G(st) = i_G(t)$. In either case,

$i_G(st) = \mathrm{Inf}(i_G(t), m)$. Applying prop. 3 of §1 gives

$$i_{G/H}(\sigma) = \frac{1}{e_{L/K'}} \sum_{t \in H} \mathrm{Inf}(i_G(t), m).$$

But $i_G(t) = i_H(t)$, and $e_{L/K'} = \mathrm{Card}(H_0)$. Applying lemma 3 to the group H, we find

$$i_{G/H}(\sigma) = 1 + \varphi_{L/K'}(m - 1). \quad \square$$

Lemma 5. *If* $v = \phi_{L/K'}(u)$ *then* $G_uH/H = (G/H)_v$.

(This statement is usually called "Herbrand's theorem".)

Keep the notation of the preceding lemma. Then

$$\sigma \in G_uH/H \Leftrightarrow j(\sigma) - 1 \geq u \Leftrightarrow \varphi_{L/K'}(j(\sigma) - 1) \geq \varphi_{L/K'}(u)$$
$$\Leftrightarrow i_{G/H}(\sigma) - 1 \geq \varphi_{L/K'}(u) \Leftrightarrow \sigma \in (G/H)_v. \quad \square$$

PROOF OF PROP. 15. Let $u > -1$ not be an integer. The derivative of the composite function $\varphi_{K'/K} \circ \varphi_{L/K'}$ is equal to

$$\varphi'_{K'/K}(v) . \varphi'_{L/K'}(u), \quad \text{with } v = \varphi_{L/K'}(u).$$

It can therefore be written in the form

$$(\mathrm{Card}(G/H)_v/e_{K'/K}) . (\mathrm{Card}(H_u)/e_{L/K'}).$$

Applying lemma 5, we see that this expression is equal to $\mathrm{Card}(G_u)/e_{L/K}$, i.e., to the derivative $\varphi'_{L/K}(u)$, which gives the desired equality. The formula for ψ follows. $\square$

PROOF OF PROP. 14. We have

$$(G/H)^v = (G/H)_x \quad \text{with } x = \psi_{K'/K}(v).$$

By lemma 5, $(G/H)_x = G_wH/H$, with $w = \psi_{L/K'}(x) = \psi_{L/K}(v)$ according to prop. 15. Thus $G_w = G^v$. $\square$

Remarks. 1) Let L/K be an infinite Galois extension, G its Galois group. One can define G^v as $\varprojlim G(L'/K)^v$, L' running through the set of finite Galois subextensions of L. The G^v again form a filtration of G; this filtration is *left continuous*: $G^v = \bigcap_{w<v} G^w$. One says that v is a "jump" for the filtration if $G^v \neq G^{v+\varepsilon}$ for all $\varepsilon > 0$. Even when L/K is finite, *a jump need not be an integer* (cf. exer. 2).

2) Let E be a subextension of the Galois extension L/K (the latter having separable residue extension, as always). Define $\varphi_{E/K}$ to be $\varphi_{L/K} \circ \psi_{L/E}$; prop. 15 shows that this definition is independent of the extension L chosen and that the $\varphi_{E/K}$ satisfy the same transitivity formulas as the $\varphi_{L/K}$.

3) The double numbering of the ramification groups can also be interpreted as follows:

Let $\Gamma_L = L^*/U_L$ be the value group of L, and let $\mathbf{R}_L = \Gamma_L \otimes \mathbf{R}$. The canonical isomorphism $\Gamma_L = \mathbf{Z}$ identifies $\mathbf{R}_L$ with $\mathbf{R}$; let T_L be the image of

$[-1, +\infty[$ under this isomorphism. It is natural to consider that the groups G_u are *indexed by elements* u *of* T_L (the valuation v_L of L was used to define them). Similarly, $\varphi_{L/K}$ can be interpreted as an *isomorphism of the ordered set* T_L *onto the ordered set* T_K and $\psi_{L/K}$ as the inverse isomorphism. From this point of view, the upper numbering becomes simply the indexing of the ramification groups *by the elements of* T_K. Prop. 15 shows that the $(T_L, \varphi_{L/K})$ form a transitive system of isomorphisms. We can use this to define the ordered set:

$$T = \varprojlim (T_L, \varphi_{L/L'}) = \varinjlim (T_L, \psi_{L/L'}),$$

where L runs through the set of finite subextensions of a given Galois extension E/K. An element t of T is by definition a system $(t_L)_{L \subset E}$, with $t_L \in T_L$ and $\varphi_{L/L'}(t_L) = t_{L'}$ if $L \supset L'$. If now G is the Galois group of an extension L/L′, put

$$G(t) = G_{t_L} = G^{t_{L'}}, \quad \text{for all } t \in T.$$

If H is a subgroup of G, then $H(t) = H \cap G(t)$, and if H is normal, $(G/H)(t) = G(t)H/H$: *the indexing of the ramification groups by* T *is compatible both with passage to a subgroup and with passage to a quotient group.*

The upper numbering is particularly interesting in the abelian case, because of the next result.

Theorem (Hasse-Arf). *If* G *is an abelian group, and if* v *is a jump in the filtration* G^v, *then* v *is an integer.*

[In other words, if $G_i \neq G_{i+1}$, then $\varphi(i)$ is an integer.]

A proof will be given in §7, Chap. V.

Remark. This theorem was first proved by Hasse (cf. [32] and [33]) for the case of a finite residue field (see also Chap. XV, §2). The general case is due to Arf [2]; a different proof, based on the proalgebraic structure of the group of units, can be found in [59]; see also Sen [106].

Example. Suppose G is cyclic of order p^n, where p is the characteristic of $\bar{K}$, and let $G(i)$ be the subgroup of G of order p^{n-i}. The theorem of Hasse-Arf means that there exist strictly positive integers

$$i_0, i_1, \ldots, i_{n-1},$$

such that the ramification groups of G are given by the formulas:

$$\begin{aligned}
G_0 = \cdots = G_{i_0} &= G = G^0 = \cdots = G^{i_0},\\
G_{i_0+1} = \cdots = G_{i_0+pi_1} &= G(1) = G^{i_0+1} = \cdots = G^{i_0+i_1},\\
G_{i_0+pi_1+1} = \cdots = G_{i_0+pi_1+p^2i_2} &= G(2) = G^{i_0+i_1+1}\\
&= \cdots = G^{i_0+i_1+i_2},\\
&\;\vdots\\
G_{i_0+pi_1+\cdots+p^{n-1}i_{n-1}+1} &= \{1\} = G^{i_0+\cdots+i_{n-1}+1}.
\end{aligned}$$

EXERCISES

1. As an application of Herbrand's theorem, recover the formulas
$$G_1H/H = (G/H)_1, \qquad G_0H/H = (G/H)_0,$$
as well as the corollary to prop. 3 of §1.

2. Let $G = \{\pm 1, \pm i, \pm j, \pm k)$ be the group of quaternions, $C = \{\pm 1\}$ its center. Suppose that G is the Galois group of a totally ramified extension L/K and that $G_4 = \{1\}$ (this is actually possible—cf. [57], n° 5). Show that then $G = G_0 = G_1$, $C = G_2 = G_3$. Deduce the formulas
$$G^v = G \quad \text{for } v \leq 1,$$
$$G^v = C \quad \text{for } 1 < v \leq \tfrac{3}{2},$$
$$G^v = \{1\} \quad \text{for } v > \tfrac{3}{2}.$$

3. Show that, if the theorem of Hasse-Arf is true for every cyclic group of prime power order, then it is true for every group (use cor. 2 to prop. 9, together with Herbrand's theorem).

§4. Example: Cyclotomic Extensions of the Field $\mathbf{Q}_p$

We are going to study the fields obtained by adjoining to $\mathbf{Q}_p$ the roots of unity. We begin with those of order prime to p.

Proposition 16. *Let K be a field complete under a discrete valuation, and having finite residue field $k = \mathbf{F}_q$, with $q = p^f$. Given an integer n prime to p. Let K_n (resp. k_n) be the field obtained by adjoining to K (resp. k) the nth roots of unity. The field K_n is an unramified extension of K and has residue field k_n. If ζ is a primitive nth root of unity, then $A_{K_n} = A_K[\zeta]$. The Galois group $G(K_n/K)$ can be identified with the group $G(k_n/k)$; it is cyclic, generated by an automorphism s such that $s(z) = z^q$ for every nth root of unity z.*

Let L be the unramified extension of K which has k_n as residue field (cf. Chap. III, §5), and let S be the set of multiplicative representatives in L (cf. Chap. II, §4). Let $\bar{\zeta}$ be a primitive nth root of unity in k_n, and let ζ be its multiplicative representative. Clearly ζ is a primitive nth root of unity; as $\bar{\zeta}$ generates k_n/k, ζ generates the A_K-algebra A_L, *a fortiori* the extension L/K. Thus $L = K_n$, and we see that $G(K_n/K) = G(k_n/k)$ is generated by an element s such that $s(a) \equiv a^q$ for all $a \in A_L$. If $z \in S$, then also $s(z) \in S$, so $s(z) \equiv z^q$ implies $s(z) = z^q$. □

Corollary 1. *The degree $[K_n : K]$ is equal to the smallest integer $r \geq 1$ such that $q^r \equiv 1 \bmod. n$.*

Indeed, r is the smallest integer such that $s^r = 1$.

Corollary 2. *The maximal unramified extension K_{nr} of K is obtained by adjoining to K all the roots of unity of order prime to p. Its Galois group can* *be identified with $\hat{\mathbf{Z}}$; it admits a generator s such that $s(z) = z^q$ for every root of unity z of order prime to p.*

This results from prop. 16 by passage to the limit on n, once we note that the union of the k_n is the algebraic closure of k.

Remark. The element s is none other than the *Artin symbol* of $\mathfrak{p}_K$, in the sense of Chap. I, §8.

We next consider the roots of unity of order $n = p^m$, $m \geq 1$; this time we limit ourselves to the ground field $\mathbf{Q}_p$.

Proposition 17. *Let K_n be the field obtained from $K = \mathbf{Q}_p$ by adjoining a primitive nth root of unity ζ, with $n = p^m$. Then*

i) $[K_n : K] = \varphi(n) = (p-1)p^{m-1}$.
ii) *The Galois group $G(K_n/K)$ can be identified with the group $G(n)$ of invertible elements in the ring $\mathbf{Z}/n\mathbf{Z}$.*
iii) *K_n is a totally ramified extension of K. The element $\pi = \zeta - 1$ is a uniformizer of K_n, and $A_{K_n} = A_K[\zeta]$.*

We know *a priori* that $G(K_n/K)$ can be identified with a subgroup of $G(n)$ (cf. Bourbaki, *Alg.*, Chap. V, §11); as the order of $G(n)$ is $\varphi(n)$, we see that assertions i) and ii) are equivalent.

On the other hand, let $u = \zeta^{p^{m-1}}$; as this is a primitive pth root of unity, we have $u^{p-1} + u^{p-2} + \cdots + 1 = 0$, whence

$$\zeta^{(p-1)p^{m-1}} + \zeta^{(p-2)p^{m-1}} + \cdots + 1 = 0.$$

If F denotes the polynomial on the left side of this equation, we see that π is a root of the equation $F(1 + X) = 0$. That is an *Eisenstein equation* of degree $\varphi(n)$: indeed, its constant term is $F(1) = p$, and the reduction mod. p of this equation is $X^{\varphi(n)} = 0$, as is easily seen. Applying prop. 18 of Chap. I, we see simultaneously that $[K_n : K] = \varphi(n)$, that π is a uniformizer of K_n, and that A_{K_n} is generated by π (or, what amounts to the same, by ζ). □

We now determine the ramification groups of $G = G(K_n/K)$. If v is an integer, $0 \leq v \leq m$, denote by $G(n)^v$ the subgroup of $G(n)$ consisting of all elements a such that $a \equiv 1 \bmod. p^v$. The group $G(n)/G(n)^v$ can be identified with the group $G(p^v)$, i.e., with the Galois group of the extension K_{p^v}/K. Thus $G(n)^v = G(K_n/K_{p^v})$.

Proposition 18. *The ramification groups G_u of $G(K_n/K)$ are:*

$$\begin{array}{ll} & G_0 = G, \\ \text{if } 1 \leq u \leq p-1, & G_u = G(n)^1 \\ \text{if } p \leq u \leq p^2-1, & G_u = G(n)^2 \\ \vdots & \vdots \\ \text{if } p^{m-1} \leq u, & G_u = G(n)^m = \{1\}. \end{array}$$

Let $a \neq 1$ be an element of $G(n)$, and let s_a be the corresponding element of G. Let v be the largest integer such that $a \equiv 1 \bmod. p^v$; then $a \in G(n)^v$ and $a \notin G(n)^{v+1}$. On the other hand,

$$i_G(s_a) = v_{K_n}(s_a(\zeta) - \zeta) = v_{K_n}(\zeta^q - \zeta) = v_{K_n}(\zeta^{q-1} - 1).$$

As ζ^{q-1} is a primitive p^{m-v}-th root of unity, $\zeta^{q-1} - 1$ is a uniformizer for the field $K_{p^{m-v}}$. Hence

$$i_G(s_a) = [K_n : K_{p^{m-v}}] = \mathrm{Card}(G(n)^{m-v}) = p^v.$$

If $p^{k-1} \leq u \leq p^k - 1$, then we see that s_a belongs to G_u if and only if $v \geq k$, which shows that indeed $G_u = G(n)^v$. □

Corollary *The jumps in the filtration* (G^v) *are integers. Moreover,*

$$G^v = G(n)^v \quad \textit{for } 0 \leq v \leq m,$$

and

$$G^v = \{1\} \quad \textit{for } v \geq m.$$

The jumps in the filtration (G_u) occur when $u = p^k - 1$, with $0 \leq k \leq m-1$ (the case $p = 2$ being exceptional: 0 is not a jump). It all comes down to proving that $\varphi_{L/K}(p^k - 1) = k$ for $k = 0, 1, \ldots, m-1$, which is easy.

Remarks. 1) The above result can be stated more suggestively after *passing to the limit on m*, i.e., after introducing the field K_{p^∞} equal to the union of the K_{p^m}. The Galois group of K_{p^∞} over K is the projective limit of the groups

$$G(p^m) = (\mathbf{Z}/p^m\mathbf{Z})^*;$$

as the projective limit of the rings $\mathbf{Z}/p^m\mathbf{Z}$ is the ring $\mathbf{Z}_p$ of p-adic integers, the limit of the $G(p^m)$ can naturally be identified with *the group* $U_p = \mathbf{Z}_p^*$ *of invertible elements of* $\mathbf{Z}_p$. If $\alpha \in U_p$, the element $s_\alpha \in G(K_{p^\infty}/K)$ associated to α transforms a p^m-th root of unity z into z^α, where this exponential has the obvious meaning. The group U_p is filtered by the U_p^n, as was mentioned in §2; this filtration can be extended to non-integral values v of the index by setting $U_p^v = U_p^n$ for n equal to the smallest integer $\geq v$. The above corollary then shows that *the canonical isomorphism of* U_p *onto* $G(K_{p^\infty}/K)$ *transforms the filtration* U_p^v *of* U_p *into the filtration of* $G(K_{p^\infty}/K)$ *obtained from the ramification groups* (with the upper numbering).

2) If we adjoin to $K = \mathbf{Q}_p$ *all* the roots of unity, we get the compositum of the extensions K_{nr} and K_{p^∞}; as these extensions are linearly disjoint over K (one being totally ramified and the other unramified), the Galois group of their compositum $K_{nr}K_{p^\infty}$ over K is isomorphic to the product of the Galois groups $G(K_{nr}/K)$ and $G(K_{p^\infty}/K)$, i.e., to $\hat{\mathbf{Z}} \times U_p$. We will see in Chap. XIV, §7, that $K_{nr}K_{p^\infty}$ is in fact *the maximal abelian extension of* K.

CHAPTER V

The Norm

Let L/K be a Galois extension satisfying the hypotheses of Chap. IV. The norm $N = N_{L/K}$ is a homomorphism of the multiplicative group L^* of L into the multiplicative group K^*; it maps U_L into U_K, and $v_K(Nx) = fv_L(x)$, where $f = [\bar{L}:\bar{K}]$. Thus we get a commutative diagram

$$\begin{array}{ccccccccc} 0 & \to & U_L & \to & L^* & \to & \mathbf{Z} & \to & 0 \\ & & \downarrow{\scriptstyle N} & & \downarrow{\scriptstyle N} & & \downarrow{\scriptstyle f} & & \\ 0 & \to & U_K & \to & K^* & \to & \mathbf{Z} & \to & 0. \end{array}$$

Following Hasse [33], we propose to determine the effect of N on the filtration of the U_L^n (resp. of the U_K^n); we restrict ourselves to those results that are independent of any hypotheses on the residue fields. The case of a finite (or, more generally, quasi-finite) residue field will be treated in Chap. XV.

Notation. If v is a non-negative real number, U_L^v denotes the group U_L^n, where n is the smallest integer $\geq v$.

§1. Lemmas

The next two lemmas are useful in comparing filtered groups.

Lemma 1. *Let*

$$\begin{array}{ccccccccc} 0 & \to & A' & \to & A & \to & A'' & \to & 0 \\ & & \downarrow{\scriptstyle f'} & & \downarrow{\scriptstyle f} & & \downarrow{\scriptstyle f''} & & \\ 0 & \to & B' & \to & B & \to & B'' & \to & 0 \end{array}$$

be a commutative diagram with exact rows. Then there is an exact sequence

$$0 \to \operatorname{Ker} f' \to \operatorname{Ker} f \to \operatorname{Ker} f'' \xrightarrow{\varphi} \operatorname{Coker} f' \to \operatorname{Coker} f \to \operatorname{Coker} f'' \to 0.$$

[Recall that $\operatorname{Ker} f = f^{-1}(0)$ is the *kernel* of f, and $\operatorname{Coker} f = B/f(A)$ is its *cokernel*.]

If $a'' \in \operatorname{Ker} f''$, choose an element $a \in A$ projecting onto a''; $f(a)$ is an element of B with image zero in B″; hence there is an element $b' \in B'$ whose image in B is equal to $f(a)$; the class of b' does not depend on the choice of a, and if it is denoted $\varphi(a'')$, then φ is a homomorphism of $\operatorname{Ker} f''$ into $\operatorname{Coker} f'$. The other maps appearing in the exact sequence are defined in the obvious way; as for the exactness itself, it is easy (cf. Cartan-Eilenberg [13], p. 40, or Bourbaki, *Alg. comm.*, Chap. I, §1). □

Lemma 2. *Let* A (resp. A′) *be an abelian group provided with a decreasing sequence of subgroups* A_n (resp. A'_n). *Suppose that* $A_0 = A$, $A'_0 = A'$, *and that* A *and* A′ *are complete Hausdorff spaces in the topologies defined by* A_n *and* A'_n (in other words, the canonical homomorphisms $A \to \lim A/A_n$ and $A' \to \lim A'/A'_n$ are bijective). *Let* $u: A \to A'$ *be a homomorphism sending* A_n *into* A'_n *for all n. If the homomorphisms*

$$u_n : A_n/A_{n+1} \to A'_n/A'_{n+1}$$

defined by u are all injective (resp. *surjective*), *then so is u.*

We quickly recall the proof of this result (cf. Bourbaki, *Alg. comm.*, Chap. III, §2):

If the u_n are injective, then $\operatorname{Ker}(u) \cap A_n = \operatorname{Ker}(u) \cap A_{n+1}$, whence by induction on n, $\operatorname{Ker}(u) \subset A_n$; as $\bigcap A_n = \{0\}$ (A being Hausdorff), we see that u is injective.

Suppose the u_n surjective, and let $a' \in A' = A'_0$. There exist $a_0 \in A_0$ and $a'_1 \in A'_1$ such that $u(a_0) = a' - a'_1$. Similarly, using the surjectivity of u_1, we see that there exist $a_1 \in A_1$ and $a'_2 \in A'_2$ such that $u(a_1) = a'_1 - a'_2$. Continuing this process, we construct sequences (a_n) and (a'_n). As A is complete, the series $a_0 + a_1 + \cdots$ converges to an element $a \in A$. Then $u(a) - a' \in A'_n$ for all n, whence $u(a) = a'$ by the Hausdorff condition. □

§2. The Unramified Case

Proposition 1. *If* L/K *is unramified, then* N *maps* U_L^n *into* U_K^n *for all n.*

Let $x = 1 + y$, with $y \in \mathfrak{p}_L^n$. Then $s(x) = 1 + s(y)$ for all $s \in G$, and $s(y) \in \mathfrak{p}_L^n$. Hence

$$(*) \qquad Nx = \prod_{s \in G} (1 + s(y)) \equiv 1 + \sum_{s \in G} s(y) \mod. \mathfrak{p}_L^{2n}.$$

But since L/K is unramified, $\mathfrak{p}_L^n \cap K = \mathfrak{p}_K^n$, and thus $Nx \equiv 1 \mod. \mathfrak{p}_K^n$. □

The map N defines by passage to the quotient maps

$$N_n : U_L^n/U_L^{n+1} \to U_K^n/U_K^{n+1}$$

that we want to determine. To do this recall (cf. Chap. IV, §2) that U_L/U_L^1 can be identified with the multiplicative group $\bar{L}^*$ of the residue field $\bar{L}$, and that U_L^n/U_L^{n+1} $(n \geq 1)$ can be identified with $\mathfrak{p}_L^n/\mathfrak{p}_L^{n+1}$, which is a one-dimensional vector space over $\bar{L}$ that we will also denote Ω_L^n. As L/K is unramified, Ω_L^n can be identified with $\bar{L} \otimes_K \Omega_K^n$. Taking these identifications into account, we get:

Proposition 2. *Suppose* L/K *is unramified. Then*

i) *The map* $N_0: \bar{L}^* \to \bar{K}^*$ *is just the norm in the residue extension* $\bar{L}/\bar{K}$.
ii) *For* $n \geq 1$, *the map* $N_n: \bar{L} \otimes_K \Omega_K^n \to \Omega_K^n$ *is just the map* $\mathrm{Tr}_{\bar{L}/\bar{K}} \otimes 1$.

Assertion i) is trivial. Assertion ii) follows from formula (∗) proved above. □

Proposition 3

a) $N(U_L^n) = U_K^n$ *for all* $n \geq 1$.
b) U_K/NU_L *is isomorphic to* $\bar{K}^*/N\bar{L}^*$.
c) K^*/NL^* *is isomorphic to* $\mathbf{Z}/f\mathbf{Z} \times \bar{K}^*/N\bar{L}^*$, *where* $f = [L:K] = [\bar{L}:\bar{K}]$.

We know that the trace is surjective in every separable extension; prop. 2 then shows that N_n is surjective for $n \geq 1$, and applying lemma 2 to $N: U_L^n \to U_K^n$, we get a).

We next apply lemma 1 to the commutative diagram below to get b):

$$\begin{array}{ccccccccc} 0 & \to & U_L^1 & \to & U_L & \to & \bar{L}^* & \to & 0 \\ & & \downarrow N & & \downarrow N & & \downarrow N & & \\ 0 & \to & U_K^1 & \to & U_K & \to & \bar{K}^* & \to & 0. \end{array}$$

Finally, the choice of a uniformizer π of K allows the identification of K^* with $\mathbf{Z} \times U_K$ and of L^* with $\mathbf{Z} \times U_L$, these identifications being compatible with the operations of G; c) then follows. □

Corollary. *The three following conditions are equivalent:*

(1) $(K^*: NL^*) = f$.
(2) $U_K = NU_L$.
(3) $\bar{K}^* = N\bar{L}^*$.

Obvious.

Remarks. 1) Condition (3) is satisfied when $\bar{K}$ is a finite field (or, more generally, quasi-finite—cf. Chap. XIII).

2) If v is a non-negative real number, prop. 1 shows that $N(U_L^v)$ is contained in U_K^v, with equality holding if $v > 0$.

Exercise

Extend propositions 1, 2, 3 to the case of an unramified extension that is not Galois.

§3. The Cyclic of Prime Order Totally Ramified Case

We suppose in this section that G is a cyclic group of prime order l, and that L/K is totally ramified (thus $\bar{L} = \bar{K}$). Denote by π a uniformizer of L.

Let s be a generator of G, and put $t = i(s) - 1$ (cf. Chap. IV, §1). The ramification groups of G are as follows:

$$G = G_0 = \cdots = G_t$$
$$\{1\} = G_{t+1} = \cdots.$$

According to Chap. IV, §2, we have $t \neq 0$ if and only if l is equal to the characteristic p of $\bar{K}$.

The function ψ of Chap. IV, §3, here becomes:

$$\psi(x) = \begin{cases} x & x \leq t \\ t + l(x - t) & x \geq t. \end{cases}$$

Lemma 3. *The different* $\mathfrak{D}$ *of the extension* L/K *is* $\mathfrak{p}_L^m$, *where* $m = (t + 1)(l - 1)$.

This is a consequence of prop. 4 of Chap. IV. □

Lemma 4. *For every integer* $n \geq 0$, $\mathrm{Tr}(\mathfrak{p}_L^n) = \mathfrak{p}_K^r$, *where* $r = [(m + n)/l]$ *and* $m = (t + 1)(l - 1)$.

[Recall that the symbol $[x]$ denotes the largest integer $\leq x$.]

Since the trace is A_K-linear, $\mathrm{Tr}(\mathfrak{p}_L^n)$ is an ideal in A_K. If r is any integer, prop. 7 of Chap. III shows that $\mathrm{Tr}(\mathfrak{p}_L^n) \subset \mathfrak{p}_K^r$ if and only if

$$\mathfrak{p}_L^n \subset \mathfrak{p}_K^r . \mathfrak{D}^{-1} = \mathfrak{p}_L^{lr-m}$$

i.e., if $r \leq (m + n)/l$. □

Lemma 5. *If* $x \in \mathfrak{p}_L^n$, *then*

$$N(1 + x) \equiv 1 + \mathrm{Tr}(x) + N(x) \bmod. \mathrm{Tr}(\mathfrak{p}_L^{2n}). \qquad (**)$$

For the computation to follow, it is convenient to use the exponential notation for the group G, i.e., to denote by x^s the transform of x by $s \in G$. Then

$$N(1 + x) = \prod_{s \in G} (1 + x^s)$$

and, expanding this product,

$$N(1 + x) = \sum x^u,$$

where u runs through those elements of the group algebra $\mathbf{Z}[G]$ which have the form $u = s_1 + \cdots + s_k$, the s_i being distinct elements of G. Put $k = n(u)$ (the *augmentation* of u). Those u of augmentation 0, 1 and l give respectively the terms 1, $\mathrm{Tr}(x)$ and $N(x)$ of the formula we seek. So it all comes down to proving that the sum of the other terms belongs to $\mathrm{Tr}(\mathfrak{p}_L^{2n})$. Let s generate G. If $u = us$, then u must be a multiple of the norm N, therefore of augmentation 0 or l. If $2 \leq n(u) \leq l - 1$, then $u \neq us$. Collecting together the $s^i u$, $0 \leq i \leq l - 1$, we get $\mathrm{Tr}(x^u)$; as $n(u) \geq 2$, we have $x^u \in \mathfrak{p}_L^{2n}$, whence $\mathrm{Tr}(x^u) \in \mathrm{Tr}(\mathfrak{p}_L^{2n})$. □

Proposition 4. *For every integer* $n \geq 0$, $\mathrm{N}(U_L^{\psi(n)}) \subset U_K^n$ *and* $\mathrm{N}(U_L^{\psi(n)+1}) \subset U_K^{n+1}$.

We will prove this proposition below. For the time being, notice that it allows, by passage to the quotient, the definition of homomorphisms

$$\mathrm{N}_n : U_L^{\psi(n)}/U_L^{\psi(n)+1} \to U_K^n/U_K^{n+1} \qquad (n \geq 0).$$

Specifically, let us identify as usual U_K/U_K^1 (resp. U_L/U_L^1) with $\bar{K}^*$ (resp. with $\bar{L}^* = \bar{K}^*$), as well as U_K^n/U_K^{n+1} (resp. U_L^n/U_L^{n+1}) with $\bar{K}$ (resp. with $\bar{L} = \bar{K}$), the latter isomorphisms coming from the choice of a uniformizer π' (resp. π) of K (resp. L). Taking these identifications into account, we have:

Proposition 5. *i) For* $n = 0$, *the map* $\mathrm{N}_0 : \bar{K}^* \to \bar{K}^*$ *is given by* $\mathrm{N}_0(\xi) = \xi^l$. *If* $t \neq 0$ (where t is the last index for which the corresponding ramification group is non-trivial), *this map is injective*; *if* $t = 0$, *its kernel is cyclic of order* l, *and is equal to the image of* G *under the map* $\theta_0 : G \to U_L/U_L^1$ (defined in Chapter IV, §2, prop. 7).

ii) For $1 \leq n < t$, *the map* $\mathrm{N}_n : \bar{K} \to \bar{K}$ *is given by* $\mathrm{N}_n(\xi) = \alpha_n \xi^p$, *for some* $\alpha_n \in \bar{K}^*$; *it is injective.*

iii) For $n = t \geq 1$, *the map* $\mathrm{N}_t : \bar{K} \to \bar{K}$ *is given by* $\mathrm{N}_t(\xi) = \alpha\xi^p + \beta\xi$, *for some* $\alpha, \beta \in \bar{K}^*$. *Its kernel is cyclic of order* $p = l$, *and is equal to* $\theta_t(G)$ (where θ_t is the map defined in Chapter IV, §2, prop. 7).

iv) For $n > t$, *the map* $\mathrm{N}_n : \bar{K} \to \bar{K}$ *is given by* $\mathrm{N}_n(\xi) = \beta_n \xi$, *for some* $\beta_n \in \bar{K}^*$; *it is bijective.*

Propositions 4 and 5 will be proved together. There are four cases to consider:

i) *The case* $n = 0$.

It is easy to see that $\mathrm{N}(U_L) \subset U_K$, $\mathrm{N}(U_L^1) \subset U_K^1$, and that $\mathrm{N}_0(\xi) = \xi^l$. If $t \neq 0$, then $l = p$, and N_0 is injective. If $t = 0$, then $l \neq p$, and the kernel of N_0 has order $\leq l$; but $\theta_0(G)$ consists of the classes in $\bar{K}^*$ of the $s(\pi)/\pi$, and evidently $\mathrm{N}(s(\pi)/\pi) = 1$; this shows that $\theta_0(G)$ is contained in $\mathrm{Ker}(\mathrm{N}_0)$, hence is equal to it.

ii) *The case* $1 \leq n < t$.

Since $t \geq 1$, we have $l = p$. Also $\psi(n) = n$. Let $x \in \mathfrak{p}_L^n$, then $\mathrm{N}(x) \in \mathfrak{p}_K^n$ since $v_L = v_K \circ \mathrm{N}$. By lemma 4, $\mathrm{Tr}(x) \in \mathfrak{p}_K^r$, where

$$r = \left[\frac{(t+1)(l-1)+n}{l}\right] = \left[n + 2 - \frac{2}{l}\right] \geq n + 1.$$

An analogous computation shows that $\mathrm{Tr}(\mathfrak{p}_L^{2n}) \in \mathfrak{p}_K^{n+1}$. Thus, by lemma 5,

$$N(1 + x) \equiv 1 + N(x) \ \mathrm{mod.}\, \mathfrak{p}_K^{n+1}.$$

As $N(x) \in \mathfrak{p}_K^n$, this formula shows that N does map U_L^n into U_K^n and U_L^{n+1} into U_K^{n+1}. Furthermore, if $x = u\pi^n$, then $N(x) \equiv u^p N(\pi)^n \ \mathrm{mod.}\, \mathfrak{p}_K^{n+1}$, and putting $N(\pi)^n = a_n \pi'^n$, this gives

$$N(1 + u\pi^n) \equiv 1 + a_n u^p \pi'^n \ \mathrm{mod.}\, \mathfrak{p}_K^{n+1}$$

whence $N_n(\xi) = \alpha_n \xi^p$, α_n denoting the image of a_n in $\bar{K}^*$.

iii) *The case* $n = t \geq 1$.

Here again $l = p$, and $\psi(t) = t$. Let $x \in \mathfrak{p}_L^t$. A computation analogous to the above shows that

$$N(1 + x) \equiv 1 + \mathrm{Tr}(x) + N(x) \ \mathrm{mod.}\, \mathfrak{p}_K^{t+1}.$$

By lemma 4, $\mathrm{Tr}(x) \in \mathfrak{p}_K^t$, and $\mathrm{Tr}(x) \in \mathfrak{p}_K^{t+1}$ if $x \in \mathfrak{p}_L^{t+1}$. It follows that N maps U_L^t into U_K^t and U_L^{t+1} into U_K^{t+1}. To determine N_t, it suffices to calculate $N(1 + u\pi^t)$, with $u \in A_K$; if we set $\mathrm{Tr}(\pi^t) = b\pi'^t$ and $N(\pi)^t = a\pi'^t$, we get

$$N(1 + u\pi^t) \equiv 1 + (bu + au^p)\pi'^t \ \mathrm{mod.}\, \mathfrak{p}_K^{t+1}$$

i.e., $N_t(\xi) = \beta\xi + \alpha\xi^p$, α and β denoting the images of a and of b in $\bar{K}$. Clearly $\alpha \neq 0$. If β were zero, N_t would be injective; but its kernel contains $\theta_t(G)$ which is cyclic of order p (that can be seen as in case i)); thus $\beta \neq 0$, and as the kernel of N_t is of order $\leq p$, we must have equality.

iv) *The case* $n > t$.

Here $\psi(n) = t + l(n - t)$. If $x \in \mathfrak{p}_L^{\psi(n)}$, lemma 4 shows that $\mathrm{Tr}(x) \in \mathfrak{p}_K^n$, and since $N(x) \in \mathfrak{p}_K^{\psi(n)}$ which is contained in $\mathfrak{p}_K^{n+1}$, we get the formula

$$N(1 + x) \equiv 1 + \mathrm{Tr}(x) \ \mathrm{mod.}\, \mathfrak{p}_K^{n+1}.$$

It follows, as before, that N maps $U_L^{\psi(n)}$ into U_K^n, $U_L^{\psi(n)+1}$ into U_K^{n+1}, and that $N_n(\xi) = \beta_n \xi$. If we had $\beta_n = 0$, we would have $\mathrm{Tr}(\mathfrak{p}_L^{\psi(n)}) \subset \mathfrak{p}_K^{n+1}$, which would contradict lemma 4. Thus $\beta_n \neq 0$. □

Corollary 1. *The homomorphism* N_n *is injective for all* n *except for* $n = t$, *in which case there is an exact sequence:*

$$0 \to G \xrightarrow{\theta_t} U_L^t/U_L^{t+1} \xrightarrow{N_t} U_K^t/U_K^{t+1}.$$

Obvious.

Corollary 2. *The homomorphism* N_n *is surjective for* $n > t$, *and, if* $\bar{K}$ *is perfect, for* $n < t$. *If* $\bar{K}$ *is algebraically closed, it is surjective for all* n.

Obvious.

Corollary 3. *We have* $N(U_L^{\psi(n)}) = U_K^n$ *for* $n > t$, *and* $N(U_L^{\psi(n)+1}) = U_K^{n+1}$ *for* $n \geq t$. *When* $\bar{K}$ *is algebraically closed, these equalities hold for all* n.

Filter $U_L^{\psi(n)}$ by the $U_L^{\psi(m)}$, and U_K^n by the U_K^m; by passage to the quotient, we get homomorphisms

$$U_L^{\psi(m)}/U_L^{\psi(m+1)} \to U_K^m/U_K^{m+1}$$

which are composites of N_m and of the canonical projection

$$U_L^{\psi(m)}/U_L^{\psi(m+1)} \to U_L^{\psi(m)}/U_L^{\psi(m)+1}.$$

If $m > t$, cor. 2 shows that these homomorphisms are surjective, and lemma 2 then shows that $N: U_L^{\psi(n)} \to U_K^n$ is surjective. A similar reasoning applies if $\bar{K}$ is algebraically closed and n is arbitrary. As for the formula $N(U_L^{\psi(n)+1}) = U_K^{n+1}$, it is a simple consequence of prop. 4 and of the fact that $U_L^{\psi(n)+1}$ contains $U_L^{\psi(n+1)}$.

Corollary 4. *We have* $N(U_L^{\psi(v)}) = U_K^v$ *if v is a real number $>t$ or if $\bar{K}$ is algebraically closed.*

Indeed, suppose that $n < v \leq n + 1$, where n is an integer. Then

$$\psi(n) < \psi(v) \leq \psi(n+1).$$

If m is the smallest integer $\geq \psi(v)$, then

$$\psi(n) + 1 \leq m \leq \psi(n+1).$$

We have $U_K^v = U_K^{n+1}$, $U_L^{\psi(v)} = U_L^m$, and cor. 3 yields $NU_L^m = U_K^{n+1}$.

Corollary 5. *If* $t = 0$, $\operatorname{Coker}(N_t) = \bar{K}^*/\bar{K}^{*l}$. *If* $t \neq 0$, $\operatorname{Coker}(N_t)$ *is isomorphic to* $\bar{K}/\wp(\bar{K})$, *with* $\wp(\xi) = \xi^p - \xi$.

If $t = 0$, we have seen that $N_t(\xi) = \xi^l$, whence the first assertion. If $t \neq 0$, we must show that $\operatorname{Coker}(N_t)$ is isomorphic to $\operatorname{Coker}(\wp)$. Now we have

$$N_t(\xi) = \alpha\xi^p + \beta\xi, \quad \text{with } \alpha, \beta \neq 0$$

and furthermore, there exists a non-zero element η in the kernel of N_t. We can therefore write

$$N_t(\xi) = \alpha\eta^p((\xi/\eta)^p - \xi/\eta) = \gamma\wp(\xi/\eta) \qquad (\gamma \neq 0),$$

whence $\operatorname{Im}(N_t) = \gamma \operatorname{Im}(\wp)$.

Corollary 6. *If $\bar{K}$ is perfect,* $N: U_L/U_L^n \to U_K/U_K^n$ *is an isomorphism for all* $n \leq t$.

By cors. 1 and 2, N_n is surjective for $n < t$. The result follows by induction on n, using lemma 1.

Corollary 7. *If $\bar{K}$ is perfect, the following three canonical homomorphism are isomorphisms:*

$$\operatorname{Coker}(N_t) \leftarrow U_K^t/N(U_L^t) \to U_K/NU_L \to K^*/NL^*$$

a) Apply lemma 1 to the diagram

$$\begin{array}{ccccccccc} 0 & \to & U_L^{t+1} & \to & U_L^t & \to & U_L^t/U_L^{t+1} & \to & 0 \\ & & \downarrow & & \downarrow & & \downarrow & & \\ 0 & \to & U_K^{t+1} & \to & U_K^t & \to & U_K^t/U_K^{t+1} & \to & 0. \end{array}$$

As $N(U_L^{t+1}) = U_K^{t+1}$, it follows that $U_K^t/N(U_L^t) \to \mathrm{Coker}(N_t)$ is an isomorphism.

b) Apply lemma 1 to the diagram

$$\begin{array}{ccccccccc} 0 & \to & U_L^t & \to & U_L & \to & U_L/U_L^t & \to & 0 \\ & & \downarrow & & \downarrow & & \downarrow & & \\ 0 & \to & U_K^t & \to & U_K & \to & U_K/U_K^t & \to & 0. \end{array}$$

Taking cor. 6 into account, it follows that $U_K^t/N(U_L^t) \to U_K/NU_L$ is an isomorphism.

c) Apply lemma 1 to the diagram

$$\begin{array}{ccccccccc} 0 & \to & U_L & \to & L^* & \to & \mathbf{Z} & \to & 0 \\ & & \downarrow & & \downarrow & & \wr\downarrow & & \\ 0 & \to & U_K & \to & K^* & \to & \mathbf{Z} & \to & 0. \end{array}$$

As $f = 1$, it follows that $U_K/NU_L \to K^*/NL^*$ is an isomorphism.

Remark. When $\bar{K}$ is a finite field, it is easy to show that $\bar{K}^*/\bar{K}^{*l}$ (resp. $\bar{K}/\wp(\bar{K})$) is cyclic of order l if l is prime to p (resp. if $l = p$). Combining cors. 5 and 7, we deduce that K^*/NL^* *is cyclic of order* l; we will return to this point in Chap. XIII.

§4. Extension of the Residue Field in a Totally Ramified Extension

Let L/K be a finite totally ramified extension. We suppose that its residue field $\bar{L} = \bar{K}$ is a *perfect* field.

Let $\bar{K}'$ be a finite extension of $\bar{K}$, and let K′ be the corresponding unramified extension of K (cf. Chap. III, §5). The extensions L/K and K′/K are *linearly disjoint*: indeed, if L′ denotes their compositum, then

$$e(L'/K) \geq e(L/K) = [L:K]$$
$$f(L'/L) \geq f(K'/K) = [K':K]$$

whence

$$[L':K] \geq [L:K].[K':K].$$

We may therefore identify L′ with $K' \otimes_K L$, and we see that L′/L is unramified, with residue extension $\bar{K}'/\bar{K}$. We will say that L′/K′ *is deduced from* L/K *by extension of the residue field from* $\bar{K}$ *to* $\bar{K}'$. This is an operation analogous to *the extension of the base field* in algebraic geometry (in fact, it is more than just an analogy—cf. Greenberg [25]). Let π_K (resp. π_L) be a uniformizer of K (resp. L); it is also a uniformizer of K′ (resp. L′). We assume henceforth that L/K is *Galois*, with Galois group G; then the same holds for L′/K′, and the ramification groups of G, as well as the homomorphisms θ_i of Chap. IV are *the same* for L/K and L′/K′ (one is tempted to say that these are "geometric" invariants).

Consider more particularly the case where G is *cyclic of prime degree l*. Then for all $n \geq 0$, we have the homomorphisms

$$\mathrm{N}_n : \mathrm{U}_{\mathrm{L}}^{\psi(n)}/\mathrm{U}_{\mathrm{L}}^{\psi(n)+1} \to \mathrm{U}_{\mathrm{K}}^n/\mathrm{U}_{\mathrm{K}}^{n+1}$$
$$\mathrm{N}'_n : \mathrm{U}_{\mathrm{L}'}^{\psi(n)}/\mathrm{U}_{\mathrm{L}'}^{\psi(n)+1} \to \mathrm{U}_{\mathrm{K}'}^n/\mathrm{U}_{\mathrm{K}'}^{n+1}.$$

We will study only the case $n \geq 1$ (the case $n = 0$ is treated similarly). By means of the uniformizers π_{K} and π_{L}, we can identify the four groups above with $\bar{\mathrm{K}}$, $\bar{\mathrm{K}}$, $\bar{\mathrm{K}}'$ and $\bar{\mathrm{K}}'$ respectively. The homomorphisms N_n and N'_n are thereby transformed into homomorphisms

$$\mathrm{N}_n : \bar{\mathrm{K}} \to \bar{\mathrm{K}}$$
$$\mathrm{N}'_n : \bar{\mathrm{K}}' \to \bar{\mathrm{K}}'.$$

These homomorphisms are given by polynomials (of degree 1 or p, according to the value of n) whose coefficients have been determined in the course of the proof of props. 4 and 5. This determination shows that the coefficients in question *are the same for* N_n *and* N'_n [in the language of algebraic geometry, that means we have for each $n \geq 1$ a homomorphism $\nu_n : \mathrm{G}_a \to \mathrm{G}_a$ rational over $\bar{\mathrm{K}}$, and that N'_n is the restriction of ν_n to the points of G_a rational over $\bar{\mathrm{K}}'$—for $n = 0$, the additive group G_a is replaced by the multiplicative group G_m]. Here is a simple application of this remark.

Proposition 6. *With the hypotheses above, let* $x \in \mathrm{K}^*$. *Then there exists an extension* $\bar{\mathrm{K}}'/\bar{\mathrm{K}}$, *of degree* $\leq l$, *such that* x *is a norm in the corresponding extended extension* L′/K′.

By cor. 7 to prop. 5, we may assume that $x \in \mathrm{U}_{\mathrm{K}}^t$. For x to be a norm in this case, it is necessary and sufficient that its class ξ in $\mathrm{U}_{\mathrm{K}}^t/\mathrm{U}_{\mathrm{K}}^{t+1}$ have the form $\mathrm{N}_t(\eta)$, where $\eta \in \mathrm{U}_{\mathrm{L}}^t/\mathrm{U}_{\mathrm{L}}^{t+1}$. Now, as the degree of N_t is l, there exists such an η in an extension $\bar{\mathrm{K}}'/\bar{\mathrm{K}}$ of degree $\leq l$; the image of x in $\mathrm{Coker}(\mathrm{N}'_t)$ is then zero, and applying once again cor. 7 to prop. 5, we see that x is a norm in L′/K′. □

Remark. The proof just given shows, in fact, that if x is not a norm, then the extension $\bar{\mathrm{K}}'/\bar{\mathrm{K}}$ is unique and is cyclic of degree l.

When we pass to the limit on $\bar{\mathrm{K}}'$, we obtain as the direct limit of the fields K′ (resp. L′) the field K_{nr} (resp. L_{nr}) which is *the maximal unramified extension of* K (resp. L); here again, $\mathrm{L}_{nr} = \mathrm{L} \otimes_{\mathrm{K}} \mathrm{K}_{nr}$.

Corollary. *With the hypotheses above, we have* $\mathrm{N}(\mathrm{L}_{nr}^*) = \mathrm{K}_{nr}^*$.

Indeed, if $x \in K_{nr}^*$, there is a finite extension $\bar{K}_0/\bar{K}$ such that x belongs to the corresponding extension K_0. Applying prop. 6 to L_0/K_0 and to x, we see that $x \in N(L_{nr}^*)$.

[*Variant.* As the residue fields of the completions $\hat{L}_{nr}$ and $\hat{K}_{nr}$ are algebraically closed, cor. 3 to prop. 5 shows that there exists $y \in \hat{L}_{nr}^*$ such that $Ny = x$. As y is the limit of elements of L_{nr}^*, it follows that, for every integer v, there exists $z_v \in L_{nr}^*$ such that $Nz_v = xu_v$, with $u_v \in U_{\hat{K}_{nr}}^v$. If $v \geq t + 1$, u_v is a norm (cor. 3 to prop. 5), hence x is also a norm.] □

The extensions of K_{nr} of the form L_{nr} are "essentially all" the extensions of K_{nr}. To make this precise, we first prove the following general lemma:

Lemma 6. *Suppose that* L *is a field that is the union of an increasing directed family of subfields* $\{L_i\}_{i \in I}$, *and let* M *be a finite extension of* L *of degree n. Then there exists an index* $i \in I$ *and an extension* M_i *of* L_i *that has degree n and is linearly disjoint from* L *over* L_i, *such that* $M_iL = M$. *If* M_i *and* M_j *both satisfy these conditions, then there is an index* $k \geq i, j$ *such that* $M_iL_k = M_jL_k$. *If* M *is separable* (resp. *Galois*), M_i *can be chosen to be separable* (resp. *Galois*) over L_i.

(In other words, the finite extensions of L are direct limits of finite extensions of the L_i.)

Let $\{m_\alpha\}$, $\alpha = 1, \ldots, n$, be a basis of M over L; then

$$m_\alpha . m_\beta = \sum c_{\alpha\beta}^\gamma m_\gamma, \quad \text{with } c_{\alpha\beta}^\gamma \in L.$$

Choose i so large that the $c_{\alpha\beta}^\gamma$ belong to L_i. The algebra M_i defined by these structure constants is clearly such that $M_i \otimes_{L_i} L = M$; it must therefore be a field, satisfying the desired conditions. Uniqueness is proved similarly. If M/L is separable, the m_α are linearly independent over $L^{1/p}$, *a fortiori* over $L_i^{1/p}$; this shows that M_i/L_i is separable. Finally, if M/L is Galois, the transforms $s(M_i)$ of M_i by the elements $s \in G(M/L)$ are such that $s(M_i)L = M$; hence there is an index $j \geq i$ such that $s(M_i)L_j = M_iL_j$, and setting $M_j = M_iL_j$, we obtain a Galois extension M_j/L_j as desired. □

Lemma 7. *Let* K *be a field complete under a discrete valuation, having perfect residue field* $\bar{K}$. *Let* K_{nr} *be the maximal unramified extension of* K, *and let* E *be a finite extension of* K_{nr} *of degree n. Then there exist a finite subextension* K′ *of* K_{nr} *and an extension* E′/K′ *of degree n, linearly disjoint from* K_{nr} *over* K′, *such that* $E = E'K_{nr}$. *The extension* E′/K′ *is then totally ramified, and* E *can be identified with* E'_{nr}. *If* E *is separable* (resp. Galois), *then* E′/K′ *can be chosen likewise.*

The residue field $\bar{K}_{nr}$ of K_{nr} is an algebraic closure of $\bar{K}$. Let $\{\bar{K}_i\}_{i \in I}$ be the family of finite subextensions of $\bar{K}_{nr}$, and let $\{K_i\}_{i \in I}$ be the corresponding family of subextensions of K_{nr}. The family $\{K_i\}$ satisfies the hypotheses of lemma 6; the existence of E'/K' follows. The fact that E' is linearly disjoint from K_{nr} shows that its residue field is equal to $\bar{K}'$ (otherwise E' would contain an unramified extension K''/K' distinct from K' and would not be linearly disjoint from K_{nr}); hence $E'_{nr} = E$. The other assertions result from lemma 6. □

We are now able to prove the result we wanted:

Proposition 7. *Let* K *be a field complete under a discrete valuation, with perfect residue field* $\bar{K}$. *Let* K_{nr} *be the maximal unramified extension of* K, *and let* $F \supset E \supset K_{nr}$ *be two finite extensions of* K_{nr}, *with* F/E *separable. Then* $N(F^*) = E^*$.

We may assume F/E to be Galois (enlarging F if necessary). By lemma 7, there is a finite subextension K' of K_{nr} such that $F = F'_{nr}$, $E = E'_{nr}$, where $F' \supset E' \supset K'$ are linearly disjoint from $K_{nr} = K'_{nr}$ over K', and where F'/E' is Galois. As $\bar{F}' = \bar{E}' = \bar{K}'$, the extension F'/E' is totally ramified, hence its Galois group is solvable (Chap. IV, §2); by "dévissage", we are reduced to the case where it is cyclic of prime degree, and in that case the corollary to prop. 6 applies. □

EXERCISE

Show that prop. 7 remains valid when the hypothesis of separability on F/E is dropped. (Treat the purely inseparable case directly, using the structure theorem of Chap. II, §4.)

§5. Multiplicative Polynomials and Additive Polynomials

Let k be a field, and let p be its exponent characteristic. A polynomial $P \in k[X]$ is called *multiplicative* if $P(XY) = P(X)P(Y)$ and if $P(1) = 1$. Such a polynomial is necessarily a *monomial* X^h; let $d(P) = h$. If $h = h_0 p^r$, with h_0 prime to p, we will say that h_0 is the *separable degree* of P, and write $d_s(P) = h_0$. The kernel of the homomorphism $P: k^* \to k^*$ is the group of $d_s(P)$-th roots of unity contained in k; the order of this group divides $d_s(P)$. If P and Q are two multiplicative polynomials, so is their composite $P \circ Q$, and

$$d(P \circ Q) = d(P) . d(Q), \qquad d_s(P \circ Q) = d_s(P) . d_s(Q).$$

A polynomial P is called *additive* if $P(X + Y) = P(X) + P(Y)$. If k has characteristic zero, then $P(X) = aX$, for some $a \in k$; if k has positive characteristic p, it is easily verified that P is a linear combination of monomials X^{p^h} having powers of p as exponents. If $P \neq 0$, it can be written uniquely in the form

$$P = P'^{p^h} = X^{p^h} \circ P' \quad \text{(assuming } k \text{ is perfect)},$$

with

$$P' = a_0 + \cdots + a_k X^{p^k}, \qquad a_0, a_k \neq 0.$$

The degree $d(P)$ of P is equal to p^{h+k}; the integer p^k is called the *separable degree* of P, and is denoted $d_s(P)$.

The kernel of the homomorphism $P: k \to k$ is equal to that of P′; it is an additive subgroup of k of order dividing $d_s(P)$. Furthermore, as P′ is separable, it can be written (over the algebraic closure of k) as

$$P' = a_k \prod_{P(\xi)=0} (X - \xi)$$

which shows that the kernel of P has order $d_s(P)$ if k contains the roots of P′. If P and Q are two additive polynomials, so is $P \circ Q$, and

$$d(P \circ Q) = d(P) . d(Q), \qquad d_s(P \circ Q) = d_s(P) . d_s(Q).$$

[The additive polynomials are just the k-endomorphisms of the algebraic group G_a; they form a ring whose structure has been studied by Ore [50]; see also Whaples [69, 70].]

§6. The Galois Totally Ramified Case

In this section, we assume the extension L/K to be Galois and totally ramified. Thus $\bar{L} = \bar{K}$. Denote by ψ the function defined in Chap. IV, §3.

Proposition 8. *For every integer* $n \geq 0$, $N(U_L^{\psi(n)}) \subset U_K^n$ *and* $N(U_L^{\psi(n)+1}) \subset U_K^{n+1}$.

We will prove this result below.

Just as in the cyclic of prime degree case, prop. 8 allows the definition of homomorphisms

$$N_0 : \bar{K}^* \to \bar{K}^*$$
$$N_n : \bar{K} \to \bar{K}, \qquad n \geq 1.$$

Proposition 9. *For* $n = 0$ (resp. $n \neq 0$), *the homomorphism* N_n *is induced by a multiplicative* (resp. *additive*) *non-constant polynomial* P_n *such that*

$$d(P_n) = \mathrm{Card}(G_{\psi(n)}) \quad \textit{and} \quad d_s(P_n) = (G_{\psi(n)} : G_{\psi(n)+1}) = \psi'_d(n)/\psi'_g(n).$$

The sequence

$$0 \to G_{\psi(n)}/G_{\psi(n)+1} \xrightarrow{\theta} U_L^{\psi(n)}/U_L^{\psi(n)+1} \xrightarrow{N_n} U_K^n/U_K^{n+1}$$

is exact.

[Recall that the homomorphism θ assigns to $s \in G_{\psi(n)}$ the class of $s(\pi)/\pi$, where π is a uniformizer of L—cf. Chap. IV, §2.]

Proof of Props. 8 and 9. We argue by induction on the order of G. The case $G = \{1\}$ is trivial. If $G \neq \{1\}$, then since it is a solvable group (Chap. IV,

§2), it has a quotient that is cyclic of prime order. Hence there is a subextension K′/K of L/K that is cyclic of prime degree l. By inductive hypothesis (resp. props. 4 and 5), the assertions are valid for L/K′ (resp. K′/K). Put

$$n' = \psi_{K'/K}(n), \qquad n'' = \psi_{L/K'}(n'),$$

Then

$$N_{L/K'}(U_L^{n''}) \subset U_{K'}^{n'} \quad \text{and} \quad N_{K'/K}(U_{K'}^{n'}) \subset U_K^n$$

whence, since $N_{L/K} = N_{K'/K} \circ N_{L/K'}$, $N(U_L^{n''}) \subset U_K^n$, which proves the first of the inclusions in prop. 8, since $\psi_{L/K} = \psi_{L/K'} \circ \psi_{K'/K}$. The second inclusion is proved similarly. □

As for prop. 9, note that N_n has the factorization

$$U_L^{n''}/U_L^{n''+1} \xrightarrow{N''} U_{K'}^{n'}/U_{K'}^{n'+1} \xrightarrow{N'} U_K^n/U_K^{n+1}$$

where N″ and N′ are defined respectively by $N_{L/K'}$ and $N_{K'/K}$. If N″ and N′ are induced by the multiplicative (resp. additive) polynomials P″ and P′, N_n is induced by their composite $P_n = P' \circ P''$, which is of the same type. The separable degree d of P_n is equal to the product-of the separable degrees d' and d'' of P′ and P″. The inductive hypothesis, applied to the extension L/K′, shows that

$$d'' = (\psi_{L/K'})'_{d/g}(n')$$

agreeing to denote by $f'_{d/g}$ the quotient f'_d/f'_g of the right derivative by the left derivative. Similarly,

$$d' = (\psi_{K'/K})'_{d/g}(n).$$

As $\psi_{L/K} = \psi_{L/K'} \circ \psi_{K'/K}$, the chain rule implies

$$d = (\psi_{L/K})'_{d/g}(n).$$

Taking into account properties of ψ (cf. Chap. IV, §3, props. 12 and 13), this can also be written as

$$d = (G_{\psi(n)} : G_{\psi(n)+1}).$$

In the same way, one can prove the formula

$$d(P_n) = \mathrm{Card}(G_{\psi(n)}).$$

On the other hand, by what was explained in §5, the kernel of N_n has order dividing $d_s(P_n)$; as $N(s(\pi)/\pi) = 1$, this kernel contains the image of θ, whose order is precisely $d_s(P_n)$ as we just saw. Thus $\mathrm{Im}(\theta) = \mathrm{Ker}(N_n)$. □

Corollary 1. N_n *is injective if and only if* $G_{\psi(n)} = G_{\psi(n)+1}$.

Obvious.

Corollary 2. N_n *is surjective in each of the following three cases*:

i) $\bar{K}$ *is algebraically closed,*
ii) $\bar{K}$ *is perfect, and* $G_{\psi(n)} = G_{\psi(n)+1}$,
iii) $G_{\psi(n)} = \{1\}$.

Indeed, in case i), P_n is a non-constant polynomial; in case ii), $P_n = cX^{p^r}$; and in case iii), $P_n = cX$. In all three cases, these polynomial maps are surjective.

Corollary 3. *We have* $N(U_L^{\psi(n)}) = U_K^n$ *if* $G_{\psi(n)} = \{1\}$, *and* $N(U_L^{\psi(n)+1}) = U_K^{n+1}$ *if* $G_{\psi(n+1)} = \{1\}$. *When* $\bar{K}$ *is algebraically closed, these equalities hold for all* $n \geq 0$.

Same proof as for cor. 3 of §3.

Corollary 4. *Let* v *be a non-negative real number. Then* $N(U_L^{\psi(v)}) = U_K^v$ *if either* $G_{\psi(v)} = \{1\}$ *or if* $\bar{K}$ *is algebraically closed.*

Same proof as for cor. 4 of §3.

Remarks. 1) We have proceeded by the method of "dèvissage", beginning with the cyclic of prime degree case; one can also work directly, as does Hasse [33].

2) If $\bar{K}$ is finite, the conditions of prop. 9 do not suffice to determine the polynomials P_n. Nevertheless, there is a choice of P_n which is invariant under residue extensions (that follows from the proof), and this choice is clearly unique; the corresponding polynomials P_n will be called *canonical.*

Proposition 6 extends without change to the case considered here, so we need only re-state it:

Proposition 10. *Suppose* $\bar{K}$ *is perfect, and let* $x \in K^*$. *Then there exists an extension* $\bar{K}'/\bar{K}$ *of degree* $\leq [L:K]$, *such that* x *is a norm in the extended extension* L'/K' *corresponding to* $\bar{K}'$.

(This proposition brings additional precision to prop. 7 of §4.)

Exercise

Assuming $\bar{K}$ perfect, extend the results of this § to the case of an arbitrary finite extension L/K.

§7. Application: Proof of the Hasse-Arf Theorem

It is the following theorem (cf. Chap. IV, §3):

Theorem 1. *Let* K *be a field complete under a discrete valuation, and let* L *be a finite abelian extension of* K, *with Galois group* G; *suppose that the residue extension* $\bar{L}/\bar{K}$ *is separable. If* v *is a jump in the filtration* $\{G^v\}$ *of* G, *then* v *is an integer.*

(Recall that to say v is a *jump* means that $G^{v+\varepsilon} \neq G^v$ for all $\varepsilon > 0$.)

The above statement uses the *upper* numbering for the ramification groups. If translated in terms of the *lower* numbering, it becomes:

Theorem 1′. *With the hypotheses of theorem 1, if μ is an integer such that $G_\mu \neq G_{\mu+1}$ then $\varphi_{L/K}(\mu)$ is an integer.*

The next proposition is a special case of th. 1′.

Proposition 11. *Let L/K be a cyclic, totally ramified extension, with Galois group G, and let μ be the largest integer such that $G_\mu \neq \{1\}$. Then $\phi_{L/K}(\mu)$ is an integer.*

Conversely, this proposition *implies* th. 1: indeed, let L/K satisfy the hypotheses of th. 1, and let v be a jump in the filtration $\{G^v\}$. We may assume L/K totally ramified (otherwise replace G by its inertia subgroup). Put $G' = G^v$ and let G'' be the next ramification group (thus $G'' = G^{v+\varepsilon}$ for all $\varepsilon > 0$ sufficiently small). By definition, $G' \neq G''$. Splitting G/G'' into a product of cyclic groups, we see that there exists a cyclic quotient group H of G/G'' such that the image H' of G' in H is $\neq \{1\}$. The group H is the Galois group of a subextension L'/K of L/K. By Herbrand's theorem, $H^v = H'$ and $H^{v+\varepsilon} = \{1\}$ for all $\varepsilon > 0$; prop. 11 then shows that v is an integer.

It remains to prove prop. 11. Denote by w the discrete valuation on L, and let π be a uniformizer of L. Put

$$r = \mathrm{Card}(G), \qquad r' = \mathrm{Card}(G_\mu), \qquad k = r/r'.$$

Choose a generator s of G; the group G_μ is then generated by $\sigma = s^k$. We again use the exponential notation x^s for $s(x)$. Finally, we assume that the residue field $\bar{K} = \bar{L}$ *is not the prime field*: we can always reduce to this case by making an extension of the residue field—cf. §4.

Let V be the set of $x \in L^*$ such that $Nx = 1$. By a classical result of Hilbert (cf. Bourbaki, *Alg.*, Chap. V, §11, th. 3), V is the set of all y^{s-1}, $y \in L^*$. Let W be the subgroup of V consisting of those y^{s-1} for which $y \in U_L$.

Lemma 8. *The group V/W is cyclic.*

Indeed, the map $y \mapsto y^{s-1}$ defines by passage to the quotient a homomorphism of $L^*/U_L = \mathbf{Z}$ *onto* V/W.

[In fact, V/W is isomorphic to G–cf. exer.]

Let m be a non-negative integer. Set

$$V_m = V \cap U_L^m, \qquad W_m = W \cap U_L^m.$$

Then $W_m \subset V_m$; the group V_m/W_m can be identified with a subgroup of V/W, and the V_m/W_m form a decreasing filtration of V/W.

Lemma 9. *For m sufficiently large, $V_m = W_m$.*

Let $t \in \mathrm{L}$ be such that $\mathrm{Tr}(t) = 1$, and let $m_0 = -w(t)$. Let $x \in \mathrm{V}_m$, with $m > m_0$; we will show that $x \in \mathrm{W}_m$. Form the "Lagrange-Hilbert resolvent" (cf. Bourbaki, *loc. cit.*):

$$y = \sum_{i=0}^{i=r-1} x^{1+s+\cdots+s^{i-1}} . t^{s^i}.$$

Since $\mathrm{Tr}(t) = 1$, one can write

$$y - 1 = \sum_{i=0}^{i=r-1} (x^{1+s+\cdots+s^{i-1}} - 1) . t^{s^i}$$

whence $w(y-1) > 0$ and $y \in \mathrm{U}_\mathrm{L}^1$. As $\mathrm{N}x = 1$, we get $y^{1-s} = x$, so that $x \in \mathrm{W}_m$. □

We now study the successive quotients $\mathrm{V}_m/\mathrm{V}_{m+1}\mathrm{W}_m$ of the filtration $\{\mathrm{V}_m/\mathrm{V}_m\}$.

Lemma 10. *If $\varphi(m)$ is an integer and if $\mathrm{G}_m = \mathrm{G}_{m+1}$, then $\mathrm{V}_m = \mathrm{V}_{m+1}$.*

Put $n = \varphi(m)$, so that $m = \psi(n)$. Let $x \in \mathrm{V}_m$, and let $\bar{x}$ be the image of x in $\mathrm{U}_\mathrm{L}^m/\mathrm{U}_\mathrm{L}^{m+1}$. Clearly x belongs to the kernel of the homomorphism

$$\mathrm{N}_n : \mathrm{U}_\mathrm{L}^m/\mathrm{U}_\mathrm{L}^{m+1} \to \mathrm{U}_\mathrm{K}^n/\mathrm{U}_\mathrm{K}^{n+1}$$

defined in the preceding section. By prop. 9, the kernel of N_n is isomorphic to $\mathrm{G}_m/\mathrm{G}_{m+1}$, therefore is trivial. Hence $\bar{x} = 0$, i.e., $x \in \mathrm{V}_{m+1}$. □

Lemma 11. *Let m be a positive integer. If the image of W_m in $\mathrm{U}_\mathrm{L}^m/\mathrm{U}_\mathrm{L}^{m+1}$ is non-trivial, then that image is equal to all of $\mathrm{U}_\mathrm{L}^m/\mathrm{U}_\mathrm{L}^{m+1}$.*

Let x be an element of W_m not belonging to $\mathrm{U}_\mathrm{L}^{m+1}$. Then $x = y^{s-1}$, with $y \in \mathrm{U}_\mathrm{L}$. We may assume $y \in \mathrm{U}_\mathrm{L}^1$ (otherwise multiply y by an element of U_K, which does not change y^{s-1}); thus $y = 1 + z$, with $w(z) \geq 1$. For any $a \in \mathrm{A}_\mathrm{K}$, set $y_a = 1 + az$ and $x_a = y_a^{s-1}$. Then

$$x_a - 1 = \frac{s(y_a) - y_a}{y_a} = \frac{a(s(y) - y)}{y_a} = a\frac{y}{y_a}(x - 1).$$

As $y/y_a \in \mathrm{U}_\mathrm{L}^1$, we see that $x_a \in \mathrm{W}_m$, and that its image $\bar{x}_a$ in $\mathrm{U}_\mathrm{L}^m/\mathrm{U}_\mathrm{L}^{m+1}$ is equal to $\bar{a}.\bar{x}$, where $\bar{a}$ denotes the image of a in $\bar{\mathrm{K}} = \bar{\mathrm{L}}$; but every element of $\mathrm{U}_\mathrm{L}^m/\mathrm{U}_\mathrm{L}^{m+1}$ has the form $\bar{a}.\bar{x}$. □

Lemma 12. *Let n be an integer such that $\mathrm{G}_{\psi(n+1)} = \{1\}$. Let m be an integer such that*

$$n < \varphi(m) < n + 1.$$

Then the images of V_m and W_m are both equal to all of $\mathrm{U}_\mathrm{L}^m/\mathrm{U}_\mathrm{L}^{m+1}$.

Let $x \in U_L^m$ represent $\bar{x} \in U_L^m/U_L^{m+1}$. Then $\psi(n) < m < \psi(n+1)$, whence $m \geq \psi(n) + 1$. By prop. 8, $Nx \in U_K^{n+1}$, and cor. 3 to prop. 9 shows that there exists $y \in U_L^{\psi(n+1)}$ such that $Ny = Nx$. Setting $x' = xy^{-1}$, we find a representative of $\bar{x}$ that belongs to V. Thus the group V_m maps *onto* U_L^m/U_L^{m+1}. As for W_m, its image in U_L^m/U_L^{m+1} is a subgroup H_m; as V_m/W_m is cyclic (being a subgroup of V/W—cf. lemma 8), the quotient of U_L^m/U_L^{m+1} by H_m is also cyclic. But U_L^m/U_L^{m+1} is isomorphic to the additive group $\bar{K}$, which is *not cyclic* because we took the precaution of insuring that $\bar{K}$ is not the prime field. Hence $H_m \neq 0$, and the preceding lemma shows that H_m must be the whole group. □

Lemma 13. *Let m be an integer, and let n + 1 be the smallest integer* $\geq \varphi(m)$. *If* $G_{\psi(n+1)} = \{1\}$, *then* $V_m = W_m$.

Let us first show that $V_m = V_{m+1}W_m$. If $\varphi(m)$ is integral, $\varphi(m) = n + 1$, $\psi(n+1) = m$, and our assertion results from lemma 10. If $\varphi(m)$ is not integral, then $n < \varphi(m) < n + 1$, and lemma 12 shows that V_m and W_m have the same image in U_L^m/U_L^{m+1}, whence $V_m = V_{m+1}W_m$.

Applying the same argument to $m + 1$, we find $V_{m+1} = V_{m+2}W_{m+1}$, whence $V_m = V_{m+2}W_m$, and by iteration, $V_m = V_{m+k}W_m$ for all $k \geq 0$. Taking k large enough gives $V_{m+k} = W_{m+k}$ (lemma 9), hence $V_m = W_{m+k}W_m = W_m$. □

We can now prove proposition 11. Suppose $\varphi(\mu)$ not integral, and let $\nu + 1$ be the smallest integer $\geq \varphi(\mu)$. Then $\mu < \psi(\nu + 1)$, whence $G_{\psi(\nu+1)} = \{1\}$, and by the preceding lemma, $V_\mu = W_\mu$. Let $\sigma = s^k$ be the generator of G_μ defined at the beginning of this (long) proof, and set $x = \pi^{\sigma - 1}$. Clearly $x \in V$; since $V_\mu = W_\mu$, there exists $y \in U_L$ such that $\pi^{\sigma-1} = y^{s-1}$. But

$$\sigma - 1 = (s - 1)(1 + s + \cdots + s^{k-1})$$

and by setting $z = y^{-1}\pi^{1+s+\cdots+s^{k-1}}$, we find $z^{s-1} = 1$, whence $z \in K^*$. As L/K is totally ramified, it follows that $w(z) \equiv 0 \bmod. r$ (recall that $r = [L:K]$); but as $w(y) = 0$ and $w(\pi) = 1$, that implies $k \equiv 0 \bmod. r$, which is absurd. □

Exercise

Keep the notation and hypotheses of the proof of prop. 11. Prove:

a) If $\varphi(m)$ is an integer, then $W_m \subset V_{m+1}$, and V_m/V_{m+1} is isomorphic to G_m/G_{m+1}.

b) If $\varphi(m)$ is not an integer, then $V_m = V_{m+1}W_m$.

c) The map $t \mapsto \pi^{t-1}$ defines by passage to the quotient an isomorphism of the group G filtered by the G_m onto the group V/W filtered by the V_m/W_m.

CHAPTER VI

Artin Representation

§1. Representations and Characters

(We recall well-known definitions and results. The reader could consult, for example, M. Hall [30], Chap. XVI or [114], for the proofs.)

Let G be a finite group, of order g. A *class function* on G is a complex valued function such that $f(sts^{-1}) = f(t)$ for all $s, t \in G$. Let V be a finite dimensional vector space over **C**, and let **GL**(V) be the group of automorphisms of V; a *linear representation* of G in V is a homomorphism $\rho: G \to \mathbf{GL}(V)$. If $s \in G$, $\rho(s)$ is an endomorphism of V, and its *trace* is defined. Set

$$\chi_\rho(s) = \operatorname{Tr}(\rho(s)).$$

The function χ_ρ is a class function on G called the *character* of the representation ρ; it determines ρ up to isomorphism. We have $\chi_\rho(s^{-1}) = \overline{\chi_\rho(s)}$. The integer $\chi(1)$, equal to the dimension of V, is called the *degree* of ρ.

The constant function with value 1 will be denoted 1_G or simply 1. It is the character of the *unit* representation of G ($\dim V = 1$ and $\rho(s) = 1$ for all $s \in G$).

The character of the *regular* representation of G will be denoted r_G; one has $r_G(1) = g$ and $r_G(s) = 0$ for $s \neq 1$. The unit representation embeds in the regular representation; the quotient is called the *augmentation* representation of G, and its character will be denoted u_G. Then $r_G = u_G + 1_G$.

A character χ is called *irreducible* if the corresponding representation is irreducible (i.e., if V is a simple **C**[G]-module). Every class function can be uniquely expressed as a linear combination of irreducible characters:

$$\varphi = \sum c_\chi \chi, \qquad c_\chi \in \mathbf{C}$$

For φ to be the character of a linear representation of G, it is necessary and sufficient that all the c_χ be non-negative integers.

The c_χ may be computed as follows: if φ and ψ are two class functions on G, put

$$(\varphi, \psi) = \frac{1}{g} \sum_{s \in G} \varphi(s)\overline{\psi(s)}.$$

Then $(\chi, \chi) = 1$, $(\chi, \chi') = 0$ if χ and χ' are distinct irreducible characters, whence

$$c_\chi = (\varphi, \chi).$$

(In other words, the irreducible characters form an orthonormal basis for the Hilbert space of class functions.) For example,

$$r_G = \sum_\chi \chi(1)\chi, \qquad u_G = \sum_{\chi \neq 1} \chi(1)\chi.$$

Next let $\alpha : H \to G$ be a homomorphism of a finite group H into G. If φ is a class function on G, the function $\alpha^*(\varphi) = \varphi \circ \alpha$ is a class function on H; if φ is the character of a linear representation $\rho : G \to \mathbf{GL}(V), \alpha^*(\varphi)$ is the character of the representation $\rho \circ \alpha : H \to \mathbf{GL}(V)$. Conversely, let ψ be a class function on H. It can be shown that there is a unique class function $\alpha_*(\psi)$ on G such that the "Frobenius identity"

$$(\varphi, \alpha_*(\psi)) = (\alpha^*(\varphi), \psi)$$

holds for all class functions φ on G. If ψ is the character of a linear representation of H in V, then $\alpha_*(\psi)$ is the character of the representation $\mathbf{C}[G] \otimes_{\mathbf{C}[H]} V$, obtained from V by extension of scalars (the *induced* representation).

Our principal application of these constructions will be to the following two special cases:

a) H *is a subgroup of* G *and* α *is the inclusion of* H *into* G

In this case we write $\varphi|H$ (or even simply φ) instead of $\alpha^*(\varphi)$, and ψ^* instead of $\alpha_*(\psi)$ (*induced* function). If $s \in G$, then

$$\psi^*(s) = \sum_{t \in G/H} \psi(tst^{-1})$$

with the convention that

$$\psi(tst^{-1}) = 0 \quad \text{if } tst^{-1} \notin H.$$

b) G *is a quotient* H/N *of* H, *and* α *is the canonical projection of* H *on* H/N

In this case we write φ instead of $\alpha^*(\varphi)$, and $\psi^\natural$ instead of $\alpha_*(\psi)$. If $s \in H/N$, then

$$\psi^\natural(s) = \frac{1}{\mathrm{Card}(N)} \sum_{t \to s} \psi(t).$$

We will need the following result:

Brauer's Theorem. *Every character of a finite group is a* **Z**-*linear combination of characters* χ_i^* *induced by characters* χ_i *of degree* 1 *of subgroups* H_i *of* G.

(A character of degree 1 of a group H is just a homomorphism of H into the multiplicative group $\mathbf{C}^*$.)

See Brauer-Tate [11] for the proof.

§2. Artin Representation

Let L/K be a finite Galois extension, with Galois group G, satisfying the hypotheses of Chaps. IV and V. Put $f = [\bar{L}:\bar{K}]$. If s is an element of G distinct from 1, we have defined in Chap. IV, §1 the positive integer $i_G(s)$. Set

$$a_G(s) = -f\,.\,i_G(s) \quad \text{if} \quad s \neq 1$$

$$a_G(1) = f \sum_{s \neq 1} i_G(s).$$

Thus

$$\sum_{s \in G} a_G(s) = 0, \quad \text{i.e., } (a_G, 1_G) = 0$$

Theorem 1. *The function* a_G *is the character of a linear representation of* G.

It is clear that a_G is a class function. Thus we can write

$$a_G = \sum c_\chi \chi,$$

where χ runs through the set of irreducible characters of G. Then

$$c_\chi = (a_G, \chi) = \frac{1}{g} \sum_{s \in G} a_G(s)\chi(s^{-1})$$

$$= \frac{1}{g} \sum_{s \in G} a_G(s^{-1})\chi(s)$$

and as $a_G(s^{-1}) = a_G(s)$, we see that $c_\chi = (\chi, a_G)$.

More generally, if φ is any class function on G, set

$$f(\varphi) = (\varphi, a_G).$$

Theorem 1 can then be reformulated as follows:

Theorem 1′. $f(\chi)$ *is a non-negative integer for every character* χ *of* G.

Before proving these two theorems, we give several properties of $f(\chi)$ and a_G:

Proposition 1. *The function a_G is equal to the function $(a_{G_0})^*$ induced by the corresponding function for the inertia group* G_0.

As G_0 is normal in G, we have $(a_{G_0})^*(s) = 0 = a_G(s)$ if $s \notin G_0$. If $s \in G_0$, $s \neq 1$, then

$$(a_{G_0})^*(s) = \sum_{t \in G/G_0} a_{G_0}(tst^{-1}) = -\sum_{t \in G/G_0} i_{G_0}(tst^{-1})$$
$$= -f \,.\, i_G(s) = a_G(s).$$

The case $s = 1$ is handled similarly. □

[This proposition would permit the reduction of theorems 1 and 1′ to the *totally ramified* case $G = G_0$, if so desired.]

Proposition 2. *Let G_i be the* i*th ramification group of* G, *let u_i be the character of the augmentation representation of G_i, and let u_i^* be the character of* G *induced by u_i. Then*

$$a_G = \sum_{i=0}^{\infty} \frac{1}{(G_0 : G_i)}\, u_i^*.$$

Put $g_i = \mathrm{Card}(G_i)$. We have $u_i^*(s) = 0$ if $s \notin G_i$, while $u_i^*(s) = -g/g_i = -f \,.\, g_0/g_i$ if $s \in G_i$, $s \neq 1$, and $\sum_{s \in G} u_i^*(s) = 0$. Then for $s \in G_k - G_{k+1}$, the sum on the right side of the equation has value $-f(k+1)$, and it is the same for $a_G(s)$. For $s = 1$, the equality follows from the orthogonality of both sides with 1_G. □

If φ is a class function on G, set

$$\varphi(G_i) = \frac{1}{g_i} \sum_{s \in G_i} \varphi(s), \qquad g_i = \mathrm{Card}(G_i).$$

Corollary 1. *If φ is a class function on* G, *then*

$$f(\varphi) = \sum_{i=0}^{\infty} \frac{g_i}{g_0} (\varphi(1) - \varphi(G_i)).$$

This follows from prop. 2, taking into account that

$$(\varphi, u_i^*) = (\varphi | G_i, u_i) = \varphi(1) - \varphi(G_i).$$

Corollary 1′. *If χ is the character of a representation of* G *in* V, *then*

$$f(\chi) = \sum_i \frac{g_i}{g_0} \operatorname{codim} V^{G_i},$$

where V^{G_i} *is the subspace of* V *fixed by* G_i, *and* $\operatorname{codim} V^{G_i} = \dim V - \dim V^{G_i}$.

This follows from cor. 1, since $\chi(1) = \dim V$ and $\chi(G_i) = \dim V^{G_i}$.

Corollary 2. *If* χ *is a character of* G, *then* $f(\chi)$ *is a non-negative rational number.*

Indeed, prop. 2 shows that $g_0 a_G$ is the character of a linear representation of G; hence $g_0 . f(\chi)$ is an integer ≥ 0.

Proposition 3. *If* N *is a normal subgroup of* G, *then*

$$a_{G/N} = (a_G)^\natural.$$

That results from prop. 3 of Chap. IV.

Corollary. *If* φ *is a class function on* G/N, *and if* φ' *is the corresponding class function on* G, *then* $f(\varphi) = f(\varphi')$.

Indeed, $f(\varphi) = (\varphi, a_G^\natural) = (\varphi', a_G) = f(\varphi')$.

Proposition 4. *Let* H *be a subgroup of* G *corresponding to the subextension* K′/K, *and let* $\mathfrak{d}_{K'/K}$ *be the discriminant of* K′/K. *Then*

$$a_G|H = \lambda r_H + f_{K'/K} . a_H, \quad \text{with } \lambda = v_K(\mathfrak{d}_{K'/K}).$$

(Recall that r_H denotes the character of the *regular* representation of H.)
If $s \neq 1$ is an element of H, then

$$a_G(s) = -f_{L/K} i_G(s), \qquad a_H(s) = -f_{L/K'} i_H(s), \qquad r_H(s) = 0,$$

and as $i_G(s) = i_H(s)$, this gives

$$a_G(s) = \lambda r_H(s) + f_{K'/K} a_H(s).$$

Now take $s = 1$. First

$$a_G(1) = f_{L/K} v_L(\mathfrak{D}_{L/K}) = v_K(\mathfrak{d}_{L/K})$$

by prop. 4 of Chap. IV. Similarly, $a_H(1) = v_{K'}(\mathfrak{d}_{L/K'})$. The formula to be proved can therefore be written

$$v_K(\mathfrak{d}_{L/K}) = [L:K'] v_K(\mathfrak{d}_{K'/K}) + f_{K'/K} v_{K'}(\mathfrak{d}_{L/K'}),$$

and from the formula for transitivity of the discriminant (Chap. III, prop. 8):

$$\mathfrak{d}_{L/K} = (\mathfrak{d}_{K'/K})^{[L:K']} . N_{K'/K}(\mathfrak{d}_{L/K'}). \quad \square$$

Corollary. *If* ψ *is a character of* H, *and* ψ^* *the character induced on* G, *then*

$$f(\psi^*) = v_K(\mathfrak{d}_{K'/K})\psi(1) + f_{K'/K} f(\psi).$$

Indeed:

$$\begin{aligned} f(\psi^*) &= (\psi^*, a_G) = (\psi, a_G|H) \\ &= (\psi, r_H) + f_{K'/K}(\psi, a_H) \\ &= \lambda\psi(1) + f_{K'/K} f(\psi), \quad \text{with } \lambda = v_K(\mathfrak{d}_{K'/K}). \end{aligned}$$

Proposition 5. *Let χ be a character of degree 1 on G. Let c_χ be the largest integer for which the restriction of χ to the ramification group G_{c_χ} is not the unit character* (if $\chi = 1_G$, take $c_\chi = -1$). *Then*

$$f(\chi) = \varphi_{L/K}(c_\chi) + 1.$$

(See Chap. IV, §3 for the definition of the function $\varphi_{L/K}$.)

If $i \leq c_\chi$, then $\chi(G_i) = 0$, whence $\chi(1) - \chi(G_i) = 1$; if $i > c_\chi$, then $\chi(G_i) = 1$, whence $\chi(1) - \chi(G_i) = 0$. Applying cor. 1 to prop. 2 gives

$$f(\chi) = \sum_{i=0}^{i=c_\chi} \frac{g_i}{g_0} = \varphi_{L/K}(c_\chi) + 1. \quad \square$$

Corollary. *Let* H *be the kernel of* χ, *let* K′ *be the subextension of* L/K *corresponding to* H, *and let* c'_χ *be the largest integer for which* $(G/H)_{c'_\chi} \neq 1$. *Then* $f(\chi) = \varphi_{K'/K}(c'_\chi) + 1$, *and this is an integer* ≥ 0.

(If H = G, put $c'_\chi = -1$.)

Herbrand's theorem (Chap. IV, lemma 5) shows that $c'_\chi = \varphi_{L/K'}(c_\chi)$. The formula $f(\chi) = \varphi_{K'/K}(c'_\chi) + 1$ then follows simply from the transitivity of the functions φ. The fact that $\varphi_{K'/K}(c'_\chi)$ is an integer comes from the theorem of Hasse-Arf (Chap. V, §7), since G/H is abelian. $\square$

Proof of Theorems 1 and 1′. We must show that $f(\chi)$ is a non-negative integer for every character χ of G. We already know (cor. 2 to prop. 2) that it is a non-negative rational number. According to Brauer's theorem (recalled in §1), $\chi = \sum n_i \chi_i^*$, $n_i \in \mathbf{Z}$, the χ_i being characters of degree 1 of subgroups H_i of G. This reduces us to showing $f(\chi^*)$ is an integer if χ is a character of degree 1; but in that case, $f(\chi)$ is integral (cor. to prop. 5), and the corollary to prop. 4 then implies that $f(\chi^*)$ is also. $\square$

Remark. 1) Theorem 1 and its proof are due to Artin [6], except that he had to assume the residue field $\bar{K}$ *finite*, the theorem of Hasse-Arf having then been proved only in this case. Moreover, as he did not have Brauer's theorem available, Artin began by reducing to the case $G = G_1$, thanks to a theorem of Speiser (cf. Chap. IV, cor. 2 to prop. 9); the group G then being a p-group, it is easy to prove that every irreducible character of G is induced by a character of degree 1 of a subgroup, which gives a more precise version of Brauer's theorem in this special case.

2) As we have seen, the proof of Artin's theorem makes essential use of the Hasse-Arf theorem; conversely, the Hasse-Arf theorem can easily be deduced from Artin's.

Terminology. The representation given in theorem 1 is called the *Artin representation* of the group G attached to the extension L/K. Notice that it has only been defined in terms of its character, i.e., only up to an isomorphism; it would be very intersting to have a direct description of it; we will return to this point in §4.

If χ is a character of G, the ideal $\mathfrak{p}_K^{f(\chi)}$ is called the *conductor* of χ, and denoted $\mathfrak{f}(\chi)$. When χ has degree 1, corresponding to a cyclic subextension K′/K, and when $\bar{K}$ is finite, $\mathfrak{f}(\chi)$ is actually the conductor of the extension K′/K, in the sense of class field theory (cf. Chap. XV); the same holds when $\bar{K}$ is algebraically closed (cf. [59], n⁰ 3.7). When χ is an irreducible character of degree >1, no such interpretation is known for $\mathfrak{f}(\chi)$.

Exercises

1. Let G^i be the ith ramification group of G with the upper numbering, and let v_i^* be the character of G induced by the augmentation representation of G^i.

   a) Show that

$$a_G = \frac{1}{g_0}\sum_{n=0}^{\infty} v_{n/g_0}^*.$$

   b) When G is abelian, show that

$$a_G = \sum_{n=0}^{\infty} v_n^*.$$

2. The following generalization of prop. 5 has recently been useful (Tunnell, Henniart): let χ be the character of an *irreducible* representation $\rho: G \rightarrow \mathbf{GL}(V)$ of degree d, and let c_χ be the largest n such that $\rho(G_n) \neq 1$. Show that

$$f(\chi) = d(1 + \phi_{L/K}(c_\chi)).$$

   (Note that the G_i are normal in G, hence the V^{G_i} of corollary 1′ are stable under G. Apply irreducibility and the same argument used for prop. 5.)

§3. Globalisation

This is easy. We just sketch the method:

Let L/K be a finite Galois extension, with Galois group G, and let A be a Dedekind domain with field of fractions K, and let B be the integral closure of A in L. We assume that if $\mathfrak{P}$ is a non-zero prime ideal of B over a prime ideal $\mathfrak{p}$ of A, the residue field $\bar{L}_{\mathfrak{P}} = B/\mathfrak{P}$ is *separable* over $\bar{K}_{\mathfrak{p}} = A/\mathfrak{p}$. Under these conditions, the completion $\hat{L}_{\mathfrak{P}}$ is Galois over $\hat{K}_{\mathfrak{p}}$, with Galois group being the decomposition group $D_{\mathfrak{P}}$. We can apply to the extension $\hat{L}_{\mathfrak{P}}/\hat{K}_{\mathfrak{p}}$ and the group $D_{\mathfrak{P}}$ the definitions and results of §§1 and 2. The corresponding Artin character will be denoted $a_{\mathfrak{P}}$; it is *a priori* defined on $D_{\mathfrak{P}}$, but can be extended by zero to all of G. Put

$$a_{\mathfrak{p}} = \sum_{\mathfrak{P}|\mathfrak{p}}$$

One checks that $a_{\mathfrak{p}} = (a_{\mathfrak{P}})^*$, for any choice of $\mathfrak{P}$ over $\mathfrak{p}$. It follows that $a_{\mathfrak{p}}$ is the character of a representation of G (which is called the *Artin representation* attached to $\mathfrak{p}$ and to the extension L/K), and that this representation is induced by the Artin representation of any one of the decomposition groups $D_{\mathfrak{P}}$, for $\mathfrak{P}|\mathfrak{p}$. If χ is a character of G, put

$$f(\chi, \mathfrak{p}) = (\chi, a_{\mathfrak{p}}) = f(\chi|D_{\mathfrak{P}}).$$

Clearly $f(\chi, \mathfrak{p}) = 0$ when $\mathfrak{p}$ is unramified. Hence we can form the product

$$\mathfrak{f}(\chi) = \prod_{\mathfrak{p}} \mathfrak{p}^{f(\chi, \mathfrak{p})}$$

which is called the *conductor* of the character χ and also denoted $\mathfrak{f}(\chi, L/K)$. The properties demonstrated in §2 can be translated as:

Proposition 6

a) $\mathfrak{f}(\chi + \chi') = \mathfrak{f}(\chi).\mathfrak{f}(\chi')$, $\mathfrak{f}(1) = (1)$.

b) *If* K'/K *is a subextension of* L/K, *corresponding to the subgroup* H *of* G, *and if* ψ *is a character of* H, *then*

$$\mathfrak{f}(\psi^*, L/K) = \mathfrak{d}_{K'/K}^{\psi(1)} . N_{K'/K}(\mathfrak{f}(\psi, L/K')).$$

c) *If* K'/K *is Galois, and if* χ *is a character of* G/H, *then*

$$\mathfrak{f}(\chi, L/K) = \mathfrak{f}(\chi, K'/K).$$

Let us apply b) to the case $\psi = 1_H$; the induced character ψ^* will be denoted $s_{G/H}$; it is the character of the representation of G defined by the operations of G on the homogeneous space G/H. As $\mathfrak{f}(\psi) = (1)$, we get:

Corollary 1. $\mathfrak{d}_{K'/K} = \mathfrak{f}(s_{G/H}, L/K)$.

By decomposing $s_{G/H}$ into a linear combination of irreducible characters, one gets a *decomposition of the discriminant* $\mathfrak{d}_{K'/K}$ *into a product of conductors*. For example, if $H = \{1\}$, then $s_{G/H} = r_G$, and we obtain the "Fuhrerdiskriminantenproduktformel" of Artin and Hasse:

Corollary 2. $\mathfrak{d}_{L/K} = \prod \mathfrak{f}(\chi)^{\chi(1)}$, *the product being taken over all irreducible characters* χ *of* G.

If G is *abelian*, this formula simplifies to

$$\mathfrak{d}_{L/K} = \prod \mathfrak{f}(\chi).$$

The case of a number field. Suppose that K is a number field (i.e., a finite extension of **Q**), with A the ring of integers of K. Consider the ideal of **Z** defined by the formula

$$\mathfrak{c}(\chi, L/K) = \mathfrak{d}_{K/Q}^{\chi(1)} \,.\, N_{K/Q}(\mathfrak{f}(\chi, L/K)).$$

It is generated by a positive integer $c(\chi, L/K)$. Applying prop. 6 and the transitivity of the discriminant, we see that $c(\chi, L/K)$ has the following properties:

(a) $c(\chi + \chi', L/K) = c(\chi, L/K) \,.\, c(\chi', L/K)$, $c(1, L/K) = |d_{K/Q}|$
(b) $c(\psi^*, L/K) = c(\psi, L/K')$
(c) $c(\chi, L/K) = c(\chi, K'/K)$

the notation being that of prop. 6.

These invariance properties are identical to those of the Artin L-functions (cf. [5]); in fact, $c(\chi, L/K)$ appears in the "exponential" term of the functional equation of $L(\chi, L/K)$—cf. [5], n° 7, or [79].

§4. Artin Representation and Homology (for Algebraic Curves)

Let k be an algebraically closed field of characteristic p, let Y be a projective non-singular connected algebraic curve, defined over k, and let G be a finite group of automorphisms of Y. Let $X = Y/G$ be the quotient curve. Let L (resp. K) be the function field of Y (resp. X); the extension L/K is Galois with group G. Moreover, each point Q of Y has a local ring that is a discrete valuation ring, with field of fractions L and residue field k; let v_Q be the corresponding valuation. Define v_P for $P \in X$ similarly. The subgroup D_Q of G consisting of all $s \in G$ such that $s(Q) = Q$ is called the *decomposition group* of Q; if $\hat{L}_Q$ (resp. $\hat{K}_P$) is the completion of L (resp. K) for the valuation v_Q (resp. v_P), then $\hat{L}_Q/\hat{K}_P$ is a Galois extension with group D_Q (the situation is analogous to §3, and could actually be reduced to it). We can apply to $\hat{L}_Q/\hat{K}_P$ the constructions of §2, and we get the functions i and a, which will be denoted i_Q and a_Q. If t is a local uniformizer at Q, then

$$i_Q(s) = v_Q(s(t) - t) \quad \text{if } s \in D_Q,\ s \neq 1.$$

In this case we can give a *geometric* interpretation of $i_G(s)$: if Γ_s denotes the graph of s in $Y \times Y$, then $i_Q(s)$ is equal to the *multiplicity* of $Q \times Q$ in the intersection $\Delta \,.\, \Gamma_s$ (where Δ denotes the diagonal of $Y \times Y$); the verification is straightforward.

As in §3, we extend a_Q by zero outside D_Q, and we set

$$a_P = \sum_{Q \to P} a_Q \quad \text{for all } P \in X.$$

If we choose point $Q \in Y$ over $P \in X$, it can be verified that $a_P = (a_Q)^*$; hence a_P is the character of a representation of G that is called the *Artin representation* attached to P.

Next we will *interpret homologically the direct sum of the Artin representations corresponding to the various points* $P \in X$. Let l be a fixed prime number distinct from p. Then the *l-adic homology groups* of Y can be defined (as in Weil [66]; see also [57], n^0 5), and one gets $H_0(Y) = H_2(Y) = \mathbf{Z}_l$, and $H_1(Y) = T_l(J)$ (Tate group of the Jacobian J of Y). These are free Z_l-modules of finite rank, on which the group G acts. Therefore they define representations, whose characters will be denoted h_i, $i = 0, 1, 2$. Put $h = h_0 - h_1 + h_2$. The integer $E(Y) = h(1)$ is the *Euler-Poincaré characteristic* of Y; if g_Y is the genus of Y, then $E(Y) = 2 - 2g_Y$. Define similarly $E(X) = 2 - 2g_X$.

Proposition 7. *With the notation above,*

$$h = E(X) . r_G - \sum_{P \in X} a_P.$$

For $s \neq 1$, we must show that $h(s) = -\sum a_P(s) = \sum i_Q(s)$, and since $\sum i_Q(s) = \deg(\Delta . \Gamma_s)$, the formula is just a special case of the *Lefschetz formula* proved by Weil (cf., for example, [40], p. 161). For $s = 1$, it is the *Hurwitz formula.*

Corollary. $\sum_{P \in X} a_P = h_1 + E(X) . r_G - 2 . 1_G$.

Indeed, $h_0 = h_2 = 1_G$.

Remark. The above results are due to A. Weil, except for the formulation (cf. [66], §V). They imply that the sum of the Artin representations A_P attached to the various points of X is *rational over* $\mathbf{Q}_l$ (i.e., can be realized by a matrix representation with coefficients in $\mathbf{Q}_l$); that suggested the rationality of each A_P over $\mathbf{Q}_l$, $l \neq p$, and this has in fact been proved, even without restricting to the case of equal characteristic—cf. [57], [114]. On the other hand, neither the representation H_1 nor the Artin representations are in general rational over $\mathbf{Q}$ (cf. [57], n^0 4 and 6), which shatters the hope of finding a "trivial" definition of these representaions. (See also Fontaine [78].)

PART THREE

GROUP COHOMOLOGY

CHAPTER VII

Basic Facts

This chapter and the next are devoted to recalling the basic definitions and results on the homology and cohomology of groups; for more details, refer to Cartan-Eilenberg [13] or Grothendieck [26], as well as [75], [88], [93] and [113].

§1. G-Modules

Let G be a group, written multiplicatively, and let A be an abelian group, written additively. We say that G *acts on the left on* A if we are given a homomorphism $G \to \mathrm{Aut}(A)$, where Aut(A) is the group of automorphisms of A; this is equivalent to having a map $(s, x) \mapsto s.x$ of $G \times A$ into A satisfying the identities

$$1.a = a$$
$$s.(a + a') = s.a + s.a'$$
$$(s.t).a = s.(t.a).$$

Under these conditions, if Λ denotes the algebra $\mathbf{Z}[G]$ of the group G over $\mathbf{Z}$, then A is provided with a structure of left Λ-module by setting

$$(\textstyle\sum n_s s)a = \textstyle\sum n_s(s.a).$$

Conversely, if A is a left Λ-module (we will say "Λ-module" to be brief), the group G acts by $a \mapsto s.a$ on A (cf. Bourbaki, *Alg.*, Chap. II, §7, no. 9). We will often say "G-module" instead of "Λ-module".

If A and A′ are G-modules, a map $f: A \to A'$ is called a G-*homomorphism* (or sometimes simply a *homomorphism*) if it is a homomorphism of Λ-modules. The G-homomorphisms form a group denoted $\mathrm{Hom}_G(A, A')$. With these homomorphisms, the G-modules form an *abelian category* in the sense of Grothendieck [26]. As in every abelian category, there is a notion of projective or injective object: the G-module A is called *projective* if the functor $\mathrm{Hom}_G(A, A')$ is exact in A′, *injective* if this functor is exact in A. Along with these notions (often too restrictive for applications), we are led to introduce the following:

The G-module A is said to be *induced* if it has the form $\Lambda \otimes X$, where X is an abelian group (the tensor product being taken over **Z**). This amounts to saying that A contains a subgroup X such that $A = \sum_{s \in G} s.X$, the sum being *direct*. Every G-module *is a quotient of an induced module*: more precisely, if A_0 denotes the abelian group underlying the G-module A, then every element of the induced module $\Lambda \otimes A_0$ is a sum of elements $s \otimes x$, $x \in A_0$, $s \in G$, and mapping such an element onto $s.x \in A$ defines a homomorphism

$$\pi : \Lambda \otimes A_0 \to A.$$

Clearly π is a *surjective* G-*homomorphism*; furthermore, if we set $\varphi(a) = 1 \otimes a$, then $\pi \circ \varphi = 1$ (but, of course, φ is not a G-homomorphism if $A \neq 0$). Thus we have a *canonical* way of writing a G-module as a quotient of an induced G-module. If π maps a direct factor of $\Lambda \otimes A_0$ (as a G-module) isomorphically onto A, we will say that A is *relatively projective* (called "weakly projective" in Cartan-Eilenberg [13], Chap. X, §8; these are the $(\Lambda, \mathbf{Z})$-projective modules in the relative theory of Hochschild [35]). The relatively projective modules can also be characterised as the *direct factors of induced modules*.

Remark. If A and B are G-modules, then $A \otimes B$ can be provided with a structure of G-module via the formula

$$s.(a \otimes b) = s.a \otimes s.b.$$

In particular, $\Lambda \otimes A$ is provided with such a structure; the G-module so obtained *is isomorphic to* $\Lambda \otimes A_0$: indeed, it is easy to see that $s \otimes a \to s \otimes s.a$ extends to a G-isomorphism of $\Lambda \otimes A_0$ onto $\Lambda \otimes A$. This allows us to write either $\Lambda \otimes A$ or $\Lambda \otimes A_0$ in the sequel.

Dually, a G-module A will be called *co-induced* if it has the form $\mathrm{Hom}_{\mathbf{Z}}(\Lambda, X)$, where X is an abelian group. It can be shown as above that each G-module A embeds canonically in the co-induced G-module $\mathrm{Hom}_{\mathbf{Z}}(\Lambda, A_0)$. The direct factors of co-induced modules will be called *relatively injective* ("weakly injective" is the terminology in Cartan-Eilenberg, *loc. cit.*). When G is *finite*, $\mathrm{Hom}_{\mathbf{Z}}(\Lambda, X)$ is isomorphic to $\Lambda \otimes_{\mathbf{Z}} X$, so that the notions of induced and co-induced modules coincide, as do the notions of relatively projective and relatively injective modules (cf. [13], p. 233, prop. 1.1).

Exercise

Let k be a commutative ring. In the definition of G-modules, replace the abelian group A by a k-module, and **Z**[G] by k[G]. How are induced and co-induced modules defined in this framework?

§2. Cohomology

Let A be a G-module, and let A^G be the submodule consisting of the elements fixed by G. If $f: A \to B$ is a G-homomorphism, then f maps A^G into B^G; thus we can speak of the *functor* A^G. It is an additive functor that is left exact, i.e., given an exact sequence of G-modules

$$0 \to A \to A' \to A''$$

the sequence of abelian groups

$$0 \to A^G \to A'^G \to A''^G$$

is also exact.

By definition, the *right derived functors* of the functor A^G are the cohomology groups of G with coefficients in A; they are denoted $H^q(G, A)$, $q \geq 0$. Recall briefly how they are computed:

First note that A^G can be identified with $\mathrm{Hom}_G(\mathbf{Z}, A)$, the group $\mathbf{Z}$ being considered as a G-module with trivial action ($s.n = n$ for all $s \in G$). Hence $H^q(G, A) = \mathrm{Ext}^q(\mathbf{Z}, A)$, since the Ext^q are the derived functors of the functor Hom_G. So choose a resolution of the G-module $\mathbf{Z}$ by projective G-modules, i.e., an exact sequence

$$\cdots \to P_i \to P_{i-1} \to \cdots \to P_0 \to \mathbf{Z} \to 0$$

where the P_i are projective (e.g., free). Putting $K^i = \mathrm{Hom}_G(P_i, A)$, the K^i form a cochain complex K, and

$$H^q(G, A) = H^q(K)$$

which gives a method of computing these groups; in the next section we will exhibit a free resolution, the "standard" resolution. The $H^q(G, A)$ should be considered not only a sequence of functors, but a "cohomological functor" (cf. Grothendieck [26], no. 2.1); that means for every $q \geq 0$, and every exact sequence of G-modules

$$0 \to A \to B \to C \to 0$$

there is a "connecting homomorphism"

$$\delta: H^q(G, C) \to H^{q+1}(G, A)$$

such that the sequence (called "the exact cohomology sequence")

$$\cdots \to H^q(G, B) \to H^q(G, C) \xrightarrow{\delta} H^{q+1}(G, A) \to H^{q+1}(G, B) \to \cdots$$

is exact; moreover, the connecting homomorphisms depend "functorially" on the given exact sequence (in an obvious sense). We often write d instead of δ.

The following properties characterise the cohomological functor $\{H^q(G, \ ), \delta\}$:

i) $H^0(G, A) = A^G$.
ii) *If* A *is injective,* $H^q(G, A) = 0$ *for all* $q \geq 1$.

[That is a general characterisation of derived functors.]

We can strengthen ii):

Proposition 1. *If* A *is relatively injective,* $H^q(G, A) = 0$ *for all* $q \geq 1$.

As A is a direct factor of a co-induced module, additivity reduces us to the case where A itself is co-induced, i.e., $A = \mathrm{Hom}_{\mathbf{Z}}(\Lambda, X)$ for some abelian group X. If B is a G-module, then $\mathrm{Hom}_{\Lambda}(B, A) = \mathrm{Hom}_{\mathbf{Z}}(B, X)$. Applying this formula to the complex introduced above, we see that

$$K^i = \mathrm{Hom}_{\mathbf{Z}}(P_i, X)$$

and $H^q(K)$ is none other than $\mathrm{Ext}^q_{\mathbf{Z}}(\mathbf{Z}, X)$, which is clearly zero for $q \geq 1$. □

Corollary. *If* $0 \to A \to A^* \to A' \to 0$ *is an exact sequence of* G*-modules, with* A^* *co-induced, then*

$$H^q(G, A') = H^{q+1}(G, A) \quad \textit{for } q \geq 1.$$

This follows from the exact cohomology sequence.

The above corollary allows us to "shift" the cohomology groups; for example, we could take $A^* = \mathrm{Hom}_{\mathbf{Z}}(\Lambda, A)$—cf. §1.

§3. Computing the Cohomology via Cochains

A free resolution of **Z** can be obtained by taking P_i to be the free **Z**-module L_i having basis the systems $(g_0, \ldots, g_i)$ of $i + 1$ elements of G, and making G operate on L_i by translations:

$$s.(g_0, \ldots, g_i) = (sg_0, \ldots, sg_i).$$

The homomorphism $d: L_i \to L_{i-1}$ is defined by the formula

$$(*) \qquad d(g_0, \ldots, g_i) = \sum_{j=0}^{j=i} (-1)^j (g_0, \ldots, \hat{g}_j, \ldots, g_i)$$

where the symbol ^ means that the letter under it should be omitted.

The homomorphism $L_0 \to \mathbf{Z}$ is defined by mapping each (g_0) onto $1 \in \mathbf{Z}$. The fact that the sequence $\cdots \to L_1 \to L_0 \to \mathbf{Z} \to 0$ so obtained is exact is well-known (a simplex is acyclic).

An element of $K^i = \mathrm{Hom}_{\Lambda}(L_i, A)$ can then be identified with a function $f(g_0, \ldots, g_i)$ having values in A, and satisfying the "covariance" condition:

$$f(s.g_0, \ldots, s.g_i) = s.f(g_0, \ldots, g_i).$$

The coboundary of f is defined by the transpose of formula (*) (left to the reader).

A covariant cochain f is uniquely determined by its restriction to systems of the form $(1, g_1, g_1 g_2, \ldots, g_1 \ldots g_i)$. That leads us to interpret the elements of K^i as "inhomogeneous cochains", i.e., as functions $f(g_0, \ldots, g_i)$ of i arguments, with values in A, whose coboundary is given by:

$$(**)\qquad df(g_1, \ldots, g_{i+1}) = g_1 . f(g_2, \ldots, g_{i+1}) + \sum_{j=1}^{j=i} (-1)^j f(g_1, \ldots, g_j g_{j+1}, \ldots, g_{i+1}) + (-1)^{i+1} f(g_1, \ldots, g_i).$$

The Group $H^1(G, A)$

Formula (**) shows that a 1-cocycle is a map f of G into A satisfying the identity

$$f(gg') = g . f(g') + f(g).$$

It is called also a *crossed homomorphism*. It is a coboundary if there exists $a \in A$ such that $f(g) = g . a - a$ for all $g \in G$.

When G acts trivially on A, we have $H^1(G, A) = \mathrm{Hom}(G, A)$.

The Group $H^2(G, A)$

A 2-cocycle is a map f of $G \times G$ into A satisfying the identity

$$g . f(g', g'') - f(gg', g'') + f(g, g'g'') - f(g, g') = 0.$$

It is also called a *factor set*.

Such functions are encountered in the classification of *extensions of* G *by* A:

Let E be a group containing A as a normal subgroup, the quotient being G; then G acts on A by inner automorphisms (A being assumed commutative, of course); this gives a first *invariant* of the extension. This being known, let $s: G \to E$ be a section ("system of representatives"); if $g, g' \in G$, then clearly $s(g) . s(g')$ and $s(g . g')$ are in the same coset modulo A, hence there exists an element $f(g, g')$ of A such that

$$s(g) . s(g') = f(g, g') . s(gg').$$

Clearly the knowledge of f allows us to write the law of composition of E; expressing the associativity of this law, a short computation shows that f is a *factor set*. Changing s modifies f by a coboundary, and one can construct an extension corresponding to a given factor set. One concludes (for the details see [13], Chap. XIV, §4) that $H^2(G, A)$ *is the set of isomorphism classes of extensions of* G *by* A (the action of G on A being given).

Exercise

Let f be a cochain of degree n. Put

$$Tf(g_1, \ldots, g_n) = g_1 \cdots g_n . f(g_n^{-1}, \ldots, g_1^{-1}).$$

Prove the identity $T(df) = (-1)^{n+1}d(Tf)$. Deduce that if f is a cocycle, Tf is a cocycle cohomologous to f (resp. $-f$) if $\deg(f) \equiv 0$ or $3 \bmod. 4$ (resp. if $\deg(f) \equiv 1$ or $2 \bmod. 4$). Check this result directly if $\deg(f) \leq 2$.

§4. Homology

Let A be a G-module, and let DA be the subgroup of A generated by the $s.a - a$, $a \in A$, $s \in G$. The quotient A/DA will be denoted A_G; it is the largest quotient module of A on which G acts trivially (whereas A^G is the largest submodule enjoying the same property).

The functor A_G is additive and right exact. Its left derived functors are, by definition, the homology groups of G with coefficients in A, denoted $H_q(G, A)$ They form a "homological functor". We have $H_0(G, A) = A_G = \mathbf{Z} \otimes_\Lambda A$, and $H_q(G, A) = \operatorname{Tor}_q^\Lambda(\mathbf{Z}, A)$. If $\{P_i\}$ is a projective resolution of $\mathbf{Z}$, the $H_q(G, A)$ are the homology groups of the complex formed by the $P_i \otimes_\Lambda A$.

[In a tensor product $B \otimes_\Lambda A$, we consider B as a right Λ-module by setting $b.s = s^{-1}.b$.]

The homological functor $\{H_q(G, \ ), \partial\}$ is characterised by the following two properties:

i) $H_0(G, A) = A_G$.
ii) *If* A *is projective*, $H_q(G, A) = 0$ *for all* $q \geq 1$.

Proposition 2. *If* A *is relatively projective*, $H_q(G, A) = 0$ *for all* $q \geq 1$.

The proof is analogous to that of prop. 1. In particular, an *induced* module has trivial homology, which again permits methods of "shifting" to be used.

The free resolution of $\mathbf{Z}$ given in the previous section also provides an explicit description of the complex $L \otimes_\Lambda A$ of the $L_i \otimes_\Lambda A$, thence of the $H_q(G, A)$. The result is as follows:

An element $x \in L_q \otimes_\Lambda A$ can be identified with a function $x(g_1, \ldots, g_q)$ with values in A, which is zero except for finitely many $(g_1, \ldots, g_q)$. The boundary d is given by the formula:

$$\begin{aligned} dx(g_1, \ldots, g_{q-1}) = & \sum_{g \in G} g^{-1}x(g, g_1, \ldots, g_{q-1}) \\ & + \sum_{j=1}^{j=q-1} (-1)^j \sum_{g \in G} x(g_1, \ldots, g_j g, g^{-1}, g_{j+1}, \ldots, g_{q-1}) \\ & + (-1)^q \sum_{g \in G} x(g_1, g_2, \ldots, g_{q-1}, g). \end{aligned}$$

When $A = \mathbf{Z}$, on which G acts trivially, we have $H_1(G, \mathbf{Z}) = G/G'$, where G' is the commutator subgroup of G (cf. [13], p. 190). As this isomorphism will play an important role in the sequel, we recall how it is defined:

Let $\pi: \Lambda \to \mathbf{Z}$ be the *augmentation* of the group algebra $\Lambda = \mathbf{Z}[G]$, i.e., the homomorphism that maps $\sum n_s s \in \Lambda$ onto $\sum n_s \in \mathbf{Z}$. Let I_G be the kernel of π; it is the subgroup of Λ generated by the elements $i_s = s - 1$, as s runs through G. A being any G-module, we have by definition $H_0(G, A) = A/I_G A$. Consider the exact sequence

$$0 \to I_G \to \Lambda \xrightarrow{\pi} \mathbf{Z} \to 0.$$

We have $H_0(G, I_G) = I_G/I_G^2$, and the image of $H_0(G, I_G)$ in $H_0(G, \Lambda)$ is zero. As Λ is free, we have $H_1(G, \Lambda) = 0$. Hence the exact homology sequence gives an isomorphism

$$d: H_1(G, \mathbf{Z}) \to H_0(G, I_G) = I_G/I_G^2.$$

On the other hand, it can be easily verified (cf. [13], *loc. cit.*) that $s \mapsto i_s$ defines by passage to the quotient an isomorphism θ of G/G' onto I_G/I_G^2. Thus we can identify $H_1(G, \mathbf{Z})$ with G/G' by means of the isomorphism $\theta^{-1} \circ d$, which we will do henceforth.

§5. Change of Group

Let $f: G' \to G$ be a homomorphism of groups, and let A be a G-module. If we put

$$s'.a = f(s').a, \qquad s' \in G', a \in A,$$

then A is provided with a structure of G'-module, which we denote f^*A (or simply A if that does not lead to confusion). Clearly A^G is a subgroup of $(f^*A)^{G'}$. This defines a morphism of the functor $H^0(G, A)$ into the functor $H^0(G', f^*A)$; as the $H^q(G', f^*A)$ form a cohomological functor (with respect to A), the universal property of derived functors (cf. Grothendieck [26]) shows that the morphism above extends to a morphism of the cohomological functor $\{H^q(G, \quad), \delta\}$ into the cohomological functor

$$\{H^q(G', f^* \quad), \delta\}.$$

In particular, for every $q \geq 0$ and every G-module A, we have a homomorphism

$$H^q(G, A) \to H^q(G', A)$$

often denoted f_q^*.

More generally, consider a G'-module A' and an additive map $g: A \to A'$. We say that f and g are *compatible* when $g(f(s').a) = s'.g(a)$ for all $s' \in G'$, $a \in A$; this amounts to saying that g is a *G'-homomorphism of f^*A into A'*. The map g thus defines a homomorphism

$$H^q(G', f^*A) \to H^q(G', A')$$

and composing it with the homomorphism obtained above, we get a homomorphism

$$(f, g)_q^*: H^q(G, A) \to H^q(G', A')$$

which is said to be *associated* to the pair (f, g); the expression of this homomorphism in terms of standard cochains is left to the reader.

EXAMPLES. 1) If H is a subgroup of G and f is the *inclusion* of H into G, we get homomorphisms

$$H^q(G, A) \to H^q(H, A)$$

which are called *restriction* homomorphisms, denoted Res.

2) Let H be a normal subgroup of G. The group A^H is a G/H-module, and the homomorphisms $G \to G/H$, $A^H \to A$ are compatible. Thus we get homomorphisms

$$H^q(G/H, A^H) \to H^q(G, A)$$

which are called *inflation*, denoted Inf.

Similar procedures hold for homology. There are homomorphisms

$$H_q(G', f^*A) \to H_q(G, A).$$

When H is a subgroup of G and f is the inclusion of H into G, we get homomorphisms

$$H_q(H, A) \to H_q(G, A)$$

called *corestriction* and denoted Cor.

Back to cohomology. Consider the case $G = G'$, $A = A'$, with the map $f: G \to G$ being the inner automorphisms $s \mapsto tst^{-1}$, and $g: A \to A$ being $a \to t^{-1}a$. One checks that these two maps are compatible, hence they define automorphisms σ_t of the cohomology $H^q(G, A)$. In fact:

Proposition 3. *The automorphisms σ_t are equal to the identity.*

The σ_t constitute an automorphism of the cohomological functor $\{H^q(G, \ ), \delta\}$; this automorphism is the identity in dimension zero (clear). Hence, by a general result (cf. [13], Chap. III or [26], no. 2.2) it is the identity in all dimensions.

DIRECT PROOF. Argue by induction on q, the case $q = 0$ being trivial. Embed A in a co-induced module A^*; let $B = A^*/A$. There is a commutative diagram

$$\begin{array}{ccccc} H^q(G, B) & \xrightarrow{\delta} & H^{q+1}(G, A) & \to & 0 \\ \downarrow{\scriptstyle\sigma_t} & & \downarrow{\scriptstyle\sigma_t} & & \\ H^q(G, B) & \xrightarrow{\delta} & H^{q+1}(G, A) & \to & 0. \end{array}$$

As σ_t is the identity on $H^q(G, B)$ by inductive hypothesis, it follows that it must be the identity on $H^{q+1}(G, A)$. □

EXERCISE

Let H be a subgroup of G, and let B be an H-module. Let B^* be the group of maps φ of G into B such that $\varphi(hs) = h\varphi(s)$ for all $h \in H$; show that $B^* = \mathrm{Hom}_{\mathbf{Z}[H]}(\mathbf{Z}[G], B)$. Make B^* into a G-module by setting $(s\varphi)(g) = \varphi(gs)$. Let $\theta: B^* \to B$ be the homomorphism defined by $\theta(\varphi) = \varphi(1)$. Show that θ is compatible with the inclusion $H \to G$.

Show that the homomorphisms

$$H^q(G, B^*) \to H^q(H, B)$$

associated to this pair of maps are *isomorphisms* (*Shapiro's lemma*). (Note that, if B is co-induced for H, then B* is co-induced for G; deduce that $H^q(G, B^*) = 0$ for $q \geq 1$ if B is co-induced. Conclude by remarking that $H^q(G, B^*)$ is a cohomological functor of the H-module B.)

§6. An Exact Sequence

Let H be a normal subgroup of a group G, and let A be a G-module. In the preceding section, we defined the homomorphisms

$$\text{Res}: H^q(G, A) \to H^q(H, A)$$
$$\text{Inf}: H^q(G/H, A^H) \to H^q(G, A).$$

Proposition 4. *The sequence below is exact:*

$$0 \to H^1(G/H, A^H) \xrightarrow{\text{Inf}} H^1(G, A) \xrightarrow{\text{Res}} H^1(H, A).$$

It is clear that $\text{Res} \circ \text{Inf} = 0$ (look at the cochains, for example). Thus there are two things to prove:

1. *Exactness at* $H^1(G/H, A^H)$. Let $f: G/H \to A^H$ be a cocycle equivalent to 0 in $H^1(G, A)$. There is an $a \in A$ such that $f(s) = sa - a$ (by abuse of language, we identify f with the map of G into A constant on the cosets of H that lifts f). But $f(s)$ depends only on the coset of s modulo H, so that $sa - a = sta - a$ for all $t \in H$, i.e., $ta = a$. Thus $a \in A^H$, so f is cohomologous to 0 in $H^1(G/H, A^H)$.

2. *Exactness at* $H^1(G, A)$. Let $f: G \to A$ be a cocycle whose restriction to H is cohomologous to 0, i.e., $f(t) = ta - a$ for some $a \in A$ and all $t \in H$. Subtracting from f the coboundary $g(s) = sa - a$, we are reduced to the case where $f(t) = 0$ for all $t \in H$. The formula $f(st) = f(s) + sf(t)$, in which we take $t \in H$, shows that f is constant on the cosets of H. Applying this formula again, but this time with $s \in H$, and keeping in mind that H is normal, we see that $f(st) = sf(t)$, which entails the invariance of f under H. Therefore f is a cocycle of G/H with values in A^H. □

The next result generalizes prop. 4.

Proposition 5. *Given a positive integer q. Suppose that* $H^i(H, A) = 0$ *for*

$$1 \leq i \leq q - 1.$$

Then the sequence below is exact:

$$0 \to H^q(G/H, A^H) \xrightarrow{\text{Inf}} H^q(G, A) \xrightarrow{\text{Res}} H^q(H, A).$$

(In studying the Brauer group, we will have to apply this proposition for $q = 2$; the hypothesis is then $H^1(H, A) = 0$.)

Corollary. $\mathrm{Inf}: H^i(G/H, A^H) \to H^i(G, A)$ *is an isomorphism for* $i \leq q - 1$.

Proof of Prop. 5. Argue by induction on q, the case $q = 1$ being prop. 4. Suppose $q \geq 2$. Let $B = \mathrm{Hom}(\mathbf{Z}[G], A)$ be the co-induced module canonically associated to A; an element of B can be identified with a function φ on G with values in A such that $s\varphi(t) = \varphi(ts)$. If $a \in A$, put $\varphi_a(t) = t.a$; then $a \to \varphi_a$ is an injective G-homomorphism of A into B; if $C = B/A$, there is an exact sequence of G-modules

$$0 \to A \to B \to C \to 0.$$

The module B is co-induced for H as well as for G, because $\mathbf{Z}[G]$ is $\mathbf{Z}[H]$-free, hence can be written $\mathbf{Z}[H] \otimes M$ (for some abelian group M), and $B = \mathrm{Hom}(\mathbf{Z}[H], \mathrm{Hom}(M, A))$. Further, as $H^1(H, A) = 0$ by hypothesis, there is an exact sequence

$$0 \to A^H \to B^H \to C^H \to 0$$

and $B^H = \mathrm{Hom}(\mathbf{Z}[G/H], A)$ is G/H-co-induced.

Consider next the diagram

$$\begin{array}{ccccccc} 0 \to & H^{q-1}(G/H, C^H) & \to & H^{q-1}(G, C) & \to & H^{q-1}(H, C) \\ & \delta\downarrow & & \delta\downarrow & & \delta\downarrow \\ 0 \to & H^q(G/H, A^H) & \to & H^q(G, A) & \to & H^q(H, A). \end{array}$$

This diagram is commutative, as is easily seen. The vertical arrows are the connecting homomorphisms defined by the exact sequences above; as B is co-induced for G and H, while B^H is co-induced for G/H, those arrows are isomorphisms. For the same reason, C satisfies the inductive hypothesis (with $q - 1$ in place of q). Hence the top row of the diagram is exact, whence the bottom row is too. □

Remark. By applying prop. 3, we see that G/H acts on all the $H^i(H, A)$, so that we may speak of the groups $H^j(G/H, H^i(H, A))$. The exact sequence of prop. 5 can be extended to the exact sequence

$$0 \to H^q(G/H, A^H) \to H^q(G, A) \to H^q(H, A)^{G/H} \to H^{q+1}(G/H, A^H) \to H^{q+1}(G, A).$$

This can be seen by dimension-shifting (with some trouble), or by applying the *spectral sequence of group extensions* (cf. [13], p. 351, or [37]).

§7. Subgroups of Finite Index

Let H be a subgroup of G, and let A be a G-module. In §5 we defined the restriction homomorphisms

$$\mathrm{Res}: H^q(G, A) \to H^q(H, A).$$

Suppose now that H has *finite index* in G. We will define homomorphisms in the opposite direction

$$\mathrm{Cor}: \mathrm{H}^q(\mathrm{H}, \mathrm{A}) \to \mathrm{H}^q(\mathrm{G}, \mathrm{A})$$

called *corestriction.*

We begin with the case $q = 0$. If $a \in \mathrm{A}^{\mathrm{H}}$ and $s \in \mathrm{G}$, the element sa evidently depends only on the *left coset* of s mod. H; as G/H is finite by hypothesis, we can form the sum

$$\mathrm{N}_{\mathrm{G/H}}(a) = \sum_{s \in \mathrm{G/H}} sa,$$

which is called the *norm* of a. One checks that $\mathrm{N}_{\mathrm{G/H}}(a)$ is invariant under G, hence we get a homomorphism

$$\mathrm{N}_{\mathrm{G/H}}: \mathrm{H}^0(\mathrm{H}, \mathrm{A}) \to \mathrm{H}^0(\mathrm{G}, \mathrm{A}).$$

This is the corestriction in dimension 0. It extends uniquely to a morphism of the cohomological functor $\{\mathrm{H}^q(\mathrm{H}, f^* \quad), \delta\}$ into the cohomological functor $\{\mathrm{H}^q(\mathrm{G}, \quad), \delta\}$: this is possible because the first functor is "effaçable" in dimension ≥ 1, in the sense of Grothendieck [26], no. 2.2 (indeed, if A is co-induced for G, we have seen that it is co-induced for H, hence $\mathrm{H}^q(\mathrm{H}, \mathrm{A}) = 0$ for $q \geq 1$). Thus we obtain

$$\mathrm{Cor}: \mathrm{H}^q(\mathrm{H}, \mathrm{A}) \to \mathrm{H}^q(\mathrm{G}, \mathrm{A}).$$

(For a more explicit definition, see Cartan-Eilenberg [13], p. 254, as well as Eckmann [22].)

Proposition 6. *If* $n = \mathrm{Card}(\mathrm{G/H})$, *then* $\mathrm{Cor} \circ \mathrm{Res} = n$.

For $q = 0$, this says that $\mathrm{N}_{\mathrm{G/H}}(a) = na$ if $a \in \mathrm{A}^{\mathrm{G}}$, which is clear. The general case can be reduced to the case $q = 0$ by shifting. □

Move now to homology. If $a \in \mathrm{A}$, and if s and s' are in the same *left coset* mod. H, then the images of $s^{-1}a$ and $s'^{-1}a$ in A_{H} coincide. The expression

$$\mathrm{N}'_{\mathrm{G/H}}(a) = \sum_{s \in \mathrm{G/H}} s^{-1}a$$

therefore makes sense in A_{H}, and we get a homomorphism

$$\mathrm{N}'_{\mathrm{G/H}}: \mathrm{A}_{\mathrm{G}} \to \mathrm{A}_{\mathrm{H}}$$

which we will also call *restriction* and denote Res. It extends as above to a morphism of functors $\{\mathrm{H}_q(\mathrm{G}, \quad), \partial\} \to \{\mathrm{H}_q(\mathrm{H}, f^* \quad), \partial\}$, again denoted Res. There is a proposition dual to prop. 6 (using the corestriction of §5).

Exercises

1. With the hypotheses of prop. 6, let q be such that $\mathrm{H}^q(\mathrm{H}, \mathrm{A}) = 0$. Show that $nx = 0$ for all $x \in \mathrm{H}^q(\mathrm{G}, \mathrm{A})$.

2. Let $G = PSL(2, \mathbf{Z})$, the *modular group*, and let A be a G-module. Show that for every $q \geq 2$ and for every $x \in H^q(G, A)$, we have $6x = 0$. (Apply exer. 1 and note that G contains a free subgroup of index 6.)

3. Let A be a G-module, and let $A^* = \operatorname{Hom}_{\mathbf{Z}[H]}(\mathbf{Z}[G], A)$ be the G-module defined in the exercise of §5. If $\varphi \in A^*$, show that $s^{-1}\varphi(s)$ depends only on the coset of s mod. H. Assuming H of finite index in G, form the sum $\pi(\varphi) = \sum_{s \in G/H} s^{-1}\varphi(s)$, and show that it defines a G-homomorphism $\pi: A^* \to A$. Combining the homomorphisms $H^q(G, A^*) \to H^q(G, A)$ induced by π with the isomorphism $H^q(G, A^*) = H^q(H, A)$, show that the corestriction is obtained.

§8. Transfer

We still suppose that H is a *subgroup of finite index* in G. Applying the definitions of §7 to the G-module **Z** (on which G acts trivially) and to the integer $q = 1$, we get a homomorphism

$$\operatorname{Res}: H_1(G, \mathbf{Z}) \to H_1(H, \mathbf{Z}).$$

Now we saw in §4 that $H_1(G, \mathbf{Z})$ can be identified with G/G' (the largest abelian quotient group of G), and $H_1(H, \mathbf{Z})$ with H/H', so that Res becomes a homomorphism

$$\operatorname{Ver}: G/G' \to H/H',$$

called *transfer* ("Verlagerung" in German). Let us make it explicit:

Let I_G be the kernel of the augmentation $\mathbf{Z}[G] \to \mathbf{Z}$. By the very definition of Res, there is a commutative diagram

$$\begin{array}{ccc} H_1(G, \mathbf{Z}) & \xrightarrow{\partial} & H_0(G, I_G) = I_G/I_G^2 \\ {\scriptstyle \operatorname{Res}}\downarrow & & \downarrow{\scriptstyle N} \\ H_1(H, \mathbf{Z}) & \xrightarrow{\partial} & H_0(H, I_G) = I_G/I_H I_G. \end{array}$$

On the other hand, since **Z**[H] embeds into **Z**[G] in a manner compatible with the action of H, there is a commutative diagram

$$\begin{array}{ccc} H_1(H, \mathbf{Z}) & \xrightarrow{\partial} & I_G/I_H I_G \\ {\scriptstyle \text{id.}}\uparrow & & \uparrow \\ H_1(H, \mathbf{Z}) & \xrightarrow{\partial} & I_H/I_H^2. \end{array}$$

In both of these diagrams, the ∂ operators are injective, since **Z**[H] and **Z**[G] are induced modules. If we replace $H_1(G, \mathbf{Z})$ with G/G' and $H_1(H, \mathbf{Z})$ with H/H', we get the following commutative diagram:

$$\begin{array}{ccccc} G/G' & \xrightarrow{\approx} & I_G/I_G^2 & & \\ & & & \searrow{\scriptstyle N} & \\ {\scriptstyle \operatorname{Ver}}\downarrow & & & & I_G/I_H I_G. \\ & & & \nearrow & \\ H/H' & \xrightarrow{\approx} & I_H/I_H^2 & & \end{array}$$

Now we just have to compute: let $s \in G$; the image of s in I_G/I_G^2 is $i_s = s - 1$. If the s_i form a system of representatives of the *right cosets* mod. H, then $N(i_s)$ is the coset of $\sum s_i(s-1)$ mod. $I_H I_G$. But for each i there exists an index $j(i)$ and an element $x_i \in H$ such that

$$s_i s = x_i s_{j(i)}.$$

Hence we can write

$$\begin{aligned} N(i_s) &\equiv \sum x_i s_{j(i)} - \sum s_{j(i)} \quad \text{mod. } I_H I_G \\ &\equiv \sum (x_i - 1) s_{j(i)} \quad \text{mod. } I_H I_G \\ &\equiv \sum (x_i - 1) \quad \text{mod. } I_H I_G \end{aligned}$$

Put $x = \prod x_i$. The image of x in $I_G/I_H I_G$ is equal to the coset of $\sum (x_i - 1)$. It follows that the image of x in H/H' is equal to the transfer of the image of s in G/G'. Changing the notation slightly, this gives:

Proposition 7. *Let $\theta : H\backslash G \to G$ be a system of representations for the homogeneous space $H\backslash G$ of right cosets of G mod. H. For every $s \in G$ and $t \in H\backslash G$, let $x_{t,s}$ be the element of H given by the formula*

$$\theta(t).s = x_{t,s}.\theta(t.s).$$

Then the transfer $\mathrm{Ver}: G/G' \to H/H'$ *is obtained by passing to the quotient the map $s \to \prod_t x_{t,s}$.*

We have recovered the classical definition of the transfer—cf. for example M. Hall [30], p. 202.

The above computation can be pushed further:

Let S be the subgroup of G generated by the element s, and let S act on the homogeneous space $H\backslash G$ on the right. The orbits of this action are the double cosets $W \in H\backslash G/S$. In each double coset, choose an element x, and let f_x be the order of the corresponding orbit (it is the smallest positive integer f for which $x.s^f \equiv x$ mod. H). As x runs through the representatives of the double cosets, it is clear that the elements $x, x.s, x.s^2, \ldots, x.s^{f_x - 1}$ form a system of representatives of $H\backslash G$.

To determine the transfer in terms of this system we must write each product $xs^e.s$ in the form $y.t$, with $y \in H$ and t representative. Now $x.s^{e+1}$ is itself a representative, except when $e = f_x - 1$, in which case

$$x.s^{f_x} \equiv x \quad \text{mod. H}.$$

by definition of f_x. Thus, in the computation of $\mathrm{Ver}(s)$, the only factors not necessarily equal to 1 are those which correspond to the products $x.s^{f_x - 1}.s$, and the latter are equal to $x.s^{f_x}.x^{-1}$. Hence:

Proposition 8. *Let x_i be a system of representatives of the double cosets* HxS, *and for each i, define $f_i = f_{x_i}$ as above. Then*

$$\mathrm{Ver}(s) = \prod x_i.s^{f_i}.x_i^{-1} \quad \text{mod. H}'.$$

Application (Artin [4]). Let L/K be a Galois extension with group G, and let A be a Dedekind domain with field of fractions K and *finite* residue fields. If L_a/K is the largest abelian extension of K within L, the Artin symbol $(\mathfrak{a}, L_a/K)$ is defined for each ideal $\mathfrak{a}$ of A unramified in L (cf. Chap. I, §8) and is an element of G/G′ (the Galois group of L_a/K). Further, let K′/K be a subextension of L/K, and let L'_a be the largest abelian extension of K′ within L; if $\mathfrak{a}'$ is the ideal generated by $\mathfrak{a}$ in the integral closure of A in K′, the Artin symbol $(\mathfrak{a}', L'_a/K')$ is well-defined and belongs to H/H′, where H = G(L/K′). Then we have the formula

$$\mathrm{Ver}((\mathfrak{a}, L_a/K)) = (\mathfrak{a}', L'_a/K').$$

[By linearity, we may assume $\mathfrak{a} = \mathfrak{p}$ to be prime; choose a prime $\mathfrak{P}$ of L over $\mathfrak{p}$, and set $s = (\mathfrak{P}, L/K)$; it is clear that $(\mathfrak{a}, L_a/K)$ is just the canonical image of s in G/G′, so we are reduced to computing Ver(s). That can be done by the method of prop. 8, noticing that the double cosets Hx_iS are in one-to-one correspondence with the prime ideals $\mathfrak{p}'_i$ of K′ over $\mathfrak{p}$. Each term $x_i . s^{f_i} . x_i^{-1}$ is none other than $(\mathfrak{p}'_i, L'_a/K')$, according to prop. 22 of Chapter I. Thus indeed

$$\mathrm{Ver}(s) = \prod(\mathfrak{p}'_i, L'_a/K') = (\mathfrak{a}', L'_a/K')$$

since $\mathfrak{a}' = \prod \mathfrak{p}'_i$.]

Remarks. 1) If G is finitely generated and H = G′, it can be shown that the transfer Ver: G/G′ → H/H′ is zero (cf. Artin-Tate [8], p. 189). Combined with what we just did, and with Artin's reciprocity law, this gives the "Hauptidealsatz"—cf. Artin [4].

2) The transfer was first defined by I. Schur (*Gesam. Abh.* I, p. 79–80), and then rediscovered by Artin [4]. Gauss' Lemma for quadratic reciprocity (cf. [115], p. 9) is a special case of it (for a cyclic group and a subgroup of order two).

Exercise

Let H be a subgroup of finite index of a group G, and let $\chi: H \to C$ be a linear representation of degree 1 of H. Let $s \to M_s$ be the linear representation of G *induced* by χ (cf. Chap. VI, §1). Show that

$$\det(M_s) = \varepsilon(s) . \chi(\mathrm{Ver}(s)),$$

where $\varepsilon(s)$ is the sign of the permutation of G/H defined by s.

APPENDIX

Non-abelian Cohomology

Let G be a group and A a group on which G acts on the left. Up till now we have only considered the case where A is abelian. We will abandon this hypothesis here and show that we can still define $H^0(G, A)$ and $H^1(G, A)$ and prove a "piece" of an exact sequence.

Write A multiplicatively. $H^0(G, A)$ is defined again as the group A^G of elements of A fixed by G (i.e., $s(a) = a$ for all $s \in G$). By a *cocycle* will be meant a map $s \mapsto a_s$ of G into A such that $a_{st} = a_s . s(a_t)$; we say that a_s and b_s are *cohomologous* if there exists $a \in A$ such that $b_s = a^{-1} . a_s . s(a)$ for all $s \in G$. This defines an equivalence relation for the set of cocycles, and the quotient set, provided with the structure of a distinguished element equal to the class of the unit cocycle $a_s = 1$ (structure of "pointed set"), will be called the *cohomology set of* G *with values in* A, and denoted $H^1(G, A)$. This definition coincides with the habitual definition when A is abelian, except that we only retain from the group structure of $H^1(G, A)$ its underlying structure of pointed set.

The objects $H^0(G, A)$ and $H^1(G, A)$ are functors in A: if $f: A \to B$ is a G-homomorphism (i.e., a group homomorphism that commutes with the operations of G), we define

$$f_0 : H^0(G, A) \to H^0(G, B)$$

$$f_1 : H^1(G, A) \to H^1(G, B)$$

as follows: f_0 is the restriction of f to A^G, the image of f_0 obviously being composed of fixed elements of B; composing a cocycle of A with f gives a cocycle of B, and this composition is compatible with the equivalence relation; f_0 is a group homomorphism, while f_1 is a "morphism of pointed sets", i.e., f sends the unit cocycle of A onto the unit cocycle of B.

By the *kernel* of a morphism of pointed sets we mean the pre-image of the distinguished element of the target set. This allows us to speak of an *exact sequence* of pointed sets.

Let $(*) 1 \to A \xrightarrow{i} B \xrightarrow{p} C \to 1$ be an exact sequence of non-abelian G-modules, with $i(A)$ being normal in B. Define a *coboundary* operator $\delta: H^0(G, C) \to H^1(G, A)$ as follows:

Given $c \in C^G$. Choose $b \in B$ such that $pb = c$. The element c being fixed by G and the sequence $(*)$ being exact, we have $s(b) \equiv b \bmod. i(A)$ for all $s \in G$. This allows us to define a_s by $a_s = i^{-1}(b^{-1}s(b))$. We will show successively that a_s is a cocycle of A and that its cohomology class is independent of the choice of b, so that class can be defined to be $\delta(c)$. (To simplify notation, we take i to be the inclusion map $i(a) = a$.) First

$$a_{st} = b^{-1}st(b) = b^{-1}s(b)s(b^{-1}t(b)) = a_s \,.\, s(a_t).$$

Secondly, if $pb' = pb = c$, then there exists $a \in A$ such that $b' = ba$; if a'_s denotes the cocycle defined by means of b', then

$$a'_s = a^{-1}b^{-1}s(b)s(a) = a^{-1}a_s s(a)$$

which is cohomologous to a_s.

Note that this coboundary operator coincides in the abelian case with the usual one.

Suppose now that A *is contained in the center of* B. We will define $\Delta: H^1(G, C) \to H^2(G, A)$, the latter being the pointed set underlying the group $H^2(G, A)$ defined in the usual way (A being abelian). Let c_s be a cocycle of C and choose $b_s \in B$ such that $p(b_s) = c_s$. As before, $b_{st} \equiv b_s \,.\, s(b_t) \bmod. A$ for all $s, t \in G$, which allows us to define $a_{s,t} \in A$ by $a_{s,t} = b_s \,.\, s(b_t) \,.\, b_{st}^{-1}$. We now show that $a_{s,t}$ is a 2-cocycle of A whose class in $H^2(G, A)$ depends neither on the choice of c_s within its class nor on the choice of $b_s \in p^{-1}(c_s)$, so that Δ can be defined by passage to the quotient:

The 2-cocycle condition to be verified is

$$s(a_{t,u})a_{s,tu} = a_{st,u}a_{s,t}$$

or, as A is abelian, $a_{s,t}^{-1}a_{s,tu}a_{st,u}^{-1}s(a_{t,u}) = 1$. Explicitly:

$$b_{st}s(b_t)^{-1}b_s^{-1}b_s s(b_{tu})b_{stu}^{-1}b_{stu}st(b_u)^{-1}b_{st}^{-1}s(a_{t,u})$$
$$= b_{st}s(b_t)^{-1}s(b_{tu})st(b_u)^{-1}b_{st}^{-1}s(a_{t,u}).$$

But $s(a_{t,u}) = s(b_t)st(b_u)s(b_{tu}^{-1})$ is in the center of B. Thus the formula to be proved can be written

$$b_{st}s(b_t)^{-1}s(b_t)st(b_u)s(b_{tu}^{-1})s(b_{tu})st(b_u)^{-1}b_{st}^{-1} = 1,$$

which is obvious.

Suppose now we replace c_s with a cohomologous cocycle $c'_s = c^{-1}c_s s(c)$, and lift c'_s to $b'_s = b^{-1}b_s s(b)$ where $b \in B$ is such that $pb = c$. Let us show that

$a_{s,t}$ does not change:

$$a'_{s,t} = b'_s s(b'_t) b'^{-1}_{st} = b^{-1} b_s s(b) s(b^{-1}) s(b_t) st(b) st(b)^{-1} b_{st}^{-1} b = b^{-1} a_{s,t} b = a_{s,t}.$$

Finally, if we modify the choice of b_s, we replace it with $b'_s = a_s b_s$, where $a_s \in \mathrm{A}$. The cocycle $a_{s,t}$ is then replaced by $a'_{s,t} = a_s b_s s(a_t) b_t b_{st}^{-1} a_{st}^{-1}$ which, since A is in the center of B, can be written:

$$a'_{s,t} = a_s s(a_t) a_{st}^{-1} a_{s,t}$$

and is indeed cohomologous to $a_{s,t}$.

This completes the definition of $\Delta : \mathrm{H}^1(\mathrm{G}, \mathrm{C}) \to \mathrm{H}^2(\mathrm{G}, \mathrm{A})$.

Proposition 1. *Let* $1 \to \mathrm{A} \xrightarrow{i} \mathrm{B} \xrightarrow{p} \mathrm{C} \to 1$ *be an exact sequence of non-abelian* G-*modules. Then the sequence of pointed sets below is exact:*

$$1 \to \mathrm{H}^0(\mathrm{G},\mathrm{A}) \xrightarrow{i_0} \mathrm{H}^0(\mathrm{G},\mathrm{B}) \xrightarrow{p_0} \mathrm{H}^0(\mathrm{G},\mathrm{C}) \xrightarrow{\delta} \mathrm{H}^1(\mathrm{G},\mathrm{A}) \xrightarrow{i_1} \mathrm{H}^1(\mathrm{G},\mathrm{B}) \xrightarrow{p_1} \mathrm{H}^1(\mathrm{G},\mathrm{C}).$$

Proposition 2. *In addition to the hypothesis of prop.* 1, *assume that* A *is in the center of* B. *Then the sequence of pointed sets below is exact:*

$$1 \to \mathrm{H}^0(\mathrm{G},\mathrm{A}) \xrightarrow{i_0} \mathrm{H}^0(\mathrm{G},\mathrm{B}) \xrightarrow{p_0} \mathrm{H}^0(\mathrm{G},\mathrm{C}) \xrightarrow{\delta} \mathrm{H}^1(\mathrm{G},\mathrm{A})$$
$$\xrightarrow{i_1} \mathrm{H}^1(\mathrm{G},\mathrm{B}) \xrightarrow{p_1} \mathrm{H}^1(\mathrm{G},\mathrm{C}) \xrightarrow{\Delta} \mathrm{H}^2(\mathrm{G},\mathrm{A}).$$

The proof consists of a series of checks:

1. *Exactness at* $\mathrm{H}^0(\mathrm{G},\mathrm{A})$. Trivial.

2. *Exactness at* $\mathrm{H}^0(\mathrm{G},\mathrm{B})$. We have $p_0 \circ i_0 = 1$ by functoriality (where "1" denotes the constant map equal to 1, not the identity map). Conversely, if $b \in \mathrm{B}^\mathrm{G}$ is in the kernel of p_0, then $b \in \mathrm{A} \cap \mathrm{B}^\mathrm{G} = i_0(\mathrm{A}^\mathrm{G})$.

3. *Exactness at* $\mathrm{H}^0(\mathrm{G},\mathrm{C})$. To say that $c \in \mathrm{C}^\mathrm{G}$ is in $p_0(\mathrm{B}^\mathrm{G})$ means that c can be lifted to an invariant element of B. To say that $\delta(c) = 1$ means the same, by definition of δ ($1 = a_s = b^{-1}s(b)$ for some $b \in p^{-1}(c)$ and all $s \in \mathrm{G}$).

4. *Exactness at* $\mathrm{H}^1(\mathrm{G},\mathrm{A})$. Given a cocycle a_s of A whose class is in the kernel of i_1; this means there exists $b \in \mathrm{B}$ such that $a_s = b^{-1}s(b)$, a condition that certainly holds when a_s is in the image of δ, by definition of δ. Conversely, if the condition holds, then $p(b) \in \mathrm{C}^\mathrm{G}$ and $p(b)$ is the class of a_s.

5. *Exactness at* $\mathrm{H}^1(\mathrm{G},\mathrm{B})$. By functoriality, $p_1 \circ i_1 = 1$. Conversely, it is clear that a cocycle of B that becomes cohomologous to 1 after projection to C is cohomologous in B to a cocycle of A.

6. *Exactness at* $\mathrm{H}^1(\mathrm{G},\mathrm{C})$, *when* A *is in the center of* B. The definition shows that $\Delta \circ p_1 = 1$. Conversely, let c_s be a cocycle of C in the kernel of Δ; we have $c_s = p(b_s)$ and the 2-cocycle $a_{s,t} = b_s s(b_t) b_{st}^{-1}$ is cohomologous to zero, i.e., of the form $a_s s(a_t) a_{st}^{-1}$; replacing b_s by $a_s^{-1} b_s$, we are reduced to the case $a_{s,t} = 1$, which means that b_s is a cocycle of B with image c_s. □

Remarks. 1) Using the Brauer group, examples can be given where *the image of Δ is not a subgroup of* $H^2(G, A)$.

2) The preceding proposition characterises the kernel of Δ, but does not tell when two elements of $H^1(G, C)$ have the same image in $H^2(G, A)$. The same situation holds for the theory of sheaves—cf. for example Frenkel [24] and Grothendieck [27]; Giraud has developed a "non-abelian homological algebra" that encompasses these theories (cf. [80] and [113]).

CHAPTER VIII

Cohomology of Finite Groups

§1. The Tate Cohomology Groups

Let G be a finite group. In the group algebra $\mathbf{Z}[G]$, the element $\sum_{s \in G} s$ will be called the *norm* and be denoted N. For every G-module A, N defines an endomorphism (also denoted N) of A by the formula

$$Na = \sum_{s \in G} s.a.$$

If I_G denotes the *augmentation ideal* of $\mathbf{Z}[G]$ (i.e., the set of linear combinations of the $s - 1$, $s \in G$), then obviously

$$I_G A \subset \operatorname{Ker}(N) \quad \text{and} \quad \operatorname{Im}(N) \subset A^G.$$

As $H_0(G, A) = A/I_G A$ and $H^0(G, A) = A^G$, it follows that N defines by passage to the quotient a homomorphism

$$N^*: H_0(G, A) \to H^0(G, A).$$

We define:

$$\hat{H}_0(G, A) = \operatorname{Ker}(N^*), \qquad \hat{H}^0(G, A) = \operatorname{Coker}(N^*).$$

In other words, if ${}_N A$ denotes the kernel of N acting on A, then

$$\hat{H}_0(G, A) = {}_N A/I_G A$$
$$\hat{H}^0(G, A) = A^G/NA.$$

Proposition 1. *If* A *is relatively projective,* $\hat{H}_0(G, A) = 0 = \hat{H}^0(G, A)$.

(Recall that "relatively projective" is equivalent to "relatively injective" when G is finite.)

It suffices to prove it for A induced. We give the proof for $\hat{H}^0$. By definition of "induced", $A = \sum_{s \in G} sX$, X being a subgroup of A and the sum being direct. Each $a \in A$ can then be expressed uniquely in the form $a = \sum s(x_s)$, $x_s \in X$. It is clear that a is fixed by G if and only if all the x_s are equal, i.e., if one can write $a = Nx$, with $x \in X$. Thus $A^G = NA$, whence $\hat{H}^0(G, A) = 0$. □

Suppose next that $0 \to A \to B \to C \to 0$ is an exact sequence of G-modules. It is easily checked that the diagram

$$\begin{array}{ccccccccc} H_1(G,C) & \to & H_0(G,A) & \to & H_0(G,B) & \to & H_0(G,C) & \to & 0 \\ \downarrow & & \downarrow N_A^* & & \downarrow N_B^* & & \downarrow N_C^* & & \downarrow \\ 0 & \to & H^0(G,A) & \to & H^0(G,B) & \to & H^0(G,C) & \to & H^1(G,A) \end{array}$$

is commutative (where N_A^* denotes the homomorphism N^* for A, and similarly for B and C).

It is well-know (cf. Cartan-Eilenberg [13], V. 10.1) that such a diagram defines a canonical homomorphism

$$\delta : \mathrm{Ker}(N_C^*) \to \mathrm{Coker}(N_A^*).$$

[Namely: if $c \in \mathrm{Ker}(N_C^*)$, lift c to $b \in H_0(G, B)$; the element $N_B^*(b)$ comes from an element $a \in H^0(G, A)$, and the image $\bar{a}$ of a in $\mathrm{Coker}(N_A^*)$ is equal by definition to $\delta(c)$. One checks that it does not depend on the choice of b.]

As $\mathrm{Ker}(N_C^*) = \hat{H}_0(G, C)$ and $\mathrm{Coker}(N_A^*) = \hat{H}^0(G, A)$, we have defined a homomorphism

$$\delta : \hat{H}_0(G, C) \to \hat{H}^0(G, A).$$

Moreover (cf. Cartan-Eilenberg, *loc. cit.*), the above diagram gives an exact sequence

$$\cdots \to H_1(G,C) \to \hat{H}_0(G,A) \to \hat{H}_0(G,B) \to \hat{H}_0(G,C) \to \hat{H}^0(G,A)$$
$$\to \hat{H}^0(G,B) \to \hat{H}^0(G,C) \to H^1(G,A) \to \cdots.$$

This leads us to define, with Tate, *cohomology groups with positive and negative exponents* by the formulas

$$\begin{aligned} \hat{H}^n(G,A) &= H^n(G,A) \quad \text{if } n \geq 1 \\ \hat{H}^0(G,A) &= A^G/NA \\ \hat{H}^{-1}(G,A) &= {}_NA/I_GA \\ \hat{H}^{-n}(G,A) &= H_{n-1}(G,A) \quad \text{if } n \geq 2. \end{aligned}$$

As we have seen, the family of groups $\hat{H}^n$ forms, in a natural way, a *cohomological functor*, defined in every dimension. Moreover, when A is relatively projective, $\hat{H}^n(G, A) = 0$ for all n (cf. prop. 1 for $n = -1$ or 0, as well as Chap. VII for the other values of n). It follows that this functor is effaçable and coeffaçable in every dimension, in the sense of Grothendieck

[26], no. 2.2. More precisely:

a) Every G-module A embeds in an induced G-module A*, and

$$\hat{H}^q(G, A) = \hat{H}^{q-1}(G, A^*/A) \quad \text{for all } q \in \mathbf{Z}.$$

b) Every G-module A is a quotient of an induced G-module A_* by a submodule A′, and

$$\hat{H}^q(G, A) = \hat{H}^{q+1}(G, A') \quad \text{for all } q \in \mathbf{Z}$$

Furthermore, A* and A_* can be chosen so that A is a direct factor (as **Z**-module) in A*, and A′ direct factor in A_* (cf. Chap. VII, §1).

Remark. These "dimension-shifting" properties could be used to give a recursive definition of the groups $\hat{H}^q(G, A)$; this is essentially the point of view of Chevalley [17].

§2. Restriction and Corestriction

Let H be a subgroup of the finite group G, and let A be a G-module. In Chap. VII, §5, the *restriction* homomorphisms

$$\text{Res}: H^q(G, A) \to H^q(H, A).$$

were defined. For $q = 0$, this homomorphism is just the inclusion of A^G in A^H; as $N_G = N_{G/H} \circ N_H$, we have $N_G A \subset N_H A$, so by passage to the quotient, we get a homomorphism

$$\text{Res}: \hat{H}^0(G, A) \to \hat{H}^0(H, A).$$

Consider next the $\hat{H}^{-n}$, with $n \geq 1$. If $n \geq 2$, $\hat{H}^{-n} = H_{n-1}$, and restriction was defined in Chap. VII, §7. For $n = 1$, one checks that this homomorphism passes to the quotient. Thus *we have defined for all $q \in \mathbf{Z}$ a homomorphism*

$$\text{Res}: \hat{H}^q(G, A) \to \hat{H}^q(H, A).$$

The next proposition shows that this definition is "reasonable":

Proposition 2. *The restriction maps form a morphism of cohomological functors.*

(In other words, these maps commute with the coboundary.)

Let $0 \to A \to B \to C \to 0$ be an exact sequence of G-modules. We must check the commutativity of the diagram

$$\begin{array}{ccc} \hat{H}^q(G, C) & \xrightarrow{\delta} & \hat{H}^{q+1}(G, A) \\ {\scriptstyle\text{Res}}\downarrow & & {\scriptstyle\text{Res}}\downarrow \\ \hat{H}^q(H, C) & \xrightarrow{\delta} & \hat{H}^{q+1}(H, A). \end{array}$$

When $q \geq 0$, this follows from the definition of Res given in Chap. VII, §5; when $q \leq -2$, it follows from Chap. VII, §7. The remaining case is $q = -1$, where $\hat{H}^q(G, C) = {}_NC/I_GC$, $\hat{H}^{q+1}(G, A) = A^G/NA, \ldots$ Here is an explicit computation for this case:

Let $c \in {}_NC$ represent $\bar{c} \in \hat{H}^{-1}(G, C)$. Lift c to $b \in B$, and take $N_G(b)$; it is an element of A^G whose class mod. N_GA is equal to $\delta(\bar{c})$. Thus the class of $N_G(b)$ mod. N_HA is equal to $\text{Res} \circ \delta(\bar{c})$. On the other hand, $\text{Res}(\bar{c})$ is the class of $\sum s_i c$, where the s_i are representatives of the right cosets mod. H; lift this element to $\sum s_i b$, so that $\delta \circ \text{Res}(\bar{c})$ is represented by $N_H(\sum s_i b)$, which is clearly equal to $N_G(b)$. □

Remark. This proposition shows that Res is *the unique morphism of cohomological functors that coincides in degree* 0 *with the map* $A^G/N_GA \to A^H/N_HA$ *induced by the inclusion* $A^G \to A^H$.

We proceed in exactly the same way for the *corestriction*

$$\text{Cor}: \hat{H}^q(H, A) \to \hat{H}^q(G, A).$$

In dimensions $q \geq 0$, it is defined as in Chap. VII, §7, while in dimensions $q \leq -1$, it is defined as in Chap. VII, §5. The proof of prop. 2 dualizes and gives:

Proposition 3. *The corestriction maps form a morphism of cohomological functors.*

More precisely, Cor is *the unique morphism of cohomological functors that coincides in degree* -1 *with the map* ${}_{N_H}A/I_HA \to {}_{N_G}A/I_GA$ *induced by the inclusion* ${}_{N_H}A \to {}_{N_G}A$.

In degree 0, Cor is induced by the map $N_{G/H}: A^H \to A^G$.

Proposition 4. *If* $n = \text{Card}(G/H)$, *then* $\text{Cor} \circ \text{Res} = n$.

This follows from prop. 6 of Chap. VII and the analogous proposition for homology.

Direct Proof. If f_n denotes multiplication by n, then the morphism of cohomological functors $\text{Cor} \circ \text{Res} - f_n$ is zero in dimension 0 (trivial to check), hence is zero in all dimensions. □

Corollary 1. *If g is the order of* G, *then all the groups* $\hat{H}^q(G, A)$ *are annihilated by g.*

Apply prop. 4 with $H = \{1\}$, remarking that the $\hat{H}^q(H, A)$ are all zero.

Corollary 2. *If* A *is a finitely generated abelian group (and is a* G*-module), then the* $\hat{H}^q(G, A)$ *are finite groups.*

Indeed, the definition of these groups in terms of chains and cochains shows that they are finitely generated groups; by corollary 1, they are torsion groups, hence finite.

§3. Cup Products

Let A and B be G-modules, and let $A \otimes B$ be their tensor product (over the ring $\mathbf{Z}$, as always). Make $A \otimes B$ into a G-module by setting

$$s.(a \otimes b) = s.a \otimes s.b$$

and extending by linearity.

Proposition 5. *If* G *is a finite group, there exists one and only one family of homomorphisms (called cup product)*

$$\hat{H}^p(G, A) \otimes \hat{H}^q(G, B) \to \hat{H}^{p+q}(G, A \otimes B)$$

denoted $(a, b) \to a.b$, *which are defined for every pair of integers* (p, q) *and every couple of* G*-modules* A, B, *and which satisfy the following four properties:*

i) *These homomorphism are morphisms of functors, when the two sides of the arrow are considered to be bifunctors covariant in* (A, B).
ii) *For* $p = q = 0$, *the cup product* $a.b$ *is obtained by passage to the quotient of the natural map* $A^G \otimes B^G \to (A \otimes B)^G$.
iii) *If* $0 \to A \to A' \to A'' \to 0$ *is an exact sequence of* G*-modules, and if the sequence*
$$0 \to A \otimes B \to A' \otimes B \to A'' \otimes B \to 0$$
is also exact, then for all $a'' \in \hat{H}^p(G, A'')$ *and* $b \in \hat{H}^q(G, B)$:
$$(\delta a'').b = \delta(a''.b),$$
where both sides are elements of $\hat{H}^{p+q+1}(G, A \otimes B)$.
iv) *If* $0 \to B \to B' \to B'' \to 0$ *is an exact sequence of* G*-modules, and if the sequence*
$$0 \to A \otimes B \to A \otimes B' \to A \otimes B'' \to 0$$
is also exact, then for all $a \in \hat{H}^p(G, A)$ *and* $b'' \in \hat{H}^q(G, B'')$:
$$a.(\delta b'') = (-1)^p \delta(a.b''),$$
where both sides are elements of $\hat{H}^{p+q+1}(G, A \otimes B)$.

Properties iii) and iv) allow the use of "dimension-shifting", and uniqueness of the product results from that. As for existence, it is proved by defining the product on the cochain level (more precisely, on a "complete resolution"): see Cartan-Eilenberg [13], Chap. XII.

Likewise, we refer to Cartan-Eilenberg for numerous formulas involving the cup product (especially those which bring in the maps Res and Cor). We mention only the two following formulas (immediate by dimension-shifting):

v) $(a.b).c = a.(b.c)$, *modulo the identification* $(A \otimes B) \otimes C = A \otimes (B \otimes C)$.

vi) $a.b = (-1)^{\dim(a).\dim(b)} b.a$, *modulo the identification* $A \otimes B = B \otimes A$.

We will occasionally need another type of cup product (which can actually be deduced from the preceding one):

Given G-modules A, B, C, let $\varphi: A \times B \to C$ be a **Z**-bilinear map that is *invariant under* G, i.e., such that $\varphi(s.a, s.b) = s.\varphi(a,b)$ for $s \in G$, $a \in A$, $b \in B$. The map φ defines a G-homomorphism

$$\varphi: A \otimes B \to C$$

If $a \in \hat{H}^p(G, A)$, $b \in \hat{H}^q(G, B)$, then $a.b \in \hat{H}^{p+q}(G, A \otimes B)$, and we can take the image of $a.b$ by φ. It is an element of $\hat{H}^{p+q}(G, C)$ that will still be called *the cup product of a and b relative to* φ, and will be denoted $a._\varphi b$ (or simply $a.b$ when there is no ambiguity about φ).

§4. Cohomology of Finite Cyclic Groups. Herbrand Quotient

Let G be a cyclic group of order n, and *choose a generator s* of G. In the group algebra, consider the two elements:

$$N = \sum_{t \in G} t = \sum_{i=0}^{i=n-1} s^i$$

$$D = s - 1.$$

Define a cochain complex K as follows: $K^i = \mathbf{Z}[G]$ for all i; $d: K^i \to K^{i+1}$ is multiplication by D (resp. by N) if i is even (resp. odd). For each G-module A, put $K(A) = K \otimes_{\mathbf{Z}[G]} A$. Then $K^i(A) = A$ for all i; $d: K^i(A) \to K^{i+1}(A)$ is multiplication by D (resp. N) if i is even (resp. odd). An exact sequence $0 \to A \to B \to C \to 0$ of G-modules gives rise to an exact sequences of complexes:

$$0 \to K(A) \to K(B) \to K(C) \to 0$$

whence to an exact cohomology sequence, and, in particular, to a coboundary operator δ.

Proposition 6. *The cohomological functor* $\{H^q(K(\ \ )), \delta\}$ *is isomorphic to the functor* $\{H^q(G,\ \ ), \delta\}$.

First of all, it is clear that $\hat{H}^0(G, A) = H^0(K(A))$, $\hat{H}^{-1}(G, A) = H^{-1}(K(A))$, and that the coboundary operator δ relating H^0 to H^{-1} is the same. Hence

$$H^q(K(A)) = 0$$

for $q = 0, -1$ when A is relatively projective, thence for all q (as the $H^q(K(A))$ depend only on the parity of q). That suffices to give the isomorphism. ☐

Corollary. *The groups* $\hat{H}^q(G, A)$ *depend only on the parity of* q.

(In other words, the cohomology of a finite cyclic group is *periodic of period two.*)

Explicitly, we have:

$$\hat{H}^q(G, A) = \mathrm{Ker}(D)/\mathrm{Im}(N) = A^G/NA \quad \text{for } q \equiv 0 \bmod. 2$$
$$\hat{H}^q(G, A) = \mathrm{Ker}(N)/\mathrm{Im}(D) = {}_N A/DA \quad \text{for } q \equiv 1 \bmod. 2.$$

Remark. The isomorphisms above do *depend on the choice of generator s* (for the coboundary operator of the complex K depends on it). This fact can also be seen as follows: the choice of s defines a *character* $\chi^s : G \to \mathbf{Q}/\mathbf{Z}$ such that $\chi_s(s) = 1/n$, and the coboundary of the exact sequence

$$0 \to \mathbf{Z} \to \mathbf{Q} \to \mathbf{Q}/\mathbf{Z} \to 0$$

transforms χ_s into an element $\theta_s = \delta\chi_s$ of $H^2(G, \mathbf{Z})$. The *periodicity isomorphisms* defined above

$$\hat{H}^q(G, A) \to \hat{H}^{q+2}(G, A)$$

are given by the cup product with θ_s: this follows, for example, from the formulas for the cup product given in Cartan-Eilenberg [13], p. 252. In particular, the isomorphism $A^G/NA \to H^2(G, A)$ is $a \mapsto a \,.\, \theta_s$, which clearly depends on the choice of s.

We are now going to alter slightly the definition of the complex K(A): we consider it as *graded by the integers mod.* 2. We write $H^0(A)$ and $H^1(A)$ in place of $H^0(K(A))$ and $H^1(K(A))$; then $H^0(A) = A_G/NA$, $H^1(A) = {}_N A/DA$. If

$$0 \to A \to B \to C \to 0$$

is an exact sequence of G-modules, the exact cohomology sequence can be written in the form of an *exact hexagon*:

$$\begin{array}{ccccc} & \nearrow & H^0(A) \to H^0(B) & \searrow & \\ H^1(C) & & & & H^0(C). \\ & \nwarrow & H^1(B) \leftarrow H^1(A) & \swarrow & \end{array}$$

Suppose that $H^0(A)$ and $H^1(A)$ are finite groups, and let $h_0(A)$ and $h_1(A)$ be their orders. The fraction

$$h(A) = h_0(A)/h_1(A)$$

is called the *Herbrand quotient* of A. In view of the periodicity of the cohomology, the Herbrand quotient is analogous to an *Euler-Poincaré characteristic.*

[This analogy can be made precise in the framework of abelian categories. Suppose that A runs through the G-objects of an abelian category $\mathscr{C}$, and let $\mathscr{D}$ be a subcategory of $\mathscr{C}$; if we suppose that $H^0(A)$ and $H^1(A)$ belong to $\mathscr{D}$, we can define $h(A)$ to be $H^0(A) - H^1(A)$, where this difference makes sense in the *Grothendieck group* $K(\mathscr{D})$ associated to $\mathscr{D}$. In our case, $\mathscr{D}$ is the category of finite abelian groups, and $K(\mathscr{D})$ is the multiplicative group of positive rational numbers—cf. Chap. I. §5. Evidently we could consider other categories, such as finite dimensional vector spaces, quasi-algebraic groups, etc.]

Proposition 7. *Given an exact sequence* $0 \to A \to B \to C \to 0$ *of* G*-modules for which at least two of the three Herbrand quotients* $h(A)$, $h(B)$, $h(C)$ *are defined. Then the third is also defined and we have*

$$h(B) = h(A) . h(C).$$

This follows from the exact hexagon written above. (More generally, when $2n$ finite groups form an exact $2n$-gon, the alternating product of their orders is equal to 1.)

[One could also invoke the "additivity" of Euler-Poincaré characteristics.]

□

Proposition 8. *If* A *is a finite* G*-module, then* $h(A) = 1$.

The exact sequence

$$0 \to A^G \to A \xrightarrow{D} A \to A_G \to 0$$

shows first of all that A^G and A_G have the same number of elements. The exact sequence

$$0 \to H^1(A) \to A_G \xrightarrow{N} A^G \to H^0(A) \to 0$$

then shows that $H^1(A)$ and $H^0(A)$ have the same number of elements.

[One could also remark that the Euler-Poincaré characteristic of $K(A)$ is obviously equal to 1; since Euler-Poincaré characteristic is invariant under passage to cohomology, one deduces that indeed $h(A) = 1$.] □

Corollary. *Let* A *and* B *be* G*-modules,* $f: A \to B$ *a* G*-homomorphism with finite kernel and cokernel. Then* A *and* B *have the same Herbrand quotient.*

Stating this more accurately: if one of $h(A)$, $h(B)$ is defined, then so is the other and they are equal. This follows from propositions 7 and 8.

EXERCISE

Extend the definition of Herbrand quotient and propositions 7 and 8 to the class of finite groups that have periodic cohomology (cf. Cartan-Eilenberg [13], Chap. XII, §11).

§5. Herbrand Quotient in the Cyclic Prime Order Case

Let G be a cyclic group of prime order p, and let A be an abelian group. If we make G act trivially on A, we get

$$H^1(G, A) = {}_pA \quad \text{(subgroup of those } a \in A \text{ such that } pa = 0)$$
$$H^2(G, A) = A_p = A/pA.$$

If the groups ${}_pA$ and A_p are finite, we can define the "trivial Herbrand quotient of A", denoted $\phi(A)$, to be the quotient of the order of A_p by that of ${}_pA$. The next proposition, due to Tate, generalizes a theorem of Chevalley ([17], th. 10.3).

Proposition 9. *Let A be a G-module for which $\varphi(A)$ is defined (i.e:, the kernel and cokernel of $p: A \to A$ are finite). Then $\varphi(A^G)$, $\varphi(A_Q)$ and $h(A)$ are defined, and they satisfy the equations*

$$h(A)^{p-1} = \varphi(A^G)^p/\varphi(A) = \varphi(A_G)^p/\varphi(A).$$

Tate's proof rests on the fact that $\mathbf{Z}[G]/(N)$ is isomorphic to the ring of integers of the field of pth roots of unity, and that in this ring, the ideal generated by p is the $(p-1)$st power of the ideal generated by D. This argument is reproduced in Artin-Tate [8]. We give below a different proof which is longer, but has the advantage of elucidating the structure of those G-modules for which $\varphi(A)$ is defined.

Lemma 1. *Let $0 \to A' \to A \to A'' \to 0$ be an exact sequence of G-modules such that $\varphi(A')$ and $\varphi(A'')$ are defined. If prop. 9 holds for A' and A'', then it also holds for A.*

Clearly $\varphi(A)$ and $h(A)$ are defined and satisfy $\varphi(A) = \varphi(A').\varphi(A'')$ and $h(A) = h(A').h(A'')$. In addition, there is an exact sequence

$$0 \to A'^G \to A^G \to A''^G \to H^1(G, A') \to \cdots,$$

and as $H^1(A')$ is finite (because $h(A')$ is defined), this gives

$$0 \to A'^G \to A^G \to A''^G \to N \to 0,$$

where N is a finite group.

We conclude that $\varphi(A^G) = \varphi(A'^G).\varphi(A''^G)$, whence the equation for A^G follows by multiplication. (Same argument for A_G.) □

Lemma 2. *Let A be a G-module for which $\varphi(A)$ is defined. Then there is an exact sequence of G-modules*

$$0 \to A' \to A \to A'' \to 0$$

such that A' is a finitely generated abelian group and $A'' = pA''$; moreover $\varphi(A')$ and $\varphi(A'')$ are defined.

By hypothesis, A/pA is finite. Hence there exists a subgroup of A that is finitely generated and maps onto A/pA; we may assume this subgroup to be stable under G (otherwise replace it with the sum of its transforms by the elements of G). Call it A', and set $A'' = A/A'$. There is an exact sequence

$$0 \to {}_pA' \to {}_pA \to {}_pA'' \to A'_p \to A_p \to A''_p \to 0.$$

By construction, $A'_p \to A_p$ is surjective; thus $A''_p = 0$, i.e., $A'' = pA''$; also, since A' is finitely generated over $\mathbf{Z}$, ${}_pA'$ and A'_p are finite, i.e., $\varphi(A')$ is defined; it is obvious that $\varphi(A'')$ is defined. □

Lemmas 1 and 2 reduce us to the cases where A is either finitely generated or divisible by p.

a) *Case where* A *is a finitely generated abelian group*

Note first that if A and A′ are G-modules of finite type such that $A \otimes_{\mathbf{Z}} \mathbf{Q}$ and $A' \otimes_{\mathbf{Z}} \mathbf{Q}$ are G-isomorphic, then $h(A) = h(A')$, $\varphi(A) = \varphi(A')$, $\varphi(A^G) = \varphi(A'^G)$ and $\varphi(A_G) = \varphi(A'_G)$: that results from the corollary to prop. 8. It reduces us to the case where $A_{\mathbf{Q}} = A \otimes_{\mathbf{Z}} \mathbf{Q}$ is a *simple* $\mathbf{Q}[G]$-module. But the algebra $\mathbf{Q}[G]$ is the product of $\mathbf{Q}$ with the field K of pth roots of unity (an isomorphism $\mathbf{Q}[G] \to \mathbf{Q} \times K$ is defined by sending a generator of G to a primitive pth root of unity, using the fact $[K:\mathbf{Q}] = p - 1$, cf. Chap. IV, §4). Hence there are *two* simple representations of $\mathbf{Q}[G]$ to be considered:

a1) The trivial representation of degree 1

In this case we can take A to be the group $\mathbf{Z}$ with trivial operation of G. Then $h(A) = \varphi(A) = \varphi(A^G)$, and the formula of prop. 9 is certainly satisfied.

a2) The representation of degree $p - 1$ given by K

We can take A to be the quotient of $\mathbf{Z}[G]$ by $\mathbf{Z}$. As $h(\mathbf{Z}[G]) = 1$, we get $h(A) = p^{-1}$; it can easily be seen that $\varphi(A) = p$, $\varphi(A^G) = \varphi(A_G) = 1$, so the formula of prop. 9 holds.

b) *Case where* A *is divisible by p*

Let A′ be the subgroup of A of all elements that are annihilated by a power of p. In $A'' = A/A'$, multiplication by p is bijective, hence

$$H^q(G, A'') = 0$$

for all $q \geq 1$; in particular, $h(A'') = 1$, $\varphi(A''^G) = 1$, and $\varphi(A'') = 1$. Thus we are reduced to A′ (by lemma 1), i.e., to the case where every element of A is annihilated by suitable power of p (still assuming A divisible by p and ${}_pA$ finite).

The structure of A can then be determined by arguing as in a). But it is quicker to use *Pontrjagin duality*: it transforms A into a compact group $\hat{A}$, which is a free module of finite type over the ring $\mathbf{Z}_p$ of p-adic integers, on which G acts. It can be immediately verified that

$$h(A) = h(\hat{A})^{-1}, \qquad \varphi(A) = \varphi(\hat{A})^{-1}, \qquad \varphi(A^G) = \varphi(\hat{A}_G)^{-1}.$$

This reduces us to proving the formula for $\hat{A}$, which is done as in case a), the ring $\mathbf{Z}_p$ replacing the ring $\mathbf{Z}$ (indeed, we showed in Chap. IV, §4 that the field obtained by adjoining to $\mathbf{Q}_p$ a primitive pth root of unity has degree $p - 1$, so the classification of simple representations of G over $\mathbf{Q}_p$ is the same as over $\mathbf{Q}$). □

Remarks. 1) The exact sequence $0 \to A^G \to A \xrightarrow{D} A \to A_G \to 0$ shows directly that $\varphi(A^G) = \varphi(A_G)$.

2) The proof given above shows that every G-module A for which $\varphi(A)$ is defined admits a composition series whose factors are of one of the six

following types:

i) a finite group,
ii) a group divisible by p,
iii) the group $\mathbf{Z}$ on which G acts trivially,
iv) the group $\Lambda = \mathbf{Z}[G]/\mathbf{Z}$,
v) the group $T = \mathbf{Q}_p/\mathbf{Z}_p$ on which G acts trivially,
vi) the group $T \otimes_{\mathbf{Z}} \Lambda$ on which G acts as on Λ.

EXERCISES

1. How must prop. 9 be modified when G is a cyclic group whose order is a power of p?

2. Let G_1 and G_2 be cyclic groups of order p, and let $G = G_1 \times G_2$. Let A be a G-module for which ${}_pA$ and A_p are finite. Denote by $h^1(A)$ the Herbrand quotient of A with respect to G_1, and similarly for $h^2(A)$. Using proposition 9, prove the following formula (which reduces to prop. 9 when one of the two groups acts trivially):

$$h^1(A^{G_2})^p . h^2(A) = h^2(A^{G_1}) . h^1(A).$$

CHAPTER IX

Theorems of Tate and Nakayama

§1. p-Groups

Let p be a prime number. Recall that a finite group G is called a *p-group* if its order Card(G) is a power of p.

Lemma 1. *Suppose* G *is a p-group acting on a finite set* E, *and let* E^G *be the subset of elements fixed by* G. *Then*

$$\mathrm{Card}(E^G) \equiv \mathrm{Card}(E) \quad \mathrm{mod.}\ p.$$

Indeed, $E - E^G$ is the disjoint union of orbits Gx not reduced to a single point, each having cardinality equal to the index of its stabilizer in G, which is divisible by p.

Lemma 2. *If a p-group acts on a p-group of order* >1, *then the fixed points form a subgroup of order* >1.

Indeed, the number of fixed points is divisible by p (lemma 1).

Theorem 1. *The center of a p-group of order* >1 *has order* >1.

Apply the preceding lemma, letting the group act on itself by inner automorphisms. □

Corollary. *A group* G *of order* p^n *admits a composition series*

$$\{1\} = G_n \subset G_{n-1} \subset \cdots \subset G_0 = G$$

with all the G_i *normal in* G (and the G_i/G_{i+1} cyclic of order p).

This follows from theorem 1, by induction on n.

Theorem 2. *Every linear representation* $\neq 0$ *of a p-group over a field of characteristic p contains the unit representation.*

Let E be the representation space. Let x be a non-zero element of E, H the subgroup of E generated by the $s.x$, $s \in G$; H is a finite dimensional vector space over the prime field $\mathbf{F}_p$. Applying lemma 2 to H gives the existence of $y \in H$, $y \neq 0$, such that $s.y = y$ for all $s \in G$. □

Corollary. *Let* G *be a p-group, and let k be a field of characteristic p. The kernel* I_G *of the augmentation homomorphism* $k[G] \to k$ *is the radical of* $k[G]$, *which is a nilpotent ideal.*

Indeed, the radical $\mathfrak{r}$ of kG is the intersection of the kernels of the irreducible representations of $k[G]$ (or of G—it is the same), and theorem 2 shows that the unit representation is the only irreducible representation of G over k; hence $\mathfrak{r} = I_G$. As $k[G]$ is a finite dimensional k-algebra, it is well-known that its radical is nilpotent (cf. Bourbaki, *Alg.*, Chap. VIII, §6, th. 3).

§2. Sylow Subgroups

Theorem 3 (Sylow). *Let* G *be a group of order* $n = p^m q$, *with p prime and* $(p, q) = 1$. *Then there exist subgroups of* G *having order* p^m (called Sylow p-subgroups); *they are all conjugate to one another, and every p-group contained in* G *is contained in one of them.*

PROOF (AFTER G. A. MILLER AND H. WIELANDT). Let E be the family of all subsets X of G having p^m elements. The group G operates on E by translations, and

$$\mathrm{Card}(E) = \binom{n}{p^m}.$$

Lemma 3. *If* $n = p^m q$, *with* $(p, q) = 1$, *then*

$$\binom{n}{p^m} \equiv q \quad \mathrm{mod.}\, p.$$

Indeed, let X and Y be indeterminates over a field of characteristic p. Then

$$(X + Y)^n = (X + Y)^{p^m q} = (X^{p^m} + Y^{p^m})^q = X^{p^m q} + qX^{p^m(q-1)}Y^{p^m} + \cdots + Y^{p^m q},$$

and comparing this with the binomial expansion of $(X + Y)^n$ gives the congruence. □

Back to the proof of Sylow's theorem: lemma 3 shows that Card(E) $\not\equiv 0$ mod. p. Hence there exists an $X \in E$ such that the orbit G.X of X in E satisfies Card(G.X) $\not\equiv 0$ mod. p. If H is the stabilizer of X (subgroup of all $s \in G$ such that $s.X = X$), then G.X is equipotent to G/H, whence (G : H) $\not\equiv 0$ mod. p, so that p^m divides the order of H. On the other hand, if $x \in X$, then $H \subset X.x^{-1}$, so

$$\text{Card(H)} \leq \text{Card(X)} = p^m.$$

Therefore Card(H) $= p^m$ and H is a Sylow p-subgroup of G.

Now let H′ be a p-group contained in G, and consider the action of H′ on the homogeneous space G/H, where H is a Sylow p-subgroup of G. Since Card(G/H) $= q \not\equiv 0$ mod. p, lemma 1 applied to G/H guarantees that the set of fixed points of H′ is non-empty; this means that H′ is contained in a conjugate of H. If in addition Card(H′) $= p^m$, H′ must be equal to a conjugate of H. □

The following "functorial" properties of the Sylow subgroups are immediate consequences of th. 3:

a) If G′ is any subgroup of G, then each Sylow p-subgroup of G′ is the intersection with G′ of a Sylow p-subgroup of G.
b) If G′ is any quotient group of G, then the Sylow p-subgroups of G′ are the images of the Sylow p-subgroups of G.

Sylow subgroups occur in cohomology via the next theorem.

Theorem 4. *Let* G *be a finite group, p a prime number, and* G_p *a Sylow p-subgroup of* G. *Then for every* G*-module* A *and every* $n \in \mathbf{Z}$, *the restriction homomorphism*

$$\text{Res} : \hat{H}^n(G, A) \to \hat{H}^n(G_p, A)$$

is injective on the p-primary component of $\hat{H}^n(G, A)$.

Given x in the kernel of Res. If $q = \text{Card}(G/G_p)$, then

$$q.x = \text{Cor} \circ \text{Res}(x) = 0 \quad \text{(Chap. VIII, prop. 4).}$$

But if x belongs to the p-primary component of $\hat{H}^n(G, A)$, there is an integer r such that $p^r.x = 0$. As $(q, p^r) = 1$, it follows that $x = 0$. □

Corollary. *Let* G *be a finite group,* A *a* G*-module, n an integer. Suppose that for every prime number p,* $\hat{H}^n(G_p, A) = 0$, *where* G_p *is a Sylow p-subgroup of* G. *Then* $\hat{H}^n(G, A) = 0$.

Indeed, all the primary components of $\hat{H}^n(G, A)$ are zero.

Remark. A characterisation of the *image* of $\text{Res} : \hat{H}^n(G, A) \to \hat{H}^n(G_p, A)$ is given in Cartan-Eilenberg ([13], Chap. XII, th. 10.1).

EXERCISES

1. With the notation from the proof of th. 3, let d be the number of Sylow p-subgroups of G. Show that the number of translates of these subgroups is dq; by comparing with the number of elements of E, deduce that $d \equiv 1 \bmod. p$.
2. Let G be a subgroup of a finite group $\tilde{G}$, and let $\tilde{P}$ be a p-Sylow subgroup of $\tilde{G}$.
   a) Show that there is a conjugate of $\tilde{P}$ whose intersection with G is a p-Sylow subgroup of G. (Hint: make G act on $\tilde{G}/\tilde{P}$ and note that one of the orbits has order prime to p.)
   b) Deduce from a) another proof of the existence of a p-subgroup of G (take for $\tilde{G}$ a group for which the existence of p-Sylow subgroups can be checked directly, for instance $GL_n(\mathbf{Z}/p\mathbf{Z})$.)

§3. Induced Modules; Cohomologically Trivial Modules

Let G be a finite group and A a G-module. A is called *cohomologically trivial* if, for every subgroup H of G and every $n \in \mathbf{Z}$, $\hat{H}^n(H, A) = 0$.

EXAMPLES. Every *induced* module is cohomologically trivial: indeed, such a module is also induced for every subgroup H of G, and we saw in Chap. VIII, §1 that the cohomology vanishes. The same holds for *relatively projective* modules, since they are direct factors of induced modules.

Starting with an induced module A, other examples can be constructed by the following process:

Let $\mathscr{C}$ be the category of abelian groups, $T: \mathscr{C} \times \mathscr{C} \to \mathscr{C}$ an additive bifunctor that we take to be bicovariant (to fix ideas). If A and B are G-modules, define a G-module structure on T(A, B) as follows: each $s \in G$ defines an element $s_A \in \mathrm{Hom}(A, A)$ and an element $s_B \in \mathrm{Hom}(B, B)$, hence an element $T(s_A, s_B) \in \mathrm{Hom}(T(A, B), T(A, B))$: this is the automorphism of T(A, B) associated to s.

Proposition 1. *If* A *is induced* (resp. *relatively projective*), *then* T(A, B) *is induced* (resp. *relatively projective*), *hence cohomologically trivial.*

We may assume that A is induced (passing to a direct factor otherwise); then A is the direct sum of the $s . A'$ for some subgroup A′. The group T(A, B) is then the direct sum of the $T(s . A', B)$; however, $T(s . A', B) = T(s . A', s . B) = s . T(A', B)$, so T(A, B) is induced. □

Corollary. *Suppose one of the* G*-modules* A, B *is relatively projective. Then the* G*-modules below are relatively projective* (*hence cohomologically trivial*):

$$A \otimes B, \quad \mathrm{Hom}(A, B), \quad \mathrm{Tor}(A, B), \quad \mathrm{Ext}(A, B).$$

[Of course, the functors ⊗, Hom, . . . , are relative to the ring **Z** of integers.]

§4. Cohomology of a p-Group

Lemma 4. *Let* G *be a p-group and* A *a* G*-module such that* $pA = 0$. *Then the three following conditions are equivalent*:

i) $A = 0$.
ii) $H^0(G, A) = 0$.
iii) $H_0(G, A) = 0$.

The implications i) $\Rightarrow$ ii) and i) $\Rightarrow$ iii) are trivial. The implication ii) $\Rightarrow$ i) has been proved in theorem 2. Let us show that iii) $\Rightarrow$ i). Let $A' = \mathrm{Hom}(A, \mathbf{F}_p)$ be the dual of A as an $\mathbf{F}_p$-vector space. It is easily seen that $H^0(G, A')$ is dual to $H_0(G, A)$. Hence $H^0(G, A') = 0$, whence $A' = 0$ and $A = 0$.

[Another proof that iii) $\Rightarrow$ i): Let $\mathfrak{r}$ be the augmentation ideal of $\mathbf{F}_p[G]$. The vanishing of $H_0(G, A)$ means that $A = \mathfrak{r}A$. But $\mathfrak{r}$ is nilpotent (cor. to th. 2). Hence $A = 0$.]

Lemma 5. *With the hypotheses of lemma* 4, *suppose that* $H_1(G, A) = 0$. *Then* A *is a free module over the algebra* $\Lambda = \mathbf{F}_p[G]$.

Let $\mathfrak{r}$ be the augmentation ideal of Λ. Then $A/\mathfrak{r}A = H_0(G, A)$, and this is a vector space over $\mathbf{F}_p$. Let h_λ be a basis of this vector space, and lift it to a family $a_\lambda \in A$. Since the h_λ generate $A/\mathfrak{r}A$, the a_λ generate A (apply lemma 4 to the quotient of A by the sub-Λ-module generated by the a_λ). Thus we have defined a surjective G-homomorphism of a free Λ-module L onto A; by construction, this homomorphism induces an isomorphism of $L/\mathfrak{r}L$ onto $A/\mathfrak{r}A$. Let R be the kernel of this homomorphism. Then there is an exact sequence

$$H_1(G, A) \to H_0(G, R) \to H_0(G, L) \to H_0(G, A).$$

As $H_1(G, A) = 0$ and $H_0(G, L) \to H_0(G, A)$ is bijective, it follows that $H_0(G, R) = 0$, whence $R = 0$ (lemma 4). □

Remark. The two lemmas above are special cases of general theorems on "non-commutative local rings"—cf. Bourbaki, *Alg. comm.*, Chap. II, §3.

Theorem 5. *Let* G *be a p-group and* A *a* G*-module annihilated by p. The following conditions are equivalent*:

i) *There exists an integer q such that* $\hat{H}^q(G, A) = 0$,
ii) A *is cohomologically trivial*,
iii) A *is an induced* G*-module*,
iv) A *is a free* $\mathbf{F}_p[G]$*-module.*

Obviously it suffices to prove i) $\Rightarrow$ iv). The shifting procedure already used several times enables us to construct a G-module B, annihilated by p, such that $\hat{H}^n(G, A) = \hat{H}^{n-q-2}(G, B)$ for all n. If $\hat{H}^q(G, A) = 0$, then $H_1(G, B) = 0$, hence, by lemma 5, B is Λ-free. Its cohomology groups are then zero; in particular,

$$H_1(G, A) = \hat{H}^{-2}(G, A) = \hat{H}^{-q-4}(G, B) = 0,$$

and lemma 5 concludes the proof. □

Theorem 6. *Let* G *be a p-group and let* A *be a* G*-module without p-torsion. The following conditions are equivalent:*

i) $\hat{H}^q(G, A) = 0$ *for two consecutive values of q,*
ii) A *is cohomologically trivial,*
iii) *the* $\mathbf{F}_p[G]$*-module* A/pA *is free.*

Since A has no p-torsion, there is an exact sequence

$$0 \to A \xrightarrow{p} A \to A/pA \to 0.$$

Passing to the cohomology gives the exact sequence

$$\hat{H}^q(G, A) \xrightarrow{p} \hat{H}^q(G, A) \to \hat{H}^q(G, A/pA) \to \hat{H}^{q+1}(G, A) \xrightarrow{p} \hat{H}^{q+1}(G, A).$$

If $\hat{H}^q(G, A) = \hat{H}^{q+1}(G, A) = 0$, this sequence shows that $\hat{H}^q(G, A/pA) = 0$, and A/pA is free by th. 5. Thus i) ⇒ iii). If iii) holds, the same exact sequence shows that multiplication by p is bijective on all the $\hat{H}^q(G, A)$; as this endomorphism is nilpotent, $\hat{H}^q(G, A) = 0$. The same reasoning applies to every subgroup H of G, for A/pA is $\mathbf{F}_p[H]$-free. Therefore iii) ⇒ ii). Finally, the implication ii) ⇒ i) is trivial. □

Corollary. *Let* A *be a* **Z***-free* G*-module satisfying the equivalent conditions of theorem* 6. *Then for every torsion-free* G*-module* B, *the* G*-module* $N = \mathrm{Hom}_{\mathbf{Z}}(A, B)$ *is cohomologically trivial.*

The module N is torsion-free. We will check that N/pN is cohomologically trivial; this will imply the result we seek, in view of the preceding theorems. The exact sequence

$$0 \to B \xrightarrow{p} B \to B/pB \to 0$$

gives the exact sequence

$$0 \to N \xrightarrow{p} N \to \mathrm{Hom}(A, B/pB) \to 0$$

whence an isomorphism $N/pN = \mathrm{Hom}(A/pA, B/pB)$. Now A/pA is free over $\mathbf{F}_p[G]$, hence induced, so the corollary to prop. 1 insures that N/pN is cohomologically trivial. □

Remark. This corollary is in fact only a lemma for th. 7 below; once that theorem is proved, we will know that A is projective, hence N is relatively projective.

§5. Cohomology of a Finite Group

Theorem 7. *Let* G *be a finite group,* A *a* **Z**-*free* G-*module, and* G_p *a Sylow p-subgroup of* G, *for each p. The following conditions are equivalent*:

i) *For every prime number p, the* G_p-*module* A *satisfies the equivalent conditions of theorem* 6,
ii) A *is* **Z**[G]-*projective.*

We must show that i) implies ii). Write A as a quotient of a free **Z**[G]-module L:

$$0 \to N \to L \to A \to 0.$$

The **Z**-module A being free yields the exact sequence

$$(*) \qquad 0 \to \mathrm{Hom}_{\mathbf{Z}}(A, N) \to \mathrm{Hom}_{\mathbf{Z}}(A, L) \to \mathrm{Hom}_{\mathbf{Z}}(A, A) \to 0.$$

By the corollary to theorem 6, i) implies that the G_p-module $\mathrm{Hom}_{\mathbf{Z}}(A, N)$ has cohomology zero in all dimensions, therefore that $H^1(G, \mathrm{Hom}_{\mathbf{Z}}(A, N)) = 0$ by the corollary to th. 4.

The exact cohomology sequence (∗) then shows that

$$\mathrm{Hom}_G(A, L) \to \mathrm{Hom}_G(A, A)$$

is surjective; in particular, the identity map of A extends to a G-homomorphism of A into L, so that A is a direct factor of L as G-module, i.e., projective. □

Remark. Let P, P′ be projective modules of finite type over **Z**[G]; call them *equivalent* if there exist free modules L, L′ of finite type such that P ⊕ L is isomorphic to P′ ⊕ L′. Let P(G) be the set of equivalence classes of projective **Z**[G]-modules of finite type (for this equivalence relation). The law of composition (P, P′) → P ⊕ P′ makes P(G) into an abelian group, called the *group of classes of projective* G-*modules.* When G is cyclic of prime order *p*, it has been shown by Rim [51] that P(G) is isomorphic to the group of ideal classes of the field of *p*th roots of unity; in particular, P(G) ≠ 0, which shows the existence of projective G-modules that are not free. Swan [62] has made a deeper study of P(G), showing in particular that it is a *finite* group.

Lemma 6. *Let* $0 \to X_1 \to X_2 \to \cdots \to X_n \to 0$ *be an exact sequence of* G-*modules. If all but one of the* X_i *are cohomologically trivial, then that one also is.*

Put $N_i = \mathrm{Ker}(X_i \to X_{i+1})$, $N_0 = N_{n+1} = 0$. Then there are $n + 1$ exact sequences

$$(E_i) \qquad 0 \to N_i \to X_i \to N_{i+1} \to 0, \qquad 0 \le i \le n.$$

If X_i is cohomologically trivial for $i \neq q$, then the sequences $E_0, \ldots, E_{q-1}$ show that $N_1, \ldots, N_q$ are cohomologically trivial, and $E_{q+1}, \ldots, E_n$ show that $N_{q+1}, \ldots, N_n$ are too. Conclude by using sequence E_q. □

Theorem 8. *Let* A *be any* G-*module. The following are equivalent*:

i) *For every prime* p, $\hat{H}^q(G_p, A) = 0$ *for two consecutive values of* q (that may depend on p),
ii) A *is cohomologically trivial*,
iii) *There exists an exact sequence* $0 \to P_k \to P_{k-1} \to \cdots \to P_0 \to A \to 0$, *where the* P_i *are projective* $\mathbf{Z}[G]$-*modules*,
iv) *There exists an exact sequence* $0 \to P_1 \to P_0 \to A \to 0$, *where the* P_i *are* $\mathbf{Z}[G]$-*projective*.

(In the terminology of Cartan-Eilenberg, condition iii) means that the *projection dimension* of A is finite, and condition iv) that it is ≤ 1.)

We have iv) $\Rightarrow$ iii) trivially, iii) $\Rightarrow$ ii) by lemma 6, and ii) $\Rightarrow$ i) trivially. Let us show i) $\Rightarrow$ iv): Let $0 \to R \to L \to A \to 0$ be an exact sequence of G-modules, with L free over $\mathbf{Z}[G]$: *a fortiori*, L is $\mathbf{Z}$-free, hence also R. On the other hand, R satisfies hypothesis i) of th. 7, so is $\mathbf{Z}[G]$-projective. □

Theorem 9. *Let* A, B *be* G-*modules, with* A *cohomologically trivial. In order that* $A \otimes B$ (resp. Hom(A, B), resp. Hom(B, A)) *be cohomologically trivial, it is necessary and sufficient that* Tor(A, B) (resp. Ext(A, B), resp. Ext(B, A)) *be*.

(Once more, the functors $\otimes$, Tor, etc. are taken over the ring $\mathbf{Z}$.)

By th. 8, iv), A has a resolution by projective modules

$$0 \to P_1 \to P_0 \to A \to 0.$$

Hence there is an exact sequence

$$0 \to \mathrm{Tor}(A, B) \to P_1 \otimes B \to P_0 \otimes B \to A \otimes B \to 0.$$

The corollary to prop. 1 shows that $P_1 \otimes B$ and $P_0 \otimes B$ are cohomologically trivial; applying lemma 6, we see that $A \otimes B$ is cohomologically trivial if and only if Tor(A, B) is. Same proof for Hom(A, B) and Ext(A, B). For Hom(B, A), use the six term exact sequence

$$0 \to \mathrm{Hom}(B, P_1) \to \mathrm{Hom}(B, P_0) \to \mathrm{Hom}(B, A) \to$$
$$\mathrm{Ext}(B, P_1) \to \mathrm{Ext}(B, P_0) \to \mathrm{Ext}(B, A) \to 0.$$

The corollary to prop. 1 shows that the four modules

$$\mathrm{Hom}(B, P_1), \quad \mathrm{Hom}(B, P_0), \quad \mathrm{Ext}(B, P_1), \quad \mathrm{Ext}(B, P_0)$$

are cohomologically trivial; conclude, as before, by applying lemma 6. □

Corollary. *If* A *is cohomologically trivial, and if* A *or* B *is torsion-free, then* $A \otimes B$ *is cohomologically trivial*.

Indeed, Tor(A, B) is zero.

EXERCISES

1. Suppose that A is cohomologically trivial, and that for every prime p dividing the order of G, one of the groups A or B is without p-torsion. Show that $A \otimes B$ is cohomologically trivial. (Show that Tor(A, B) is cohomologically trivial and then apply th. 9.)

2. Let G be the cyclic group of order 6. Show that there exists a G-module A which is not cohomologically trivial, yet for which $\hat{H}^n(G, A) = 0$ for all n. (Take $A = \mathbf{Z}/3\mathbf{Z}$, on which G operates by $x \to -x$, and show that $\hat{H}^0(g, A) \neq 0$ if g denotes the subgroup of G of order 3.)

3. Let G be the cyclic group of order 2 and let $A = \mathbf{Z}/8\mathbf{Z}$; make G operate on A by $x \to 3x$. Let G operate trivially on $B = \mathbf{Z}/2\mathbf{Z}$. Show that A is cohomologically trivial but $A \otimes B$ is not. Deduce that A is not relatively projective (cf. [13], p. 263, exer. 3). (Note that A is isomorphic to the multiplicative group $\mathbf{F}_9^*$ of the field with 9 elements, on which the Galois group of the extension $\mathbf{F}_9/\mathbf{F}_3$ acts.)

§6. Dual Results

Lemma 7. *Let G be a finite group, and let A be an injective $\mathbf{Z}[G]$-module. Then A is $\mathbf{Z}$-injective, i.e., divisible.*

We must show that the functor $\mathrm{Hom}_{\mathbf{Z}}(C, A)$ is exact in C. If $\Lambda = \mathbf{Z}[G]$, there is a functorial isomorphism

$$\mathrm{Hom}_{\mathbf{Z}}(C, A) = \mathrm{Hom}_{\Lambda}(C \otimes_{\mathbf{Z}} \Lambda, A).$$

As Λ is $\mathbf{Z}$-free, the functor $C \otimes_{\mathbf{Z}} \Lambda$ is exact, and since A is Λ-injective, so is the functor $\mathrm{Hom}_{\Lambda}(\ \ , A)$, and the result follows. □

Theorem 10 (Dual to Theorem 7). *Let G be a finite group, and let A be a $\mathbf{Z}$-injective G-module. In order that A be cohomologically trivial, it is necessary and sufficient that A be $\mathbf{Z}[G]$-injective.*

Sufficiency is trivial. To see the necessity, embed A in a $\mathbf{Z}[G]$-injective module I, obtaining an exact sequence

$$0 \to A \to I \to R \to 0.$$

Since A is $\mathbf{Z}$-injective, this yields the exact sequence

$$0 \to \mathrm{Hom}_{\mathbf{Z}}(R, A) \to \mathrm{Hom}_{\mathbf{Z}}(I, A) \to \mathrm{Hom}_{\mathbf{Z}}(A, A) \to 0.$$

If A is cohomologically trivial, theorem 9 tells us that $\mathrm{Hom}_{\mathbf{Z}}(R, A)$ is also; the exact cohomology sequence then shows that the map

$$\mathrm{Hom}_G(I, A) \to \mathrm{Hom}_G(A, A)$$

is surjective, hence A is a direct factor of I, and is thus $\mathbf{Z}[G]$-injective. □

Theorem 11 (Dual to Theorem 8). *In order that a G-module A be cohomologically trivial, it is necessary and sufficient that there be an exact sequence $0 \to A \to I_0 \to I_1 \to 0$, where the I_i are injective $\mathbf{Z}[G]$-modules.*

As before, there is an exact sequence

$$0 \to A \to I_0 \to R \to 0$$

with I_0 $\mathbf{Z}[G]$-injective. Since A is cohomologically trivial, so is R; on the other hand, I_0 is $\mathbf{Z}$-injective (by lemma 7), hence R is too. Theorem 10 then guarantees that R is $\mathbf{Z}[G]$-injective. □

Note. The results of the three preceding sections are essentially due to Nakayama ([47], [48]). For the presentation, I have followed the paper [51] of Dock Sang Rim, who has greatly simplified the proofs of Nakayama and generalised some of his results. See also Lang [93] and Tate [118].

§7. A Comparison Theorem

Theorem 12. *Let G be a finite group, A and A′ G-modules, and $f: A' \to A$ a G-homomorphism. For each prime number p, let G_p be a Sylow p-subgroup of G, and suppose there is an integer n_p such that the homomorphism*

$$f_i^*: \hat{H}^i(G_p, A') \to \hat{H}^i(G_p, A)$$

is surjective for $i = n_p$, bijective for $i = n_p + 1$, injective for $i = n_p + 2$.

If B is a G-module such that $\mathrm{Tor}(A, B) = 0 = \mathrm{Tor}(A', B)$, then the homomorphism

$$\hat{H}^i(g, A' \otimes B) \to \hat{H}^i(g, A \otimes B)$$

is bijective for every subgroup g of G and every integer i. In particular, $\hat{H}^i(g, A') \to \hat{H}^i(g, A)$ is bijective for all i.

We will use a construction analogous to the "mapping-cylinder" in topology. Let $\bar{A}'$ be the induced module canonically defined by A′, $i: A' \to \bar{A}'$ the canonical injection (cf. Chap. VII, §6). Put $A^* = A \oplus \bar{A}'$. The pair (f, i) defines an injection $\theta: A' \to A^*$; if A″ denotes the cokernel of θ, we have the exact sequence

$$0 \to A' \to A^* \to A'' \to 0.$$

As $\bar{A}'$ is cohomologically trivial, the cohomology of A^* can be identified with that of A. The hypothesis on the f_i^*, together with the exact cohomology sequence, gives

$$\hat{H}^q(G_p, A'') = 0 \quad \text{for } q = n_p, n_p + 1.$$

By theorem 8, A″ *is cohomologically trivial.* On the other hand, A′ is a direct factor in $\bar{A}'$ (as $\mathbf{Z}$-module, of course), hence also in A^*; as A^* is the direct sum of A and a number of copies of A′, the hypothesis on B implies $\mathrm{Tor}(A^*, B) = 0$, whence $\mathrm{Tor}(A'', B) = 0$, and theorem 9 tells us that $A'' \otimes B$ is cohomologically trivial. The exact sequence

$$0 \to A' \otimes B \to A^* \otimes B \to A'' \otimes B \to 0$$

enables us to deduce the bijectivity of $\hat{H}^q(g, A' \otimes B) \to \hat{H}^q(g, A^* \otimes B)$. As the same holds for $\hat{H}^q(g, A^* \otimes B) \to \hat{H}^q(g, A \otimes B)$, the theorem is proved. □

Remark. Suppose A and A′ are **Z**-free. The G-modules $\bar{A}'$ and A″ are then projective (the first is even free). In other words, f factors into

$$A' \xrightarrow{i} A' \oplus P' \xrightarrow{F} A \oplus P \xrightarrow{\pi} A$$

with P and P′ projective, F an isomorphism, and i (resp. π) denoting the obvious injection (resp. projection). When A and A′ are finitely generated, P and P′ can be taken to be finitely generated; in the terminology of Eckmann-Hilton ([23], and see also [58]), f is a *homotopy equivalence.*

Exercise

With the notation and hypotheses of the above remark, show that the element $(f) = P' - P$ of the group P(G) depends only on f, not on the choice of P and P′. Show that $(fg) = (f) + (g)$, and that $(f) = 0$ if and only if P and P′ can be chosen free of finite rank over **Z**[G].

§8. The Theorem of Tate and Nakayama

Theorem 13. *Let* G *be a finite group,* A, B, C *three* G*-modules, and let* $\varphi: A \times B \to C$ *be a* G*-invariant bilinear map. Let* $q \in \mathbf{Z}$, $a \in \hat{H}^q(G, A)$. *Given any subgroup* g *of* G *and any* G*-module* D, *denote by*

$$f(n, g, D): \hat{H}^n(g, B \otimes D) \to \hat{H}^{n+q}(g, C \otimes D)$$

the homomorphism defined by cup product with the class $a_g = \mathrm{Res}_{G/g}(a)$ (relative to the obvious bilinear map of $A \times (B \otimes D)$ into $C \otimes D$).

Suppose that for every prime p and Sylow p-subgroup G_p *of* G, *there is an integer* n_p *for which* $f(n, G_p, \mathbf{Z})$ *is surjective for* $n = n_p$, *bijective for* $n = n_p + 1$, *and injective for* $n = n_p + 2$.

Then $f(n, g, D)$ *is bijective for all n, all g, and every* G*-module* D *such that*

$$\mathrm{Tor}(B, D) = \mathrm{Tor}(C, D) = 0.$$

We first treat the case $q = 0$. The class $a \in \hat{H}^0(G, A)$ can be represented by an element $a \in A^G$. Putting $f(b) = \varphi(a, b)$, we obtain a homomorphism of G-modules

$$f: B \to C.$$

It is easy to check that the homomorphism

$$f(n, g, \mathrm{D}) : \hat{\mathrm{H}}^n(g, \mathrm{B} \otimes \mathrm{D}) \to \hat{\mathrm{H}}^n(g, \mathrm{C} \otimes \mathrm{D})$$

is merely the homomorphism induced by $f \otimes 1 : \mathrm{B} \otimes \mathrm{D} \to \mathrm{C} \otimes \mathrm{D}$, and we are reduced to theorem 12.

The general case is handled by dimension-shifting. Let us show how to pass from $q - 1$ to q: embed A in its canonical induced module $\bar{\mathrm{A}}$, and set $\mathrm{A}_1 = \bar{\mathrm{A}}/\mathrm{A}$. Define similarly $\mathrm{C}_1 = \bar{\mathrm{C}}/\mathrm{C}$ and $\varphi_1 : \mathrm{A}_1 \times \mathrm{B} \to \mathrm{C}_1$. The class $a \in \hat{\mathrm{H}}^q(\mathrm{G}, \mathrm{A})$ can be written $a = \delta(a_1)$, $a_1 \in \hat{\mathrm{H}}^{q-1}(\mathrm{G}, \mathrm{A}_1)$. This class a_1 defines, by cup product, homomorphisms

$$f_1(n, g, \mathrm{D}) : \hat{\mathrm{H}}^n(g, \mathrm{B} \otimes \mathrm{D}) \to \hat{\mathrm{H}}^{n+q-1}(g, \mathrm{C}_1 \otimes \mathrm{D}).$$

Combining f_1 with the isomorphism

$$\delta : \hat{\mathrm{H}}^{n+q-1}(g, \mathrm{C}_1 \otimes \mathrm{D}) \to \hat{\mathrm{H}}^{n+q}(g, \mathrm{C} \otimes \mathrm{D}),$$

we obtain $f(n, g, \mathrm{D})$ (using the fact that cup products commute with the coboundaries—cf. Chap. VIII, §3). If the theorem holds for the class a_1, it also holds for a. □

The most important special case of this theorem is the following:

Theorem 14. *Let* G *be a finite group,* A *a* G*-module, and* $a \in \mathrm{H}^2(\mathrm{G}, \mathrm{A})$. *Let* G_p *be a Sylow p-subgroup of* G, *for each prime p, and suppose that*

1) $\mathrm{H}^1(\mathrm{G}_p, \mathrm{A}) = 0$;
2) $\mathrm{H}^2(\mathrm{G}_p, \mathrm{A})$ *is generated by* $\mathrm{Res}_{\mathrm{G}/\mathrm{G}_p}(a)$, *and has order equal to that of* G_p.

Then for any G*-module* D *such that* $\mathrm{Tor}(\mathrm{A}, \mathrm{D}) = 0$, *the cup product with* $a_g = \mathrm{Res}_{\mathrm{G}/g}(a)$ *induces isomorphisms*

$$\hat{\mathrm{H}}^n(g, \mathrm{D}) \to \hat{\mathrm{H}}^{n+2}(g, \mathrm{A} \otimes \mathrm{D})$$

for every integer n and every subgroup g of G.

Apply th. 13 with $\mathrm{B} = \mathbf{Z}$, $\mathrm{C} = \mathrm{A}$, $q = 2$, $\varphi : \mathrm{A} \times \mathbf{Z} \to \mathrm{A}$ being the obvious map. Take $n_p = -1$. For $n = -1$, hypothesis 1) shows that the cup product is surjective; for $n = 0$, hypothesis 2) shows that it is bijective; for $n = 1$, it is injective, since $\mathrm{H}^1(\mathrm{G}_p, \mathbf{Z}) = 0$. Thus all the hypotheses hold. □

Corollary (Tate [63]). *For every integer n and every subgroup g of* G, *the cup product with* a_g *defines an isomorphism*

$$\hat{\mathrm{H}}^n(g, \mathbf{Z}) \to \hat{\mathrm{H}}^{n+2}(g, \mathrm{A}).$$

This is the special case $\mathrm{D} = \mathbf{Z}$.

In Chap. XI we will see how this result applies to class formations.

CHAPTER X

Galois Cohomology

§1. First Examples

Let K/k be a finite Galois extension with Galois group G. The group G acts both on the additive group of K and on the multiplicative group K^*; hence we can investigate the corresponding cohomology groups.

Proposition 1. *For every integer n,* $\hat{H}^n(G, K) = 0$.

Indeed, the *normal basis* theorem (Bourbaki, *Alg.*, Chap. V, §10) shows that K is an *induced* module, and we know that the cohomology of such a module is trivial.

[If one wishes to avoid the normal basis theorem, one can simply remark that K contains an element of trace 1, which implies that K is relatively projective (Cartan-Eilenberg [13], p. 233, prop. 1.1), hence cohomologically trivial.] □

Proposition 2. $H^1(G, K^*) = 0$.

Let $s \mapsto a_s$ be a 1-cocycle. If $c \in K$, form the "Poincaré series"

$$b = \sum_{s \in G} a_s . s(c).$$

It follows from the linear independence of automorphisms (Bourbaki, *loc. cit.*, §7, no. 5) that c can be chosen so that $b \neq 0$. On the other hand,

$$\begin{aligned} s(b) &= \sum s(a_t) . st(c) \\ &= \sum a_s^{-1} a_{st} . st(c) = a_s^{-1} . b \end{aligned}$$

which shows that a_s is a coboundary. □

Corollary. *If* G *is cyclic, and* s *is a generator, and if* $x \in \mathrm{K}^*$ *has norm* 1, *then there exists* $y \in \mathrm{K}^*$ *such that* $x = y/s(y)$.

That follows from the determination of $\mathrm{H}^1(\mathrm{G}, \mathrm{K}^*)$ when G is cyclic.

Remarks. 1) This corollary is none other than the famous "Hilbert theorem 90"; in the literature, prop. 2 is often referred to by that name.

2) The higher cohomology groups $\mathrm{H}^q(\mathrm{G}, \mathrm{K}^*)$ are not in general zero. We will return later to the case $q = 2$ (the Brauer group).

Prop. 2 generalises as follows (cf. Speiser [60]): let $\mathbf{GL}(n, \mathrm{K})$ be the group of invertible matrices of degree n with entries from K; the group G acts in an obvious way on $\mathbf{GL}(n, \mathrm{K})$, which allows us to define the cohomology set $\mathrm{H}^1(\mathrm{G}, \mathbf{GL}(n, \mathrm{K}))$—cf. Chap. VII, appendix.

Proposition 3. $\mathrm{H}^1(\mathrm{G}, \mathbf{GL}(n, \mathrm{K})) = \{1\}$.

The proof is analogous to that of prop. 2. Let a_s be a 1-cocycle, $c \in \mathrm{M}_n(\mathrm{K})$ any matrix. Again form the Poincaré series

$$b = \sum_{s \in \mathrm{G}} a_s \,.\, s(c)$$

and check that $s(b) = a_s^{-1} \,.\, b$; this formula shows that a_s is a coboundary, provided that c can be chosen so that b is an *invertible* matrix. When K is infinite, the existence of such c results simply from the *algebraic independence of automorphisms* (Bourbaki, *Alg.*, Chap. V, §10, th. 4). Unfortunately this argument is not applicable when K is finite; that is why we use another procedure, shown to me by Cartier:

Let x be a vector in K^n, and form $b(x) = \sum_{s \in \mathrm{G}} a_s(s(x))$. The $b(x)$ generate K^n as a vector space over K, as x runs through K^n: indeed, if a linear form u vanishes on all the $b(x)$, then for every $h \in \mathrm{K}$,

$$0 = u(b(hx)) = \sum a_s \,.\, u(s(h)s(x)) = \sum s(h) u(a_s(s(x))).$$

As h varies, a linear relation among the $s(h)$ is obtained. By the linear independence theorem of Dedekind already cited, each of the $u(a_s(s(x))) = 0$, and since the a_s are invertible, that implies $u = 0$.

With this point settled, let $x_1, \ldots, x_n$ be vectors in K^n such that the $y_i = b(x_i)$ are linearly independent over K. Let c be the matrix of the map that sends the canonical basis e_i to the x_i. A computation of the corresponding matrix b shows that $b(e_i) = y_i$, hence b is invertible. □

Corollary. $\mathrm{H}^1(\mathrm{G}, \mathbf{SL}(n, \mathrm{K})) = \{1\}$.

The exact sequence

$$\{1\} \to \mathbf{SL}(n, \mathrm{K}) \to \mathbf{GL}(n, \mathrm{K}) \xrightarrow{\det} \mathrm{K}^* \to \{1\}$$

gives rise (cf. Chap. VII, appendix) to the exact sequence

$$\mathrm{H}^0(\mathrm{G}, \mathbf{GL}(n, \mathrm{K})) \to \mathrm{H}^0(\mathrm{G}, \mathrm{K}^*) \to \mathrm{H}^1(\mathrm{G}, \mathbf{SL}(n, \mathrm{K})) \to \{1\},$$

which is

$$\mathbf{GL}(n, k) \to k^* \to \mathrm{H}^1(\mathrm{G}, \mathbf{SL}(n, \mathrm{K})) \to \{1\}.$$

As the determinant map $\mathbf{GL}(n, k) \to k^*$ is surjective, the corollary follows.

Remark. Let A be a *group scheme over k* (or, in the terminology of Weil, a "group variety defined over k"). For any commutative k-algebra K whatever, the set A_K of points of A with values in K is a *group*, depending functorially on K. In particular, if K/k is a finite Galois extension with Galois group G, then G acts on A_K, and the cohomology $H^q(G, A_K)$ is defined (if A is non-abelian, only H^0 and H^1 are defined, and H^1 is merely a "pointed set"). There is a simple geometric interpretation for $H^1(G, A_K)$: it is the set of classes of principal homogeneous spaces for A, defined over k, which have a rational point in K (cf. [41], [113]). The cases treated above are $A = G_a$, G_m, $\mathbf{GL}(n)$, $\mathbf{SL}(n)$; in §2 we will examine the orthogonal and symplectic groups.

Of course, there is no reason to restrict ourselves to linear groups. There are extremely interesting problems related to abelian varieties, for which we can only refer the reader to Tate's exposé [64] and the papers of Cassels [14].

EXERCISES

1. Extend prop. 3 to the group of automorphisms of an infinite dimensional vector space.

2. Let K/k be a finite Galois extension with group G, and let M be a unitary k-algebra, finite dimensional over k. Let $M_K = M \otimes_k K$, and let M_K^* be the group of inverible elements of M_K: it is a G-module. Show that $H^1(G, M_K^*) = \{1\}$. (When k is infinite, use the Poincaré series. When k is finite, reduce to the case where M is semi-simple, and apply prop. 3 (or use a general result of Lang [39]).)

§2. Some Examples of "Descent"

Let V be a vector space over k, provided with a fixed tensor x of type (p, q) (i.e., $x \in \bigotimes^p V \otimes \bigotimes^q V^*$, where V^* is the dual of V). Two pairs (V, x) and (V', x') are called *k-isomorphic* if there is a k-linear isomorphism

$$f: V \to V'$$

such that $f(x) = x'$.

Now let K/k be a finite Galois extension with group G. Let $V_K = V \otimes_k K$ be the vector space over K obtained by extending scalars; the tensor x defines a tensor x_K on V_K in the obvious way, and we will often denote it simply as x. We say that (V, x) and (V', x') are K-*isomorphic* (or "become isomorphic over K") if (V_K, x_K) and (V'_K, x'_K) are isomorphic. Denote by $E_{V,x}(K/k)$ *the set of k-isomorphism classes of pairs* (V', x') *that are* K-*isomorphic to* (V, x). We propose to interpret $E_{V,x}(K/k)$ as an H^1.

To do this, let A_K be the group of K-*automorphisms* of (V_K, x_K). The group G acts on A_K as follows: first of all, it acts on V_K by $s(x \otimes \lambda) = x \otimes s(\lambda)$; then, if $f: V_K \to V_K$ is a K-linear map, put

$$s(f)(x) = s.f(s^{-1}(x)), \quad \text{i.e., } s(f) = s \circ f \circ s^{-1}.$$

We propose to *compare* $E_{V,x}(K/k)$ *with* $H^1(G, A_K)$. To simplify, write $E(K/k)$ instead of $E_{V,x}(K/k)$.

So let $(V', x') \in E(K/k)$, and let $f: V_K \to V'_K$ be a K-isomorphism. Put

$$p_s = f^{-1} \circ s(f) = f^{-1} \circ s \circ f \circ s^{-1}, \qquad s \in G.$$

Evidently $p_s \in A_K$; furthermore, a simple computation shows that $s \mapsto p_s$ is a 1-cocycle, and that changing f has the effect of replacing p_s with an equivalent cocycle. Thus the class of p_s in $H^1(G, A_K)$ is well-determined, and we have defined a map

$$\theta: E(K/k) \to H^1(G, A_K).$$

Proposition 4. *The map θ just defined is bijective.*

We first show θ is *injective*. Let (V'_1, x'_1) and (V'_2, x'_2) correspond to the same cocycle p_s, and let f_1, f_2 be the corresponding K-isomorphisms. Then $f_1^{-1} \circ s(f_1) = f_2^{-1} \circ s(f_2)$ whence $s(f_2 f_1^{-1}) = f_2 f_1^{-1}$. The map $f = f_2 f_1^{-1}$ is a k-isomorphism of (V'_1, x'_1) onto (V'_2, x'_2), so θ is injective.

We next show θ is *surjective*. Let p_s be a 1-cocycle of G with values in A_K; as $A_K \subset \mathbf{GL}(V_K)$, proposition 3 gives us a K-automorphism f of V_K such that

$$p_s = f^{-1} \circ s(f) \quad \text{for all } s \in G.$$

Extend f to the tensor algebra of V_K, and put $x' = f(x)$. The element x' is "rational over k" (i.e., belongs to the tensor algebra of V over k): indeed,

$$s(x') = s(f)(s(x)) = s(f)(x) = f \circ p_s(x) = f(x) = x'.$$

Hence (V, x') belongs to $E(K/k)$, and clearly the image of this element by θ is equal to the class of p_s. □

Remark. We could have defined θ by noting that the set of "isomorphisms" of (V, x) onto (V', x') is a principal homogeneous space for the algebraic group A (see [113], Chap. III, §1).

EXAMPLES. Take as tensor x a non-degenerate quadratic form Φ. The set $E(K/k)$ is then the set of classes of quadratic forms that are K-isomorphic to Φ. The group A_K is the orthogonal group $\mathbf{O}_K(\Phi)$ of the form Φ over K. Thus:

Corollary 1. *The set $H^1(G, \mathbf{O}_K(\Phi))$ is in bijective correspondence with the set of classes of quadratic k-forms that are K-isomorphic to Φ.*

This interpretation of $H^1(G, \mathbf{O}_K(\Phi))$ allows the construction of examples where this set is non-trivial (e.g., $k = \mathbf{R}$, $K = \mathbf{C}$).

Instead of taking a quadratic form, one could take a non-degenerate alternating (skew-symmetric) form; the group A_K is then the symplectic group $\mathbf{Sp}(n, K)$. As two non-degenerate alternating forms of the same rank are equivalent (Bourbaki, *Alg.*, Chap. IX, §5), one gets:

Corollary 2. $H^1(G, \mathbf{Sp}(n, K)) = \{1\}$.

Remark. What we have done is a special case of the method of "Galois descent" in algebraic geometry (cf. [56], Chap. V, §4), itself included in Grothendieck's "theory of descent" (cf. [28], [29], [80] and [113]).

§3. Infinite Galois Extensions

It is often convenient to "pass to the limit" by taking larger and larger Galois extensions. Let us see how this fits with the cohomological point of view.

Let K/k be a Galois extension (not necessarily finite). Its Galois group G is a compact topological group, totally disconnected, equal to the projective limit of the Galois groups $G(K'/k)$, where K' runs through the set of finite Galois subextensions of K.

Let A be a G-module. It is called a *topological* G-module if, for all $a \in A$, the set of those $s \in G$ such that $s(a) = a$ is an open subgroup of G. It amounts to the same thing to say that

$$A = \bigcup A^H$$

when H runs through the set of open normal subgroups of G. This condition is realized when A is the group of rational points in K of a group scheme of finite type over k; for example, take $A = K$ or $A = K^*$.

If A is a topological G-module, define $H^q(G, A)$ by the formula

$$H^q(G, A) = \varinjlim H^q(G/H, A^H)$$

as H runs through the set of open normal subgroups of G (the direct limit is taken relative to the *inflation* homomorphisms—cf. Chap. VII, §5). These cohomology groups enjoy properties entirely analogous to those of the usual cohomology groups; they form a functor (the coboundary also being defined by passage to the limit); they can be computed via *continuous cochains* with values in A. Their properties have been systematically studied by Tate (see [20], [88], [93], [113]). We will give two elementary applications of them.

a) Artin-Schreier Theory

Let k be a field of characteristic $p \neq 0$, K its separable algebraic closure, and G the Galois group of K/k. The map

$$\wp : K \to K$$

defined by $\wp(x) = x^p - x$, is a G-homomorphism. It is surjective because the equation $\wp(x) = a$ is separable; its kernel is $\mathbf{Z}/p\mathbf{Z}$ (on which G acts trivially). Thus we have an exact sequence

$$0 \to \mathbf{Z}/p\mathbf{Z} \to \mathrm{K} \to \mathrm{K} \to 0.$$

Writing out the corresponding exact cohomology sequence, and taking into account that $\mathrm{H}^1(\mathrm{G}, \mathrm{K}) = 0$ by prop. 1, we obtain

$$k \to k \to \mathrm{H}^1(\mathrm{G}, \mathbf{Z}/p\mathbf{Z}) \to 0.$$

In other words, the group $\mathrm{Hom}(\mathrm{G}, \mathbf{Z}/p\mathbf{Z})$ of continuous homomorphisms of G into $\mathbf{Z}/p\mathbf{Z}$ *is isomorphic to* $k/\wp(k)$. The isomorphism assigns to $a \in k$ the homomorphism $\varphi_a : \mathrm{G} \to \mathbf{Z}/p\mathbf{Z}$ defined by solving the equation $x^p - x = a$ and setting $\varphi_a(s) = s(x) - x$.

Denote by k_p the composite of all the abelian extensions of type $(p, \ldots, p)$ of k; then we see that *the Galois group* $\mathrm{G}(k_p/k)$ *is isomorphic to the compact dual* $(k/\wp(k))\hat{}$ *of the discrete group* $k/\wp(k)$.

If one wants to pass from $\mathbf{Z}/p\mathbf{Z}$ to $\mathbf{Z}/p^n\mathbf{Z}$, one must use the exact sequence

$$0 \to \mathbf{Z}/p^n\mathbf{Z} \to \mathrm{W}_n(\mathrm{K}) \xrightarrow{\mathrm{f}-1} \mathrm{W}_n(\mathrm{K}) \to 0$$

where W_n denotes the additive group of Witt vectors of length n.

b) Kummer Theory

Let n be an integer prime to the characteristic of k, and assume that k^* contains the group μ_n of nth roots of unity. If K again denotes the separable closure of k, then we have the exact sequence of G-modules

$$0 \to \mu_n \to \mathrm{K}^* \xrightarrow{u} \mathrm{K}^* \to 0$$

with $u(x) = x^n$. Applying proposition 2, we deduce an *isomorphism* $k^*/k^{*n} \to \mathrm{Hom}(\mathrm{G}, \mu_n)$. This isomorphism assigns to an element $a \in k^*$ the homomorphism $\varphi_a : \mathrm{G} \to \mu_n$ defined by solving the equation $x^n = a$ and setting $\varphi_a(s) = s(x).x^{-1}$. Denote by k_n the composite of all the abelian extensions of k whose Galois group is annihilated by n; then we see that the Galois group $\mathrm{G}(k_n/k)$ is isomorphic to $\mu_n \otimes (k^*/k^{*n})\hat{}$, which is itself isomorphic (non-canonically) to $(k^*/k^{*n})\hat{}$.

§4. The Brauer Group

We will now concern ourselves exclusively with the *multiplicative group*. If K/k is a Galois extension (finite or infinite), with Galois group G, we write $\mathrm{H}^q(\mathrm{K}/k)$ instead of $\mathrm{H}^q(\mathrm{G}, \mathrm{K}^*)$. These groups depend functorially on the pair (K, k). More precisely, let K'/k' be a Galois extension with group G', the field k' being an extension of k. Suppose there is a k-monomorphism f of K into K′; it defines a homomorphism $\bar{f}: G' \to G$ by assigning to $s' \in G'$ the unique element $s = \bar{f}(s') \in G$ such that $f \circ s = s' \circ f$. The homomorphism $\bar{f}$ is *compatible* with $f: K^* \to K'^*$, in the sense of Chap. VII, §5. Hence, it induces homomorphisms

$$f_q: H^q(K/k) \to H^q(K'/k').$$

Proposition 5. *The homomorphisms f_q are independent of the choice of f.*

Indeed, two such choices differ by an element of G, and prop. 3 of Chap. VII. §5 can be applied to this element. □

(Of course, this proposition generalises to any group scheme over k.)

In particular, if $k' = k$, and if K and K′ are isomorphic, then there is a *canonical* isomorphism of $H^q(K/k)$ with $H^q(K'/k)$; this applies notably when K is the *separable closure* k_s of k, in which case the groups $H^q(k_s/k)$ will also be denoted $H^q(\ \ /k)$. Clearly they depend functorially on k.

The group $H^1(\ \ /k)$ is zero (prop. 2). The group $H^2(\ \ /k)$ is called the *Brauer group* of the field k, and will also be denoted B_k. By definition, it is the direct limit of the $H^2(K/k)$ as K runs through the set of finite Galois extensions of k. This direct limit is actually a union. Namely:

Proposition 6. *Let L/k be a Galois extension containing the Galois extension K/k. Then there is an exact sequence*

$$0 \to H^2(K/k) \to H^2(L/k) \to H^2(L/K).$$

Let $G = G(L/k)$, and let $H = G(L/K)$. Since $H^1(H, L^*) = 0$, we can apply prop. 5 of Chap. VII, §6, with $q = 2$; we get the exact sequence

$$0 \to H^2(G/H, K^*) \xrightarrow{\text{Inf}} H^2(G, L^*) \xrightarrow{\text{Res}} H^2(H, L^*),$$

as desired. □

Corollary. *There is an exact sequence*

$$0 \to H^2(K/k) \to B_k \to B_K.$$

Pass to the limit on L.

Remarks. 1) Prop. 6 and its corollary remain valid when K/k is an infinite Galois extension, as can be seen by passage to the limit.

2) An element $a \in B_k$ is said to be *split by* K if it is in the kernel of the map $B_k \to B_K$; when K/k is Galois, this comes to the same (by the corollary) as saying that a belongs to $H^2(K/k)$, considered as a subgroup of B_k.

EXERCISES

1. Let k be a field of characteristic p, and let $K = k^{1/p}$. Show that an element $a \in B_k$ is split by K if and only if $pa = 0$. Deduce that the p-primary component of B_k is zero if k is perfect.
2. Let K/k be an extension of degree n, and let $a \in B_k$ be split by K. Show that $na = 0$. (Treat the separable and purely inseparable cases separately; use exer. 1 for the latter, and prop. 6 of Chap. VII for the former.)

§5. Comparison with the Classical Definition of the Brauer Group

First, recall:

Proposition 7. *Let k be a field and* A *a finite dimensional k-algebra. The following conditions are equivalent:*

a) A *has no non-trivial two-sided ideal, and its center is k.*
b) *If* K *is the algebraic closure of k, then the extended algebra is isomorphic to a matrix algebra over* K.
b′) *There exists a finite Galois extension K/k such that A_K is isomorphic to a matrix algebra over* K.
c) A *is isomorphic to a matrix algebra over a division algebra with center k.*

For the proof, see Bourbaki, *Alg.*, Chap. VIII, §§5, 10. Such an algebra is called *central simple* over k. Two such algebras are said to be *equivalent* if their division algebras associated by c) are k-isomorphic; when the two central simple algebras have the same dimension, equivalence reduces to k-isomorphic.

Let A_k be the set of classes of central simple algebras (for the equivalence relation just defined); the tensor product defines by passage to the quotient a structure of abelian group on A_k. This is the group known classicially as the "Brauer group" (cf. Bourbaki, *loc. cit.*, §10). It is a covariant functor of k: if K is an extension of k, then extension of scalars from k to K defines a homomorphism

$$A_k \to A_K.$$

Denote by $A(K/k)$ the kernel of this homomorphism. Proposition 7 implies that A_k *is the union of the* $A(K/k)$ *as* K *runs through the set of finite Galois extensions of* k. To show that A_k is isomorphic to B_k, it is enough to construct *isomorphisms* $A(K/k) \to H^2(K/k)$ compatible with the injections

$$A(K/k) \to A(K'/k) \quad \text{and} \quad H^2(K/k) \to H^2(K'/k), \quad \text{for } K' \supset K.$$

We proceed by "descent".

Let $A(n, K/k)$ be the set of classes of k-algebras A such that $A \otimes_k K$ is isomorphic to the matrix algebra $\mathbf{M}_n(K)$. The group $A(K/k)$ is the union of the subsets $A(n, K/k)$ for all positive n. The procedure of §2 applies to $A(n, K/k)$: an element of $A(n, K/k)$ can be considered as a pair (V, x), where V is a vector space of dimension n^2 and where x is a tensor of type $(1, 2)$ (the law of composition), this pair being K-isomorphic to the standard pair defined by the matrix algebra $\mathbf{M}_n$. If G is the Galois group of K/k, and C_K denotes the group of K-automorphisms of $\mathbf{M}_n(K)$, we conclude that the map

$$\theta: A(n, K/k) \to H^1(G, C_K)$$

defined in §2 is a bijection.

But it is well-known that every automorphism of $\mathbf{M}_n(K)$ is inner. Hence the exact sequence

$$(*) \qquad \{1\} \to K^* \to \mathbf{GL}(n, K) \to C_K \to \{1\}.$$

This sequence allows us to *identify* C_K *with the projective group* $\mathbf{PGL}(n, K)$. Summarizing:

Proposition 8. *There is a canonical bijection*

$$\theta: A(n, K/k) \to H^1(G, \mathbf{PGL}(n, K)).$$

On the other hand, the exact sequence $(*)$ defines (cf. Chap. VII, appendix) a "coboundary" operator

$$\Delta_n: H^1(G, \mathbf{PGL}(n, K)) \to H^2(G, K^*).$$

Composing θ and Δ_n gives a map

$$\delta_n: A(n, K/k) \to H^2(G, K^*) = H^2(K/k).$$

An easy computation shows that if $C \in A(n, K/k)$ and $C' \in A(n', K/k)$, then

$$\delta_{nn'}(C \otimes C') = \delta_n(C) + \delta_{n'}(C').$$

On the other hand, $\delta_n(C) = 0$ if and only if C is a matrix algebra (as follows from prop. 2 of the appendix to Chap. VII). The compatibility of the maps δ_n results from this, and they define a homomorphism

$$\delta: A(K/k) \to H^2(K/k).$$

Proposition 9. *The homomorphism* $\delta: A(K/k) \to H^2(K/k)$ *is bijective.*

We have already seen that it is injective. The next lemma shows that it is surjective.

Lemma 1. *If* $n = [K:k]$, *then the map* $\delta_n: A(n, K/k) \to H^2(K/k)$ *is surjective.*

(Compare with Bourbaki, *Alg.*, Chap. VIII, §10, prop. 7.)

By prop. 8 and the definition of δ_n, it suffices to show that Δ_n is surjective, i.e., that every 2-cocycle $a_{s,t}$ with values in K* can be written

$$a_{s,t} = p_s s(p_t) p_{st}^{-1}, \quad \text{with } p_s \in \mathbf{GL}(n, \mathrm{K}).$$

Let V be a vector space over K having as basis a family of vectors e_s, $s \in$ G. Let $p_s \in \mathrm{Hom}_{\mathrm{K}}(\mathrm{V}, \mathrm{V})$ be the automorphism of V that sends e_t to $a_{s,t} e_{st}$. Let us check that these p_s satisfy the above equation. We have

$$p_s s(p_t)(e_u) = a_{s,tu} s(a_{t,u}) e_{stu}$$
$$a_{s,t} p_{st}(e_u) = a_{s,t} a_{st,u} e_{stu}$$

and the equation follows from the fact that $a_{s,t}$ is a cocycle. □

Thus we have obtained the desired isomorphism $\mathrm{A(K}/k) \to \mathrm{H}^2(\mathrm{K}/k)$. When K'/k is a Galois extension containing K/k, it is trivial to check that the diagram

$$\begin{array}{ccc} \mathrm{A(K}/k) & \to & \mathrm{H}^2(\mathrm{K}/k) \\ \downarrow & & \downarrow \\ \mathrm{A(K}'/k) & \to & \mathrm{H}^2(\mathrm{K}'/k). \end{array}$$

is commutative. Hence we can pass to the limit over K and get *an isomorphism* $\delta: \mathrm{A}_k \to \mathrm{B}_k$, which proves the equivalence of the two definitions of the Brauer group.

Remark. The isomorphism $\mathrm{A(K}/k) \to \mathrm{H}^2(\mathrm{K}/k)$ defined above is the *opposite of the standard one*, defined by means of "crossed products" ([7], [19], [110], [123]); see exer. 2 below.

Exercises

1. Let K/k be a finite Galois extension with group G, and let A be a finite dimensional k-algebra. Let u_s be a 1-cocycle of G with values in $\mathrm{Aut_K(A} \otimes_k \mathrm{K})$. Let B be the set of all $x \in \mathrm{A} \otimes_k \mathrm{K}$ for which $x = u_s(s(x))$ for all $s \in$ G. Show that B is a sub-k-algebra of $\mathrm{A} \otimes_k \mathrm{K}$, that the inclusion $i: \mathrm{B} \to \mathrm{A} \otimes_k \mathrm{K}$ extends to a K-isomorphism $j: \mathrm{B} \otimes_k \mathrm{K} \to \mathrm{A} \otimes_k \mathrm{K}$, and that $u_s = j \circ s(j)^{-1}$. Deduce that the cohomology class associated to B by the method of prop. 4 is equal to that of u_s.

2. Apply exercise 1 to the case $\mathrm{A} = \mathbf{M}_n(k)$, u_s being the inner automorphism defined by an element $p_s \in \mathbf{GL}(n, \mathrm{K})$. Deduce that B is the subalgebra of $\mathbf{M}_n(\mathrm{K})$ of those x such that $x \circ p_s = p_s \circ s(x)$. Determine B explicitly when the p_s are as in the proof of lemma 1 (one finds that B is the "crossed product" defined by the factor system $st(a_{t^{-1},s^{-1}})$, which is cohomologous to $-a$).

3. Let D be a division algebra with center k and rank n^2 over k. Let x be the element of the Brauer group B_k corresponding to D, and let d be the order of x.
   a) Show that d divides n. (Apply exer. 2 of §4 to a maximal commutative subfield of D.)
   b) Show that every prime factor of n divides d. (Let p be a prime not dividing d; let K/k be a Galois extension, with group G, that splits x, and let G_p be a Sylow p-

subgroup of G. Show that the image of x in $H^2(G_p, K^*)$ under restriction is zero; if K' is the fixed field of G_p, deduce that K' splits x, and since $[K':k]$ is prime to p, conclude that p does not divide n.)

§6. A Geometric Interpretation of the Brauer Group: Severi-Brauer Varieties

Let k be a ground field, and let $\mathbf{P}_{n-1}$ be the projective space of dimension $n-1$, considered as a *scheme over k*. If V is any scheme over k, V is called a *Severi-Brauer variety* if there exists a separable algebraic extension K/k such that V is isomorphic to $\mathbf{P}_{n-1}$ over K, for some n (i.e., $V \otimes_k K$ and $\mathbf{P}_{n-1} \otimes_k K$ are isomorphic K-schemes). This notion is due to Châtelet [15], [16].

Examples. 1) A non-singular conic in the projective plane is a Severi-Brauer variety of dimension 1: it is K-isomorphic to $\mathbf{P}_1$ if and only if it has a rational point in K.

2) The positive divisors belonging to a divisor class rational over k form a Severi-Brauer variety. [We leave to the reader the task of making this statement precise!]

Let V be a Severi-Brauer variety of dimension $n-1$, and let K/k be a finite Galois extension, of group G, over which V "splits", i.e., for which there is an isomorphism

$$f: \mathbf{P}_{n-1} \otimes_k K \to V \otimes_k K.$$

For every $s \in G$, $p_s = f^{-1} \circ s(f)$ is an automorphism of $\mathbf{P}_{n-1} \otimes_k K$, i.e., belongs to $\mathbf{PGL}(n, K)$. Applying the theory of descent, we conclude that *the Severi-Brauer varieties over k, of dimension n − 1, that are split by* K, *correspond bijectively* (up to isomorphism) *with the elements of* $H^1(G, \mathbf{PGL}(n, K))$, *therefore also with the elements of* $A(n, K/k)$. This correspondence between central simple algebras and Severi-Brauer varieties can be made explicit as follows:

Let A be a central simple k-algebra of degree n^2. Let Gr(A) be the Grassmannian of all linear subvarieties of A of dimension n; Gr(A) has a natural structure of scheme over k. Let V be the closed subscheme of Gr(A) whose points are the right ideals of A of dimension n over k. It can be shown that V is a Severi-Brauer variety, and that under the above correspondence, it corresponds to A.

For more details, see Châtelet (*loc. cit.*) and Amitsur [1]. These results have been generalized by Grothendieck [83] to the framework of schemes (over an arbitrary base); for the case of affine schemes, see Azumaya [10], Auslander-Goldman [9], as well as the exercises in Bourbaki, *Alg. comm.*, Chap. II, §5.

EXERCISES

1. (Châtelet). Show that a Severi-Brauer variety is trivial (i.e., isomorphic to $\mathbf{P}_{n-1}$ over k) if and only if it has a rational point in k.

2. (Amitsur). Let V and V′ be Severi-Brauer varieties of the same dimension, and let a, a' be the corresponding elements of the Brauer group. Show that if V and V′ are birationally equivalent, then a and a' generate the same subgroup. [It is not known whether the converse is true.]

§7. Examples of Brauer Groups

The Brauer group is one of the principal invariants available for measuring the "degree of complexity" of a field k. We begin by giving examples of fields for which it is *zero*:

A field k is called *quasi-algebraically closed* (or C_1) if it satisfies the following condition:

If $f(X_1, \ldots, X_n)$ is a homogeneous polynomial of degree $d \neq 0$ in n variables, with coefficients in k, whose only zero in k^n is $(0, \ldots, 0)$, then $d \geq n$.

Proposition 10. *If the field k has property C_1, then the Brauer group of every finite extension K of k is zero.*

Indeed, let D be a division algebra with center K and of degree r^2 over K. Let $s = [K:k]$. If $x \in D$, let $\mathrm{Nrd}(x) \in K$ be its *reduced norm* (Bourbaki, *Alg.*, Chap. VIII, §12, n⁰ 3), and put

$$f(x) = N_{K/k}(\mathrm{Nrd}(x)).$$

The function f vanishes only for $x = 0$. On the other hand, if we take a basis $\{e_i\}$ for D over k, and write $x = \sum x_i e_i$, then f becomes a polynomial in the n variables $x_1, \ldots, x_n$, where $n = s.r^2$, and is homogeneous of degree sr (*loc. cit.*, prop. 11). Since k is C_1, $sr^2 \leq sr$, i.e., $r = 1$ and $D = K$; this shows that the Brauer group of K is zero. □

The vanishing of the Brauer group has interesting consequences.

Proposition 11. *For a given field k, the following are equivalent:*

1) *The Brauer group of every finite separable extension of k is zero.*
2) *If K is finite separable over k, and L/K is a finite Galois extension, then the G(L/K)-module L* is cohomologically trivial.*
3) *With the same hypotheses as in 2), the norm homomorphism $N_{L/K}: L^* \to K^*$ is surjective.*

Suppose 2) holds, then $\hat{H}^0(G(L/K), L^*) = 0$, whence 3) holds, and

$$H^2(G(L/K), L^*) = 0$$

whence 1) holds by passage to the limit over L. Conversely, suppose 1) (resp. 3)) holds. If H is any subgroup of G(L/K), then

$$\hat{H}^2(H, L^*) = 0 \quad (\text{resp. } \hat{H}^0(H, L^*) = 0);$$

on the other hand, as $H^1(H, L^*) = 0$ (prop. 2), we can apply th. 8 of Chap. IX and deduce that L^* is cohomologically trivial. □

Remark. It suffices to verify 3) when L/K is *cyclic of prime degree*: indeed, we pass from there to the solvable case by dévissage, and then to the general case by using the Sylow subgroups.

Corollary. *Let* L/K *be an extension satisfying conditions* 2). *If* E *is a torsion-free* G(L/K)*-module, then the* G(L/K)*-module* $L^* \otimes E$ *is cohomologically trivial.*

That follows from the corollary to th. 9 of Chap. IX. □

Application. Let A be a "torus" over k, with character group X. Let K/k be a Galois extension with group G, and assume K large enough so that every $\chi \in X$ is rational over K. The group A_K of points of A rational over K is then isomorphic to $K^* \otimes X'$, where X' is the dual of X. If k satisfies the conditions of prop. 11, then $\hat{H}^n(G, A_K) = 0$ for all n. In particular, every principal homogeneous space of A is trivial over K.

Examples of Fields with Zero Brauer Group

a) A *finite* field. Indeed, in this case, $H^1(G, K^*)$ and $H^2(G, K^*)$ have the same order (cf. Chap. VIII, prop. 8), and as the first group is zero (prop. 2), so is the second.

[One could just as well obtain this result by proving the stronger result that a finite field is C_1; cf. [82], [115].]

b) *The maximal unramified extension* K_{nr} of a field K complete under a discrete valuation and having *perfect* residue field. Indeed, condition 3) of prop. 11 holds (cf. Chap. V, §4, prop. 7). This applies in particular to the case where the residue field is algebraically closed.

[Again the fields K_{nr} are actually C_1; the proof (which is far from trivial) can be found in Lang's thesis [38].]

c) An extension *of transcendence degree* 1 of an algebraically closed field. In fact, such a field is C_1 ("Tsen's theorem", cf. Lang, *loc. cit.*).

d) An algebraic extension of **Q** containing all the roots of unity. This follows, by passage to the limit, from the determination of the Brauer group of a number field. [It has been conjectured by Artin that such a field is C_1.]

Examples of Fields with Non-zero Brauer Group

e) The field $\mathbf{R}$. Its Brauer group is equal to $H^2(G, \mathbf{C}^*)$, where $G = G(\mathbf{C}/\mathbf{R})$ is cyclic of order 2. Hence (Chap. VIII, §4)

$$B_{\mathbf{R}} = \mathbf{C}^{*G}/N\mathbf{C}^* = \mathbf{R}^*/\mathbf{R}^*_+ = \mathbf{Z}/2\mathbf{Z}.$$

The non-zero element of $B_{\mathbf{R}}$ corresponds to the *quaternions* $\mathbf{H}$. The corresponding Severi-Brauer variety is the conic

$$x^2 + y^2 + z^2 = 0.$$

f) A field *complete* under a discrete valuation with *finite residue field.* We will show in Chap. XIII that the Brauer group is isomorphic to $\mathbf{Q}/\mathbf{Z}$.

g) *An algebraic number field k.* Let k_i be the completions of k for the various topologies defined by absolute values on k. By f), B_{k_i} is equal to $\mathbf{Q}/\mathbf{Z}$ for ultrametric absolute values, is equal to the subgroup $\{0, \frac{1}{2}\}$ of $\mathbf{Q}/\mathbf{Z}$ if $k_i = \mathbf{R}$, and is zero if $k_i = \mathbf{C}$. As k embeds in each k_i, B_k gets mapped in the product of the B_{k_i}. One can show that in fact it *embeds into the direct sum* $\bigoplus B_{k_i}$, and that the following sequence is exact:

$$0 \to B_k \to \bigoplus B_{k_i} \xrightarrow{\sigma} \mathbf{Q}/\mathbf{Z} \to 0$$

σ being the map defined by $\sigma((x_i)) = \sum x_i$.

This is proved together with the theorems of class field theory (cf. [8], [19], [75], [123]).

h) *An algebraic function field of one variable over a finite base field.* The result is the same as for a number field.

CHAPTER XI

Class Formations

The notion of class formation was introduced by Artin-Tate [8] after earlier work by Weil [67] and by Hochschild-Nakayama [36]. This notion clarifies the cohomological aspect of class field theory, both in the local and the global cases. The present chapter is restricted to the main properties of class formations; the reader will find in Artin-Tate [8] further developments (the Šafarevič theorem, Weil groups).

§1. The Notion of Formation

We begin with a group G and a family of subgroups $\{G_E\}_{E \in X}$, of *finite index* in G; we assume the indexing to be such that $G_E = G_{E'}$ implies $E = E'$ (so that one could identify X with a set of subgroups of G, but that would not be convenient for applications). We assume further that

a) For every finite family F_i of elements of X, there exists $F \in X$ such that $G_F = \bigcap G_{F_i}$.
b) Every subgroup G' of G that contains a subgroup G_F, $F \in X$, has the form $G' = G_{F'}$ for some $F' \in X$.
c) If $s \in G$ and $F \in X$, the subgroup $s . G_F . s^{-1}$ has the form $G_{F''}$, with $F'' \in X$.

Examples. 1. Let E be a field, Ω a Galois extension of E, and let X be the set of subfields of Ω that are finite extensions of E. Take G to be the *Galois topological group* $G(\Omega/E)$, and for G_F take the subgroup $G(\Omega/F)$. Then properties a), b), c) are a consequence of Galois theory.

[In practice, Ω is usually taken to be the *separable closure* of E.]

2. More generally, take G to be any topological group, and $\{G_E\}$ to be the family of all closed subgroups of finite index.

Remark. In what follows, the group G will play a role only through its *finite quotients* G/G_E, where G_E is normal in G and $E \in X$. Thus we could replace G with its completion $\hat{G}$, equal to the *projective limit* of the G/G_E, which is a compact totally disconnected group.

For G and $\{G_E\}_{E \in X}$ satisfying the above conditions, we will transfer to X the terminology of Galois theory. The elements of X will be called "fields". We write $E \subset F$ if $G_E \supset G_F$, we say "F/E is a Galois extension" if G_F is a normal subgroup of G_E, and we call the quotient group G_E/G_F (also denoted $G_{F/E}$) "the Galois group G(F/E)". In the situation of condition a), we call F "the composite" of the F_i, and write $F = \prod F_i$; in the situation of c), we write $F'' = s(F)$ or simply sF. It makes sense to speak of "abelian extensions, solvable extensions", etc. Note that every extension F/E is contained in a Galois extension: namely, since G_F has finite index in G_E, it only has finitely many conjugates $s.G_F.s^{-1}$, and their intersection has the form $G_{F'}$, with $F' \in X$; then F'/E is Galois and contains F.

Next we suppose given a G-module A satisfying the following condition:

(*) For every $a \in A$, the stabilizer of a (equal to the group of all $s \in G$ such that $s.a = a$) is one of the subgroups G_E, $E \in X$.

[This condition means that A can be considered as a *topological G-module*, in the sense of Chap. X, §3.]

A *formation* is then a system $(G, \{G_E\}_{E \in X}, A)$ satisfying the above conditions. Then for $E \in X$, we denote by A_E the subgroup of A consisting of all $a \in A$ fixed by G_E; thus $A_E = H^0(G_E, A)$, and axiom (*) means that A is the union of the A_E, as E runs through X. If F/E is Galois, the group G(F/E) acts on A_F, and

$$H^0(G(F/E), A_F) = A_E.$$

Cohomology in a Formation

We return to the *Galois* case F/E. As the group G(F/E) acts on A_F, the *cohomology* groups $H^q(G(F/E), A_F)$ are defined; they will be denoted simply $H^q(F/E)$. Similarly, $H_q(F/E)$ denotes the *homology*, and $\hat{H}^n(F/E)$ the *Tate cohomology*, $n \in \mathbf{Z}$. There are the following operations relating these groups:

i) Suppose F/E and F′/E′ are Galois, with $E' \supset E$ and $F' \supset F$. Then there is a canonical homomorphism $G_{F'/E'} \to G_{F/E}$ compatible with the inclusion $A_F \to A_{F'}$ (cf. Chap. VII, §5). It induces canonical homomorphisms

$$H^q(F/E) \to H^q(F'/E'), \qquad q \geq 0.$$

ii) Suppose $F' \supset F \supset E$, with F′/E and F/E both Galois (the special case $E = E'$ of i)). Then the canonical homomorphism of i) is called *inflation*, denoted

$$\mathrm{Inf}\colon \mathrm{H}^q(\mathrm{F/E}) \to \mathrm{H}^q(\mathrm{F'/E}), \qquad q \geq 0.$$

iii) Suppose $\mathrm{F} \supset \mathrm{E'} \supset \mathrm{E}$, with F/E Galois, hence also F/E′ (the special case $\mathrm{F} = \mathrm{F'}$ of i)). The group $\mathrm{G}_{\mathrm{F/E'}}$ is a subgroup of $\mathrm{G}_{\mathrm{F/E}}$, and $\mathrm{A}_{\mathrm{F}} = \mathrm{A}_{\mathrm{F'}}$. The canonical homomorphisms $\mathrm{H}^q(\mathrm{F/E}) \to \mathrm{H}^q(\mathrm{F/E'})$ then extend to the Tate cohomology groups, and take the name *restriction* (cf. Chap. VIII, §2), denoted

$$\mathrm{Res}\colon \hat{\mathrm{H}}^n(\mathrm{F/E}) \to \hat{\mathrm{H}}^n(\mathrm{F/E'}), \qquad n \in \mathbf{Z}.$$

Alongside these homomorphisms are the ones in the opposite direction, called *corestriction* (*loc. cit.*), denoted

$$\mathrm{Cor}\colon \hat{\mathrm{H}}^n(\mathrm{F/E'}) \to \hat{\mathrm{H}}^n(\mathrm{F/E}), \qquad n \in \mathbf{Z}.$$

iv) Suppose F/E is Galois, and let $s \in \mathrm{G}$. The map $t \mapsto s^{-1}ts$ is an isomorphism $\mathrm{G}(s\mathrm{F}/s\mathrm{E}) \to \mathrm{G}(\mathrm{F/E})$. The map $a \mapsto s.a$ is an isomorphism of A_{F} with $\mathrm{A}_{s\mathrm{F}}$. These two maps are compatible, hence they define (by transport of structure) isomorphisms

$$s^*\colon \hat{\mathrm{H}}^n(\mathrm{F/E}) \to \hat{\mathrm{H}}^n(s\mathrm{F}/s\mathrm{E}), \qquad n \in \mathbf{Z}.$$

When s "is the identity on E", i.e., $s \in \mathrm{G}_{\mathrm{E}}$, we have $s\mathrm{E} = \mathrm{E}$, $s\mathrm{F} = \mathrm{F}$, and s^* is *the identity map* of $\hat{\mathrm{H}}^n(\mathrm{F/E})$ (as can be seen by applying prop. 3 of Chap. VII, §5).

§2. Class Formations

A *class formation* consists of a *formation* $(\mathrm{G}, \{\mathrm{G}_{\mathrm{E}}\}_{\mathrm{E} \in \mathrm{X}}, \mathrm{A})$, together with a homomorphism

$$\mathrm{inv}_{\mathrm{E}}\colon \mathrm{H}^2(\mathrm{F/E}) \to \mathbf{Q}/\mathbf{Z}$$

for each Galois extension F/E, satisfying axioms I and II below:

Axiom I. *For every Galois extension* F/E, $\mathrm{H}^1(\mathrm{F/E}) = 0$.

This axiom is equivalent to the apparently weaker axiom:

Axiom I′. $\mathrm{H}^1(\mathrm{F/E}) = 0$ *for every cyclic extension of prime degree.*

Namely, if axiom I′ holds, let F/E be a Galois extension of prime power degree p^n. In view of the elementary properties of p-groups (Chap. IX, §1), there exists E′, with $\mathrm{F} \supset \mathrm{E'} \supset \mathrm{E}$, such that E′/E is cyclic of degree p. By prop. 4 of Chap. VII, §6, there is an exact sequence

$$0 \to \mathrm{H}^1(\mathrm{E'/E}) \xrightarrow{\mathrm{Inf}} \mathrm{H}^1(\mathrm{F/E}) \xrightarrow{\mathrm{Res}} \mathrm{H}^1(\mathrm{F/E'}).$$

Arguing by induction on n, we may assume $\mathrm{H}^1(\mathrm{F/E'}) = 0$. Since axiom I′ gives $\mathrm{H}^1(\mathrm{E'/E}) = 0$, the exact sequence yields $\mathrm{H}^1(\mathrm{F/E}) = 0$.

Now if F/E is an arbitrary Galois extension, and the G_p are the Sylow subgroups of $G_{F/E}$, we have just seen that $H^1(G_p, A_F) = 0$ for all p; then the corollary to theorem 4 of Chap. IX, §2 tells us that $H^1(F/E) = 0$. □

Before stating axiom II, we introduce the notation $H^q(\ /E)$ for the direct limit of the $H^q(F/E)$, as F runs through the set of Galois extensions of E; if $F' \supset F \supset E$, the homomorphisms

$$H^q(F/E) \to H^q(F'/E)$$

in this directed system are the *inflation* homomorphisms (cf. §1). For $q = 2$ these homomorphisms are *injective*: that follows from axiom I and prop. 5 of Chap. VII, §6. The group $H^2(\ /E)$ is therefore the *union* of the $H^2(F/E)$, so the situation is entirely analogous to that of the Brauer group.

The final data for a class formation is a *homomorphism* $\mathrm{inv}_E : H^2(\ /E) \to \mathbf{Q}/\mathbf{Z}$ satisfying the axiom:

Axiom II

a) *The homomorphism* inv_E *is injective, and maps* $H^2(F/E)$ *onto the unique subgroup of* $\mathbf{Q}/\mathbf{Z}$ *of order equal to* $[F:E]$.

b) *For any extension* E′/E, *we have*

$$(**) \qquad \mathrm{inv}_{E'} \circ \mathrm{Res}_{E/E'} = [E':E] \, . \, \mathrm{inv}_E.$$

Proposition 1

i) *For every extension* E′/E, *the homomorphism*

$$\mathrm{Res}_{E/E'} : H^2(\ /E) \to H^2(\ /E')$$

is surjective.

ii) *For every extension* E′/E, *the homomorphism*

$$\mathrm{Cor}_{E'/E} : H^2(\ /E') \to H^2(\ /E)$$

is injective, and $\mathrm{inv}_E \circ \mathrm{Cor}_{E'/E} = \mathrm{inv}_{E'}$.

iii) *For every* $s \in G$, *if* s^* *denotes the isomorphism* $H^2(\ /E) \to H^2(\ /sE)$ *defined by passing to the limit the homomorphisms of* §1, *then*

$$\mathrm{inv}_{sE} \circ s^* = \mathrm{inv}_E.$$

To prove i), it suffices to show that if F is a Galois extension of E containing E′, the restriction homomorphism maps $H^2(F/E)$ onto $H^2(F/E')$; let $n = [F:E]$ and $n' = [F:E']$; if $x' \in H^2(F/E')$, the element $y' = \mathrm{inv}_{E'}(x')$ of $\mathbf{Q}/\mathbf{Z}$ satisfies $n' . y' = 0$; since $\mathbf{Q}/\mathbf{Z}$ is divisible, there exists $y \in \mathbf{Q}/\mathbf{Z}$ such that

$$[E':E] \, . \, y = y'$$

and we have $n . y = 0$; thus there exists $x \in H^2(F/E)$ such that $\mathrm{inv}_E(x) = y$, and formula (**) shows that $\mathrm{Res}(x)$ and x' have the same invariant. As $\mathrm{inv}_{E'}$ is injective, we must have $x' = \mathrm{Res}(x)$, which proves i).

Since $\mathrm{Res}_{E/E'}$ is surjective, the formula $\mathrm{inv}_E \circ \mathrm{Cor}_{E'/E} = \mathrm{inv}_{E'}$ is equivalent to the formula

$$\mathrm{inv}_E \circ \mathrm{Cor}_{E'/E} \circ \mathrm{Res}_{E/E'} = \mathrm{inv}_{E'} \circ \mathrm{Res}_{E/E'}.$$

Now the left side of the latter formula is equal to $[E':E].\mathrm{inv}_E$ (by prop. 6 of Chap. VII, §7), and so is the right side, according to (**). Since $\mathrm{inv}_{E'}$ is injective, we see also that $\mathrm{Cor}_{E'/E}$ is injective, and ii) is proved.

To prove iii), denote by E_0 the "base field", corresponding to the subgroup G itself. Since $s \in G_{E_0}$, the homomorphism

$$s^*: H^2(\ \ /E_0) \to H^2(\ \ /sE_0) = H^2(\ \ /E_0)$$

is equal to the identity (§1), so we do have $\mathrm{inv}_{sE_0} \circ s^* = \mathrm{inv}_{E_0}$. Now if E is an arbitrary field, it follows from i) that every $x \in H^2(\ \ /E)$ has the form $\mathrm{Res}_{E_0/E}(x_0)$, with $x_0 \in H^2(\ \ /E_0)$. Since Res and s^* commute (transport of structure!), we get

$$\begin{aligned}\mathrm{inv}_{sE}(s^*x) &= \mathrm{inv}_{sE} \circ \mathrm{Res}_{sE_0/sE}(s^*x_0) = [sE:sE_0].\mathrm{inv}_{sE_0}(s^*x_0)\\ &= [E:E_0].\mathrm{inv}_{E_0}(x_0) = \mathrm{inv}_E(x). \quad \square\end{aligned}$$

§3. Fundamental Classes and Reciprocity Isomorphism

For the rest of this chapter, $(G, \{G_E\}_{E \in X}, A, \mathrm{inv}_E)$ will denote a class formation.

If F/E is any Galois extension, and $n = [F:E]$ is its degree, then there exists a unique element $u_{F/E} \in H^2(F/E)$ such that $\mathrm{inv}_E(u_{F/E}) = 1/n$. Moreover, by axiom II, this element *generates* the group $H^2(F/E)$, which is cyclic of order n. The properties of the operations Inf, Res, Cor, s^* with respect to the invariants translate into the following formulas for the $u_{F/E}$:

$$\mathrm{Inf}(u_{F/E}) = [F':F].u_{F'/E} \quad \text{if } F' \supset F \supset E, \text{ with } F'/E \text{ and } F/E \text{ Galois.}$$

$$\mathrm{Res}(u_{F/E}) = u_{F/E'} \quad \text{if } F \supset E' \supset E, \text{ with } F/E \text{ Galois.}$$

$$\mathrm{Cor}(u_{F/E'}) = [E':E].u_{F/E} \quad \text{if } F \supset E' \supset E, \text{ with } F/E \text{ Galois.}$$

$$u_{sF/sE} = s^*(u_{F/E}) \quad \text{if } F/E \text{ is Galois and } s \in G.$$

[In fact, we could have used the $u_{F/E}$ to begin with, instead of inv_E; we would have had to assume that $H^2(F/E)$ is cyclic of order $[F:E]$, with $u_{F/E}$ as generator, and that the above formulas hold.]

The classes $u_{F/E}$ satisfy the hypotheses of the Tate-Nakayama theorem (th. 14, Chap. IX, §8). We re-state only Tate's: Theorem:

Theorem 1. *For every Galois extension* F/E *and every integer* n, *the homomorphism*

$$\theta^n(F/E): \hat{H}^n(G_{F/E}, Z) \to \hat{H}^{n+2}(G_{F/E}, A_F) = \hat{H}^{n+2}(F/E)$$

defined by $x \to u_{F/E}.x$ *(cup product) is bijective.*

If $F \supset E' \supset E$, with F/E Galois, the homomorphisms θ^n *commute with restriction and corestriction*: that follows from the above formulas for the $u_{F/E}$ and known formulas for the cup product. For example, let us verify this for restriction. We must show that

$$\mathrm{Res}_{E/E'}(u_{F/E} \,.\, x) = u_{F/E'} \,.\, \mathrm{Res}_{E/E'}(x)$$

for all $x \in H^n(G_{F/E}, \mathbf{Z})$.

According to Cartan-Eilenberg [13], Chap. XII, p. 256, the left side is equal to $\mathrm{Res}_{E/E'}(u_{F/E}) \,.\, \mathrm{Res}_{E/E'}(x)$, and the result follows from the second formula for $u_{F/E}$.

If $F' \supset F \supset E$, with F'/E and F/E Galois, the homomorphisms θ^n do not commute with *inflation*, but we have the equation

$$\mathrm{Inf} \circ \theta^n(F/E) = [F' : F] \cdot \theta^n(F'/E) \circ \mathrm{Inf}$$

which can be checked as above.

Special cases. We have $H^1(G_{F/E}, \mathbf{Z}) = \mathrm{Hom}(G_{F/E}, \mathbf{Z}) = 0$, whence $H^3(F/E) = 0$. On the other hand, the exact sequence

$$0 \to \mathbf{Z} \to \mathbf{Q} \to \mathbf{Q}/\mathbf{Z} \to 0$$

shows that $H^2(G_{F/E}, \mathbf{Z})$ can be identified with the group $\mathrm{Hom}(G_{F/E}, \mathbf{Q}/\mathbf{Z})$ of *characters of degree* 1 of $G_{F/E}$; hence $H^4(F/E) = \mathrm{Hom}(G_{F/E}, \mathbf{Q}/\mathbf{Z})$.

The most important special case, of course, is $n = -2$. Indeed, we know that $\hat{H}^{-2}(G_{F/E}, \mathbf{Z})$ can be identified with the group $G_{F/E}$ made abelian, which group will be denoted $G^a_{F/E} = G_{F/E}/G'_{F/E}$ (an explicit isomorphism was given in Chap. VII, §4). On the other hand,

$$\hat{H}^0(F/E) = A_E/N_{F/E}A_F,$$

where $N_{F/E}$ denotes the norm operator on the $G_{F/E}$-module A_F. We see thus that θ^{-2} defines an isomorphism.

$$\theta : G^a_{F/E} \to A_E/N_{F/E}A_F.$$

If u is a 2-cocycle representing the class $u_{F/E}$, a cup product computation (Appendix, lemma 4) shows that

$$\theta(s) \equiv \sum_{t \in G} u(t,s) \ \text{mod.}\ N_{F/E}A_F, \qquad s \in G.$$

The isomorphism inverse to θ is called the *reciprocity isomorphism*, and is denoted

$$a \mapsto (a, F/E), \qquad a \in A_E,\ (a, F/E) \in G^a_{F/E}.$$

The next proposition gives a convenient characterisation of the element $(a, F/E)$.

Proposition 2. *Let $\chi \in \mathrm{Hom}(G_{F/E}, \mathbf{Q}/\mathbf{Z})$ be a character of degree* 1 *of the group $G_{F/E}$ (or, what amounts to the same thing, of the group $G^a_{F/E}$). For any $s \in G^a_{F/E}$, put*

$$\langle \chi, s \rangle = \chi(s) \in \mathbf{Q}/\mathbf{Z}.$$

On the other hand, let $d\chi \in \mathrm{H}^2(\mathrm{G}_{\mathrm{F/E}}, \mathbf{Z})$ *be the image of* χ *by the coboundary associated to the exact sequence*

$$0 \to \mathbf{Z} \to \mathbf{Q} \to \mathbf{Q}/\mathbf{Z} \to 0.$$

Finally, if $a \in \mathrm{A}_\mathrm{E}$, *denote by* $\bar{a}$ *its image in* $\hat{\mathrm{H}}^0(\mathrm{F/E})$. *Then we have the formula*

$$(***) \qquad \langle \chi, (a, \mathrm{F/E}) \rangle = \mathrm{inv}_\mathrm{E}(\bar{a} \,.\, d\chi)$$

where $\bar{a} \,.\, d\chi \in \mathrm{H}^2(\mathrm{F/E})$ *denotes the cup product of the cohomology classes* $\bar{a}$ *and* $d\chi$.

The formula (***) actually characterises $(a, \mathrm{F/E})$ in the abelian group $\mathrm{G}^a_{\mathrm{F/E}}$, since it determines its scalar product with every element of the dual group.

To simplify the notation, we set $s_a = (a, \mathrm{F/E})$; if $s \in \mathrm{G}_{\mathrm{F/E}}$, we denote by $\bar{s}$ the canonical image of s in $\hat{\mathrm{H}}^{-2}(\mathrm{G}_{\mathrm{F/E}}, \mathbf{Z}) = \mathrm{G}^a_{\mathrm{F/E}}$. By the very definition of θ, we have

$$u_{\mathrm{F/E}} \,.\, \bar{s}_a = \bar{a} \quad \text{in } \hat{\mathrm{H}}^0(\mathrm{G}_{\mathrm{F/E}}, \mathrm{A}_\mathrm{F}).$$

Using the associativity of the cup product, this gives

$$\bar{a} \,.\, d\chi = u_{\mathrm{F/E}} \,.\, (\bar{s}_a \,.\, d\chi), \quad \text{with } \bar{s}_a \,.\, d\chi \in \hat{\mathrm{H}}^0(\mathrm{G}_{\mathrm{F/E}}, \mathbf{Z}).$$

Since d commutes with cup products, we have

$$\bar{s}_a \,.\, d\chi = d(\bar{s}_a \,.\, \chi), \quad \text{with } \bar{s}_a \,.\, \chi \in \hat{\mathrm{H}}^{-1}(\mathrm{G}_{\mathrm{F/E}}, \mathbf{Q}/\mathbf{Z}).$$

If $n = [\mathrm{F}:\mathrm{E}]$, we can identify $\hat{\mathrm{H}}^{-1}(\mathrm{G}_{\mathrm{F/E}}, \mathbf{Q}/\mathbf{Z})$ with the subgroup $(1/n)\mathbf{Z}/\mathbf{Z}$ of elements of order dividing n, and then the cup product $\bar{s}_a \,.\, \chi$ becomes identified with $\langle \chi, s_a \rangle$, (cf. Appendix, lemma 3). Writing $\langle \chi, s_a \rangle = r/n$, with $r \in \mathbf{Z}$, we must compute

$$d(r/n) \in \hat{\mathrm{H}}^0(\mathrm{G}_{\mathrm{F/E}}, \mathbf{Z})$$

and the straightforward computation gives $d(r/n) = r$. We deduce

$$u_{\mathrm{F/E}} \,.\, (\bar{s}_a \,.\, d\chi) = r \,.\, u_{\mathrm{F/E}},$$

and the invariant of this cohomology class is therefore r/n, i.e., precisely $\langle \chi, s_a \rangle$. □

Functorial Properties of the Reciprocity Map

They are summarized by the four commutative diagrams below.

$$(1) \quad \begin{array}{ccc} \mathrm{A}_{\mathrm{E}'} & \xrightarrow{\text{norm}} & \mathrm{A}_\mathrm{E} \\ \downarrow & & \downarrow \\ \mathrm{G}^a_{\mathrm{F/E}'} & \xrightarrow{\text{can.}} & \mathrm{G}^a_{\mathrm{F/E}} \end{array} \qquad (2) \quad \begin{array}{ccc} \mathrm{A}_\mathrm{E} & \xrightarrow{\text{incl.}} & \mathrm{A}_{\mathrm{E}'} \\ \downarrow & & \downarrow \\ \mathrm{G}^a_{\mathrm{F/E}} & \xrightarrow{\text{Ver}} & \mathrm{G}^a_{\mathrm{F/E}'} \end{array}$$

$$(3) \quad \begin{array}{ccc} \mathrm{A}_\mathrm{E} & \xrightarrow{s^*} & \mathrm{A}_{s\mathrm{E}} \\ \downarrow & & \downarrow \\ \mathrm{G}^a_{\mathrm{F/E}} & \xrightarrow{s^*} & \mathrm{G}^a_{s\mathrm{F}/s\mathrm{E}} \end{array} \qquad (4) \quad \begin{array}{ccc} \mathrm{A}_\mathrm{E} & = & \mathrm{A}_\mathrm{E} \\ \downarrow & & \downarrow \\ \mathrm{G}^a_{\mathrm{F'/E}} & \xrightarrow{\text{proj.}} & \mathrm{G}^a_{\mathrm{F/E}}. \end{array}$$

In diagram (1), we have $F \supset E' \supset E$, with F/E Galois, and $G_{F/E'}$ is a subgroup of $G_{F/E}$; this defines a canonical homomorphism of $G^a_{F/E'}$ into $G^a_{F/E}$. The two horizontal arrows in this diagram are the homomorphisms Cor, for $n = 0$ and $n = -2$.

The situation is the same in diagram (2), where Ver denotes the *transfer* (Chap. VII, §7). The two horizontal arrows are none other than the homomorphisms Res, for $n = 0$ and $n = -2$.

In diagram (3), $s \in G$, and the homomorphism $s^*: A_E \to A_{sE}$ sends a to $s.a$; the analogous homomorphism on $G^a_{F/E}$ sends t to sts^{-1}. Commutativity is a consequence of transport of structure, taking into account that

$$s^*(u_{F/E}) = u_{sF/sE}.$$

In diagram (4), we have $F' \supset F \supset E$, with F/E and F'/E Galois. Here, commutativity does not follow from that of inflation but can be seen directly from the characterisation of $(a, F/E)$ given in prop. 2.

The symbol $(a, */E)$

The commutativity of diagram (4) allows us to pass to the limit over increasing extensions F/E, where E remains fixed. More precisely, define $\mathfrak{A}(E)$ to be the *projective limit* of the $G^a_{F/E}$ as F runs through the directed set of Galois extensions of E. In the case of the actual Galois theory, $\mathfrak{A}(E)$ is the Galois topological group of the maximal abelian extension of the field E. If $a \in A_E$, the commutativity of diagram (4) shows that the $(a, F/E)$ are coherent, and therefore define an element of $\mathfrak{A}(E)$ to be denoted $(a, */E)$. In this way we obtain a canonical homomorphism

$$A_E \to \mathfrak{A}(E)$$

If E'/E is any extension, passage to the limit in diagrams (1) and (2) yields commutative diagrams

$$(1)_* \quad \begin{array}{ccc} A_{E'} & \xrightarrow{\text{norm}} & A_E \\ \downarrow & & \downarrow \\ \mathfrak{A}(E') & \xrightarrow{\text{can.}} & \mathfrak{A}(E) \end{array} \qquad\qquad (2)_* \quad \begin{array}{ccc} A_E & \xrightarrow{\text{incl.}} & A_{E'} \\ \downarrow & & \downarrow \\ \mathfrak{A}(E) & \xrightarrow{\text{Ver}} & \mathfrak{A}(E') \end{array}$$

Diagram (3) gives the formula

$$(s.a, */sE) = s.(a, */E).s^{-1}.$$

EXERCISES

Suppose that for each integer n, there exists an extension F/E whose degree is a multiple of n. Show that $H^q(\ /E) = 0$ for all $q \geq 3$.

§4. Abelian Extensions and Norm Groups

Let E'/E be an arbitrary extension, and let s_i be representatives of the left cosets of G_E mod. $G_{E'}$. For each $a' \in A_{E'}$, put

$$N_{E'/E}(a') = \sum s_i . a'.$$

One checks that $N_{E'/E}(a')$ does not depend on the choice of the s_i and is an element of A_E. This homomorphism

$$N_{E'/E}: A_{E'} \to A_E$$

is called the *norm*. In the actual Galois case, we get the usual norm.

Definition. *A subgroup* I *of* A_E *is called a norm group if there exists an extension* E'/E *such that* $I = N_{E'/E}(A_{E'})$.

When E'/E is an *abelian* extension, the reciprocity isomorphism

$$A_E/N_{E'/E}A_{E'} \to G_{E'/E}$$

shows that $N_{E'/E}(A_{E'})$ is a subgroup *of finite index* in A_E, this index being the degree $[E':E]$. The next proposition reduces the general case to the abelian case.

Proposition 3. *Let* E'/E *be any extension, and let* E'' *be the largest abelian extension of* E *contained in* E'. *Then*

$$N_{E'/E}(A_{E'}) = N_{E''/E}(A_{E''}).$$

Let F be a Galois extension of E containing E', and let $G = G_{F/E}$ be its Galois group. Let $H = G_{F/E'}$. Then the Galois group of F/E'' is G'.H.

Let $a \in N_{E''/E}(A_{E''})$. We have $(a, E''/E) = 1$ in G/(G'.H), which means that the element $(a, F/E)$ of G/G' lies in the image of the homomorphism $H/H' \to G/G'$. The commutativity of diagram (1) of §3 and the fact that $A_{E'} \to H/H'$ is surjective show that there exists $a' \in A_{E'}$ such that

$$(N_{E'/E}a', F/E) = (a, F/E).$$

It follows that there exists $a'' \in A_F$ with $N_{F/E}a'' = N_{E'/E}a' - a$, whence

$$a = N_{E'/E}(a' - N_{F/E'}a'')$$

which shows that $N_{E''/E}(A_{E''}) \subset N_{E'/E}(A_{E'})$. The opposite inclusion results from the transitivity of the norms. ☐

Corollary. *The norm group* $N_{E'/E}A_{E'}$ *has finite index in* A_E; *this index divides* $[E':E]$, *and is equal to it if and only if* E' *is an abelian extension of* E.

Indeed, $(A_E : N_{E''/E}A_{E''}) = [E'':E]$, which divides $[E':E]$, and equality holds if and only if $E' = E''$.

Let F/E be an abelian extension; to simplify the notation, put

$$I_F = N_{F/E}A_F.$$

Proposition 4. *The map* $F \to I_F$ *is a bijection of the set of abelian extensions of* E *onto the set of norm groups of* A_E; *this correspondence reverses inclusion* *and satisfies*

$$I_{F.F'} = I_F \cap I_{F'}, \qquad I_{F \cap F'} = I_F + I_{F'}.$$

Furthermore, every subgroup of A_E *which contains a norm group is itself a norm group.*

If F and F′ are abelian extensions, so is F . F′, and $I_{F.F'} \subset I_F \cap I_{F'}$; conversely, if $a \in I_F \cap I_{F'}$, the element $(a, F.F'/E)$ of $G_{F.F'/E}$ has trivial image in $G_{F/E}$ and $G_{F'/E}$, hence is trivial, which shows that $a \in I_{F.F'}$. In particular, if $I_F \supset I_{F'}$, we see that $I_{F.F'} = I_F$, whence $[F.F':E] = [F:E]$ and $F' \subset F$. It follows that the correspondence $F \to I_F$ is bijective and reverses inclusion. The other assertions are immediate. □

One consequence of this proposition is that the norm groups define a *topology* on A_E. Denoting by $\hat{A}_E$ the Hausdorff completion of A_E in this topology, we have (by definition)

$$\hat{A}_E = \lim A_E/I \quad \text{(I running through the norm groups)}$$

and the symbol $(a, */E)$ defines an *isomorphism of* $\hat{A}_E$ *with the topological group* $\mathfrak{A}(E)$ of §3.

The topology just defined on A_E is sometimes called the "norm" topology, to distinguish it from the one which occurs in the existence theorem.

§5. The Existence Theorem

It is a characterisation of the norm groups of A_E. Toward that end, we assume given a topology on each A_E, compatible with its group structure, and such that if $E \subset F$, the topology on A_E is *induced* by that on A_F; we also assume that for every $s \in G$, the map $a \mapsto s.a$ is a *continuous* map of A_E into A_{sE}. These assumptions imply that the norm map $N_{F/E}: A_F \to A_E$ is *continuous.*

We are going to impose three axioms below on these topologies, and we will show that the norm groups are precisely the closed subgroups of finite index in A_E (th. 2).

Axiom III-1. *For every extension* F/E, *the norm homomorphism*

$$N_{F/E}: A_F \to A_E$$

has closed image and compact kernel.

[When A_E and A_F are locally compact and countable at infinity—which is the case in all the applications—axiom III-1 means that the norm map is *proper.*]

It follows that $N_{F/E}A_F$, being *closed of finite index* in A_E, is an *open* subgroup of A_E (its complement is a finite union of closed sets). In particular, the given topology on A_E is *finer* than the norm topology.

Definition. *The universal norm group of* E *(denoted* D_E*) is the intersection of all the norm groups of* A_E.

If $a \in A_E$, then $a \in D_E$ if and only if $(a, F/E) = 1$ for every abelian extension F/E; the group D_E is therefore the kernel of the reciprocity *map* $A_E \to \mathfrak{A}(E)$.

Proposition 5. *For every extension* F/E, *we have* $N_{F/E}D_F = D_E$.

The inclusion $N_{F/E}D_F \subset D_E$ follows from the transitivity of the norms. Conversely, let $a \in D_E$, and let F′ be an extension of F; denote by K(F′) the set of $b \in A_F$ whose norm in E is equal to a and which are norms of elements of F′. Then

$$K(F') = N_{F'/F}A_{F'} \cap N_{F/E}^{-1}(a)$$

which shows that K(F′) is compact; also, since $a \in D_E$, there exists $b' \in A_F$ such that $Nb' = a$, and the image of b' in A_F belongs to K(F′), which is therefore non-empty. As F′ varies, the K(F′) form a directed decreasing family of non-empty compact subspaces of K(F); their intersection is therefore non-empty, and if $b \in \bigcap K(F')$, it is clear that $b \in D_F$ and $N_{F/E}(b) = a$. □

Axiom III-2. *For every prime number p, there exists a field* E_p *such that, for* $E \supset E_p$, *the map* $\varphi_p : x \mapsto px$ *of* A_E *into itself satisfies the condition*:

i) The kernel of φ_p *is compact, and its image contains* D_E.

[This is the only axiom whose verification is difficult in practice.]

Proposition 6. *For every field* E, *the group* D_E *is divisible and equal to* $\bigcap n \,.\, A_E$.

First we show that for every E and every prime p, $D_E = p \,.\, D_E$. Let $a \in D_E$ and let F be an extension of E containing E_p (i.e., "sufficiently large"); denote by L(F) the set of $b \in A_E$ such that $pb = a$ and $b \in N_{F/E}A_F$; it is a compact subset of A_E. We claim L(F) is non-empty: indeed, by prop. 5, $a = N_{F/E}x$, with $x \in D_F$, whence $x = py$, with $y \in A_F$ (axiom III-2), and we can take $b = N_{F/E}\,y$. As before, the intersection of the L(F) must be non-empty, and if b belongs to it, we have $pb = a$ and $b \in D_E$. It follows that D_E is divisible, so that $D_E \subset \bigcap n \,.\, A_E$. Conversely, if $a \in \bigcap n \,.\, A_E$, and if F/E has degree n, then $a = n \,.\, b$, $b \in A_E$, whence $a = N_{F/E}b$: thus $a \in D_E$. □

Axiom III-3. *There exists a compact subgroup* U_E *of* A_E *such that every closed subgroup of finite index in* A_E *that contains* U_E *is a norm group.*

Theorem 2. *Suppose axioms* III-1, III-2, III-3 *hold. For a subgroup of* A_E *to be a norm group, it is necessary and sufficient that it be closed and of finite index.*

We know that these conditions are necessary. Let I be a subgroup satisfying them; if $(A_E : I) = n$, then $n . A_E \subset I$, whence $D_E \subset I$ by prop. 6. If N runs through the family of norm groups, then

$$\bigcap (N \cap U_E) = D_E \cap U_E \subset I.$$

As the $N \cap U_E$ are compact and I is open, there exists an N such that $N \cap U_E \subset I$. We then get the inclusion

$$N \cap (U_E + (N \cap I)) \subset I.$$

Indeed, if a belongs to this intersection, we can write $a = a' + a''$, with $a' \in U_E$ and $a'' \in N \cap I$; we have $a' = a - a'' \in N$, whence $a' \in U_E \cap N$ and $a' \in I$, which shows that $a \in I$.

The group $N \cap I$ is closed and of finite index (being the intersection of two such subgroups); hence $U_E + (N \cap I)$, which contains it, is closed, of finite index, and contains U_E; by III-3, it is a norm group. Then the same holds for $N \cap (U_E + (N \cap I))$ (prop. 4), hence for I (prop. 4). □

Example. In the case of local class field theory, we take $A_E = E^*$, and we provide E^* with the topology induced from E, which makes it into a locally compact group. Axiom III-1 is trivial. In axiom III-3, take U_E to be the group of *units*; the exact sequence

$$0 \to U_E \to E^* \to \mathbf{Z} \to 0$$

shows that the subgroups of finite index of E^* containing U_E are the inverse images of the subgroups $n\mathbf{Z}$ of $\mathbf{Z}$; they are norm groups for the *unramified* extensions of E. Furthermore, it can be shown that $D_E = 0$. The structure of the topological group $\mathfrak{A}(E) = \hat{A}_E$ follows: there is an exact sequence

$$0 \to U_E \to \mathfrak{A}(E) \to \hat{\mathbf{Z}} \to 0$$

where $\hat{\mathbf{Z}}$ denotes the completion of $\mathbf{Z}$ for the topology defined by the $n\mathbf{Z}$ (it is the product of the additive groups $\mathbf{Z}_p$ over all primes p). We will return to this in Chapter XIV, §6.

APPENDIX

Computations of Cup Products

Notation. G denotes a finite group, and A, B, ... are G-modules. If a is an element of A fixed by G (resp. such that $\mathrm{N}a = 0$), denote by $\bar{a}^0$ (resp. $\bar{a}_0$) its canonical image in $\hat{\mathrm{H}}^0(\mathrm{G}, \mathrm{A})$ (resp. $\hat{\mathrm{H}}^{-1}(\mathrm{G}, \mathrm{A})$).

Lemma 1. *Given G-modules A, B and $a \in \mathrm{A}^{\mathrm{G}}$, let $f_a : \mathbf{Z} \to \mathrm{A}$ be the G-homomorphism such that $f_a(1) = a$. Let $x \in \hat{\mathrm{H}}^n(\mathrm{G}, \mathrm{B})$. Then the cup product*

$$\bar{a}^0 . x \in \hat{\mathrm{H}}^n(\mathrm{G}, \mathrm{A}) \otimes \mathrm{B})$$

is equal to the image of x under the homomorphism $f_a \otimes \mathrm{I} : \mathrm{B} \to \mathrm{A} \otimes \mathrm{B}$.

We will give the proof for $n \geq 0$, for example. When $n = 0$, the very definition of cup product gives $\bar{a}^0 . x = (f_a \otimes 1)(x)$. Use induction on n: we know that there is an exact sequence of G-modules

$$0 \to \mathrm{B} \to \mathrm{B}' \to \mathrm{B}'' \to 0$$

with B′ an *induced* G-module and B a direct factor in B′ for its structure of abelian group. We can then write $x = dy$, with $y \in \hat{\mathrm{H}}^{n-1}(\mathrm{G}, \mathrm{B})$, whence

$$\bar{a}^0 . x = \bar{a}^0 . dy = d(\bar{a}^0 . y)$$

and applying the inductive hypothesis to $\bar{a}^0 . y$ gives the result. □

Lemma 2. *Given $a \in \mathrm{A}$ with $\mathrm{N}a = 0$, and let f be a 1-cocycle of G with values in B, $\bar{f} \in \mathrm{H}^1(\mathrm{G}, \mathrm{B})$ its cohomology class. Then*

$$\bar{a}_0 . \bar{f} = \bar{c}^0 \quad \textit{in } \hat{\mathrm{H}}^0(\mathrm{G}, \mathrm{A} \otimes \mathrm{B}),$$

with

$$c = -\sum_{t \in \mathrm{G}} ta \otimes f(t).$$

Consider an exact sequence

$$0 \to B \to B' \to B'' \to 0$$

as above. Since $H^1(G, B') = 0$, there exists $b' \in B'$ such that $f(t) = t.b' - b'$. Let $b'' \in (B'')^G$ be the image of b' in B''. Then

$$\bar{f} = d(\bar{b}''^0) \quad \text{in } H^1(G, B),$$

whence $\bar{a}_0.\bar{f} = -d(\bar{a}_0.\bar{b}''^0)$, since $\bar{a}_0$ is a class of *odd* degree.

By lemma 1, $\bar{a}_0.\bar{b}''^0 = a \otimes b''^0$. Remembering that the operator $d: \hat{H}^{-1}(G, A \otimes B'') \to H^0(G, A \otimes B)$ is defined by the norm, we get

$$\bar{a}_0.\bar{f} = -\bar{c}^0, \quad \text{with } c = N(a \otimes b') = \sum_{t \in G} ta \otimes tb'.$$

Now this can be written

$$c = \sum_{t \in G} ta \otimes (f(t) + b') = \sum_{t \in G} ta \otimes f(t)$$

since $Na = 0$. □

Consider next the exact sequence

$$0 \to I \to \mathbf{Z}[G] \to \mathbf{Z} \to 0$$

and, for $s \in G$, denote by i_s the element $s - 1$ of I. Then $(\bar{i}_s)_0 \in \hat{H}^{-1}(G, I)$. Let $\bar{s} \in \hat{H}^{-2}(G, \mathbf{Z})$ be such that $d\bar{s} = (\bar{i}_s)_0$. The map $s \mapsto \bar{s}$ defines by passage to the quotient *the canonical isomorphism* of G/G' onto $\hat{H}^{-2}(G, \mathbf{Z})$ (cf. Chap. VII, §4).

Lemma 3. *Let* B *be a* G*-module and* $f: G \to B$ *a* 1*-cocycle,* $\bar{f} \in H^1(G, B)$ *its cohomology class. Then for every* $s \in G$,

$$\bar{s}.\bar{f} = \overline{f(s)}_0 \quad \textit{in } \hat{H}^{-1}(G, B).$$

[We identify the G-modules $\mathbf{Z} \otimes B$ and B.]

The homomorphism $d: \hat{H}^{-1}(G, B) \to \hat{H}^0(G, I \otimes B)$ is an isomorphism. Thus it will suffice to show that the images under d of $\bar{s}.\bar{f}$ and $\overline{f}(s)_0$ coincide. In view of the definition of d, we have

$$d(\overline{f(s)}_0) = \bar{x}^0, \quad \text{with } x = \sum_{t \in G} t \otimes t.f(s).$$

On the other hand, we have

$$d(\bar{s}.\bar{f}) = (\bar{i}_s)_0.\bar{f} = \bar{y}^0, \quad \text{with } y = -\sum_{t \in G} t.i_s \otimes f(t)$$

in virtue of lemma 2. This can be written

$$y = \sum_{t \in G} (t - ts) \otimes f(t)$$

$$= \sum_{t \in G} t \otimes f(t) - \sum_{t \in G} ts \otimes f(t).$$

But $f(ts) = f(t) + t.f(s)$, and the sum $\sum ts \otimes f(t)$ can be written in the form $\sum ts \otimes f(ts) - \sum ts \otimes t.f(s)$. The first term cancels the term $\sum t \otimes f(t)$, and there remains

$$y = \sum_{t \in G} ts \otimes t.f(s).$$

Hence $x - y = \sum_{t \in G} t(1-s) \otimes t.f(s) = \mathrm{N}((1-s) \otimes f(s))$, which shows that $\bar{x}^0 = \bar{y}^0$. □

Lemma 4. *Let* B *be a* G*-module,* $u: \mathrm{G} \times \mathrm{G} \to \mathrm{B}$ *a 2-cocycle,* $\bar{u} \in \mathrm{H}^2(\mathrm{G}, \mathrm{B})$ *its cohomology class. Then for all* $s \in \mathrm{G}$,

$$\bar{s}.\bar{u} = \bar{a}^0, \quad \textit{with } a = \sum_{t \in G} u(t, s).$$

Once again introduce an exact sequence $0 \to \mathrm{B} \to \mathrm{B}' \to \mathrm{B}'' \to 0$ of G-modules, with B′ *induced.* Since $\mathrm{H}^2(\mathrm{G}, \mathrm{B}') = 0$, there exists a 1-cochain $f': \mathrm{G} \to \mathrm{B}'$ such that

$$u(x, y) = x.f'(y) - f'(xy) + f'(x), \qquad x, y \in \mathrm{G}.$$

Composing f' with $\mathrm{B}' \to \mathrm{B}''$, we get a 1-cocycle $f'': \mathrm{G} \to \mathrm{B}''$ whose cohomology class $\bar{f}''$ satisfies $d(\bar{f}'') = \bar{u}$. We deduce

$$\begin{aligned} \bar{s}.\bar{u} = \bar{s}.d(\bar{f}'') = d(\bar{s}.\bar{f}'') &= d(\overline{f''(s)_0}) \quad \text{(lemma 3)} \\ &= \overline{\mathrm{N}(f'(s))}^0. \end{aligned}$$

Thus we are reduced to computing $a = \mathrm{N}(f'(s)) = \sum_{t \in G} t.f'(s)$. The identity

$$u(t, s) = t.f'(s) - f'(ts) + f'(t)$$

gives by summation

$$\sum_{t \in G} u(t, s) = a - \sum_{t \in G} f'(ts) + \sum_{t \in G} f'(t) = a. \quad \square$$

PART FOUR

LOCAL CLASS FIELD THEORY

CHAPTER XII

Brauer Group of a Local Field

Throughout this chapter, K denotes a field complete under a valuation v, A denotes the valuation ring of v, and $\bar{K}$ the residue field.

§1. Existence of an Unramified Splitting Field

Theorem 1. *Suppose that the residue field* $\bar{K}$ *is perfect. Then each element of the Brauer group* B_K *of* K *is split by a finite unramified extension of* K.

Let K_{nr} be the maximal unramified extension of K. We know that its Brauer group is zero (Chap. X, §7, example b)). If $a \in B_K$, the image of a in $B_{K_{nr}}$ is zero; as K_{nr} is the directed union of finite unramified extensions K′ of K, it follows that the image of a in one of the $B_{K'}$ is zero. □

[We have used the following fact, whose verification is easy: if a field E is the directed union of a family of subfields E_i, then the canonical homomorphism $\lim B_{E_i} \to B_E$ is an isomorphism.]

Corollary. *The Brauer group of* K *may be identified with* $H^2(K_{nr}/K)$.

This is equivalent to theorem 1.

Remarks. 1) The assumption $\bar{K}$ perfect cannot be dropped (cf. Chap. XIV, §5, exer. 2).

2) The proof of th. 1 given above ultimately rests upon the norm computations in Chap. V. The reader will find in the next section another proof,

based on the interpretation of the Brauer group as the group of classes of central simple algebras.

Exercises

1. Show that $H^q(\quad /K) = H^q(K_{nr}/K)$ for all $q \geq 0$ (use the spectral sequence of group extensions, obtained by passage to the limit from the one for finite groups; cf. [37]).

2. Let L/K be a finite totally ramified Galois extension, with group G. Let w be the valuation on L.
   a) Show that the homomorphism $\hat{H}^q(G, L^*) \to \hat{H}^q(G, \mathbf{Z})$ induced by $w: L^* \to \mathbf{Z}$ is zero for all $q \in \mathbf{Z}$. (Construct an extension L_0/K_0 as in exer. 4 of Chap. II, §2; factor w into $L^* \to \hat{L}_0^* \to \mathbf{Z}$, and remark that $\hat{L}_0^*$ is cohomologically trivial.)
   b) Deduce an exact sequence
   $$0 \to \hat{H}^{q-1}(G, \mathbf{Z}) \xrightarrow{\delta} \hat{H}^q(G, U_L) \to \hat{H}^q(G, L^*) \to 0.$$
   Make explicit the special cases $q = 1$ and $q = 2$.
   c) Let π be a uniformizer of L, and let $\alpha \in H^1(G, U_L)$ be the class of the crossed homomorphism $s \to \pi^{s-1}$ (in exponential notation). Show that $\delta(x) = \alpha . x$ for all $x \in \hat{H}^{q-1}(G, \mathbf{Z})$.
   d) Let V (resp. W) be the subgroup of L^* consisting of the elements $\prod_{s \in G} x_s^{s-1}$, where x_s runs through L^* (resp. U_L). Let G^a be the quotient of G by its commutator subgroup. Show that the map $s \to \pi^{s-1}$ defines, by passage to the quotient, an isomorphism of G^a onto V/W. (Apply a) with $q = -1$.)

§2. Existence of an Unramified Splitting Field (Direct Proof)

Let D be a *division algebra* with center K and rank n^2 over K. We begin by defining a "discrete valuation" on D prolonging the one on K. The method is exactly the same as in the commutative case (cf. Chap. II, §2):

Let $\mathrm{Nrd}: D^* \to K^*$ be the *reduced norm* (cf. Bourbaki, *Alg.*, Chap. VIII, §12). For any $x \in D^*$, set

$$v'(x) = v(\mathrm{Nrd}(x)), \qquad v'(0) = +\infty.$$

Then the map $v': D^* \to \mathbf{Z}$ is a homomorphism, and $v'(x) = nv(x)$ if $x \in K^*$ (since $\mathrm{Nrd}(x) = x^n$ in this case); let d the positive generator of the subgroup $v'(D^*)$ of $\mathbf{Z}$, and set

$$\omega = \frac{1}{d} v'.$$

The map $w: D^* \to \mathbf{Z}$ is a surjective homomorphism.

Proposition 1

a) *For all* $x \in K^*$, $w(x) = (n/d)v(x)$.
b) $w(x + y) \geq \mathrm{Inf}(w(x), w(y))$, *and* $w(xy) = w(x) + w(y)$.

c) *Let a be any real number strictly between* 0 *and* 1. *Setting* $\|x\|_D = a^{w(x)}$ *gives a norm on* D; *the topology defined by this norm is the product topology* (*identifying* D *with* K^n).
d) *For an element* $x \in D$ *to be integral over* A, *it is necessary and sufficient that* $w(x)$ *be* ≥ 0. *The set* B *of such elements is a subring of* D.

If L is a commutative subfield of D containing K and $x \in L$, then $\mathrm{Nrd}(x)$ is a power of $N_{L/K}(x)$ (as can be seen, e.g., by reducing to the case where L is maximal, in which case it result from the definition of the reduced norm—cf. Bourbaki, *loc. cit.*). Hence the restriction of w to L is a *multiple* of the discrete valuation v_L prolonging v; applying this remark to the field $L = K(y/x)$ gives $w(1 + x^{-1}y) \geq \mathrm{Inf}(\omega(1), \omega(x^{-1}y))$, whence b) follows by adding $w(x)$ to both sides. Assertion a) is trivial. One deduces that $\|x\|_D$ is a norm on D, and makes D into a normed vector space over K; then c) follows from completeness of K (cf. Bourbaki, *Esp. Vect. Top.*, Chap. I, §2, th. 2). Finally, for $x \in D$ to be integral over A, it is necessary and sufficient (cf. Chap. II, §2) that the valuation of x in $K(x)$ be ≥ 0, which is equivalent to $w(x) \geq 0$. Formulae b) show that B is a ring. □

Lemma 1. *Suppose* $\bar{K}$ *is perfect and* $n \geq 2$. *Then there is a commutative subfield* L *of* D, *containing* K, *that is unramified over* K *and distinct from* K.

Suppose that such a field did not exist. Then for every commutative extension L of K within D, the residue field $\bar{L}$ would be equal to $\bar{K}$ (otherwise, by cor. 3 to th. 3, Chap. III, §5, L would contain an unramified extension distinct from K). Let $\pi \in D$ be such that $w(\pi) = 1$ (π is a "uniformizer" of D), and let $b \in B$. Applying the preceding argument to the field $L = K(b)$, we see that there exists $a \in A$ such that $w(b - a) \geq 1$, i.e.,

$$b = a + \pi b_1, \quad \text{with } b_1 \in B.$$

Applying the same argument to b_1 and iterating the procedure, we get

$$b = a + \pi a_1 + \cdots + \pi^{N-1} a_{N-1} + \pi^N b_N, \quad \text{with } a_i \in A,\ b_N \in B$$

for any natural number N, which shows that b is in the closure of $K(\pi)$. But $K(\pi)$, being a vector subspace of D, is closed (cf. Bourbaki, *loc. cit.*, cor. 1 to th. 2), so $b \in K(\pi)$. Thus $B \subset K(\pi)$. However, for any $x \in D$, we have $\pi^m x \in B$ for sufficiently large m. Thus $D = K(\pi)$ and D is commutative, contradicting the hypothesis. □

Proposition 2. *Suppose* $\bar{K}$ *is perfect. Then there is a maximal subfield of* D *that is unramified over* K.

Argue by induction on n, the case $n = 1$ being trivial. For $n \geq 2$, lemma 1 furnishes an unramified extension K′ of K within D. Let D′ be the centralizer of K′ in D. Then D′ is a division algebra with center K′ (Bourbaki, *Alg.*,

Chap. VIII, §10, th. 2), and its degree is $<n^2$. By inductive hypothesis, there is a maximal commutative subfield L of D′, containing K′, and unramified over K′. The field L is then unramified over K, and

$$\begin{aligned}[L:K]^2 = [L:K']^2[K':K]^2 = [D':K'][K':K]^2 = [D':K][K':K]\\ = [D:K] \quad \text{(cf. Bourbaki, } \textit{loc. cit.}\text{)}\end{aligned}$$

which shows that L is a maximal subfield of D. □

Theorem 1 is a direct consequence of this proposition, since a maximal subfield of a division algebra is a splitting field for the division algebra. □

EXERCISES

1. a) Show that B is a free A-module of rank n^2. (Either use the reduced trace or argue as in Chap. II, §2.)
   b) Show that every ideal of B is two-sided, and is equal either to 0 or to the set of all $x \in D$ such that $w(x) \geq n$ (for a fixed non-negative integer n).
   c) Deduce from b) that every torsion-free B-module of finite type is free.
2. Let $\bar{B} = B/\pi B$. Show that $\bar{B}$ is a division ring. Let $\varphi = [\bar{B}:\bar{K}]$; if e is the ramification index of w with respect to v (i.e., the integer n/d of prop. 1), show that $e\varphi = n^2$.
3. Suppose that $\bar{K}$ is perfect. Let $\bar{E}$ be the center of $\bar{B}$, and set

   $$[\bar{E}:\bar{K}] = e', \qquad [\bar{B}:\bar{E}] = f^2.$$

   Then $ee'f^2 = n^2$. Show that every subfield of $\bar{B}$ is of degree $\leq n$ over $\bar{K}$ (lift a primitive element of such a field to B); deduce that $ef \leq n$. Using prop. 2, show that there exists a subfield of $\bar{B}$ of degree n over $\bar{K}$, and deduce that $ef = n$, whence $e = e'$.
4. Keeping the above notation, define the *different* of B relative to A by means of the reduced trace. Show that it is equal to B if and only if $e = 1$.

[For more details on the structure of B, see Deuring [19], Chap. VI, or Schilling [54], Chap. V, or Weil [123].]

§3. Determination of the Brauer Group

From now on, $\bar{K}$ is assumed to be *perfect*.

Proposition 3. *The group* B_K *is the union of the subgroups* $H^2(L/K)$, *as* L *runs through the family of finite unramified Galois extensions of* K.

Indeed, these extensions correspond to the subextensions $\bar{L}$ of $\bar{K}_{nr}$ that are finite and Galois over $\bar{K}$, and the union of these $\bar{L}$ is $\bar{K}_{nr}$; the union of the L is therefore K_{nr}. □

This proposition reduces us to the determination of $H^2(L/K)$. Put

$$\mathfrak{g} = G(L/K) = G(\bar{L}/\bar{K}).$$

Consider the exact sequence

$$0 \to U_L \to L^* \to \mathbf{Z} \to 0,$$

where v denotes the valuation of L which prolongs the one on K with ramification index 1. This is an exact sequence of $\mathfrak{g}$-modules; moreover, this sequence splits: indeed, choosing a uniformizer of K allows the identification of L^* with $U_L \times \mathbf{Z}$ as $\mathfrak{g}$-modules (of course, that would not work if L/K were ramified).

Hence we get a split exact sequence of abelian groups

$$0 \to H^q(\mathfrak{g}, U_L) \to H^q(L/K) \to H^q(\mathfrak{g}, \mathbf{Z}) \to 0, \qquad q \geq 0.$$

As in Chap. IV, denote by U_L^n the subgroup of U_L consisting of those $a \in U_L$ such that $v(1 - a) \geq n$.

Lemma 2. *For all* $q \geq 1$, $H^q(\mathfrak{g}, U_L^1) = 0$.

Filter U_L^1 by the U_L^n; the quotients U_L^n/U_L^{n+1} are isomorphic as $\mathfrak{g}$-modules to the additive group $\bar{L}$; therefore they have trivial cohomology (Chap. X, §1, prop. 1), and lemma 2 is a consequence of the following more general lemma.

Lemma 3. *Let* $\mathfrak{g}$ *be a finite group,* M *a* $\mathfrak{g}$*-module filtered by a decreasing sequence* M_n, $n \geq 1$, *of submodules, with* $M_1 = M$. *Suppose that* M *is complete Hausdorff in the topology defined by the* M_n. *Fix an integer* $q \geq 0$. *If* $H^q(\mathfrak{g}, M_n/M_{n+1}) = 0$ *for all* $n \geq 1$, *then* $H^q(\mathfrak{g}, M) = 0$.

Let $\varphi(g_1, \ldots, g_q)$ be a q-cocycle of g with values in M. Since

$$H^q(\mathfrak{g}, M_1/M_2) = 0,$$

there is a $(q - 1)$-cochain ψ_1 of g with values in M_1 such that $\varphi = \delta\psi_1 + \varphi_1$, φ_1 being a q-cocycle with values in M_2. Iterating this procedure, one gets a sequence (ψ_n, φ_n), where ψ_n is a $(q - 1)$-cochain with values in M_n, and φ_n a q-cocycle with values in M_{n+1}, with

$$\begin{aligned} \varphi &= \delta\psi_1 + \varphi_1 \\ &\vdots \\ \varphi_n &= \delta\psi_{n+1} + \varphi_{n+1} \\ &\vdots \end{aligned}$$

Put $\psi = \psi_1 + \cdots + \psi_n + \cdots$. In view of the hypotheses on M, this series converges and defines a $(q - 1)$-cochain of g with values in M. By the preceding equations, $\varphi = \delta\psi$. □

Back to the cohomology of the group U_L. There is an exact sequence of $\mathfrak{g}$-modules

$$0 \to U_L^1 \to U_L \to \bar{L}^* \to 0.$$

Taking lemma 2 into account, the exact cohomology sequence yields

$$H^q(\mathfrak{g}, U_L) = H^q(\mathfrak{g}, \bar{L}^*) = H^q(\bar{L}/\bar{K}), \qquad q \geq 1.$$

Summarizing:

Proposition 4. *Let* L/K *be a finite unramified Galois extension with Galois group* $\mathfrak{g}$. *Then there is a split exact sequence*

$$0 \to H^q(\bar{L}/\bar{K}) \to H^q(L/K) \to H^q(\mathfrak{g}, \mathbf{Z}) \to 0, \qquad q \geq 1.$$

Passing to the limit over L gives:

Corollary. *Let* $\mathfrak{g}$ *be the Galois group of* K_{nr}/K. *There is a split exact sequence*

$$0 \to H^q(\quad/\bar{K}) \to H^q(K_{nr}/K) \to H^q(\mathfrak{g}, \mathbf{Z}) \to 0, \qquad q \geq 1.$$

(Note that $\mathfrak{g}$ is also the Galois group of $\bar{K}_{nr}/\bar{K}$, where $\bar{K}_{nr}$ is the algebraic closure of $\bar{K}$.)

Now set $q = 2$. The group $H^2(\quad/\bar{K})$ is none other than the *Brauer group* $B_{\bar{K}}$ of $\bar{K}$; similarly, $H^2(K_{nr}/K)$ may be identified with B_K (cor. to th. 1). It remains to compute $H^2(\mathfrak{g}, \mathbf{Z})$. Now we have the exact sequence

$$0 \to \mathbf{Z} \to \mathbf{Q} \to \mathbf{Q}/\mathbf{Z} \to 0.$$

As $\mathbf{Q}$ has trivial cohomology, we deduce the isomorphism

$$\mathrm{Hom}(\mathfrak{g}, \mathbf{Q}/\mathbf{Z}) \to H^2(\mathfrak{g}, \mathbf{Z})$$

already used in the preceding chapter. Of course the group $\mathrm{Hom}(\mathfrak{g}, \mathbf{Q}/\mathbf{Z})$ is the group of *continuous* homomorphisms of $\mathfrak{g}$ (i.e., those homomorphisms having open kernel); we denote it by $X(\mathfrak{g})$ and call it the *character group* of $\mathfrak{g}$ (it is the Pontrjagin dual of $\mathfrak{g}/\mathfrak{g}'$). Thus we finally get the following theorem, due to Witt [72]:

Theorem 2. *Let* K *be a field that is complete under a discrete valuation and has perfect residue field* $\bar{K}$. *Let* $\mathfrak{g}$ *be the Galois group of the algebraic closure of* $\bar{K}$ *over* $\bar{K}$, *and let* $X(\mathfrak{g})$ *be its character group. Then there is a split exact sequence*

$$0 \to B_{\bar{K}} \to B_K \to X(\mathfrak{g}) \to 0.$$

(As we have seen, the splitting of this sequence comes from choosing a uniformizer of K.)

Remark. The homomorphism $B_{\bar{K}} \to B_K$ has a neat interpretation in the Azumaya-Auslander-Goldman theory, where the *Brauer group* B_A *of the ring* A is defined; the functoriality of Brauer groups gives homomorphisms of B_A into $B_{\bar{K}}$ and into B_K, but since the first homomorphism $B_A \to B_{\bar{K}}$ is an isomorphism (cf. [10], th. 31, or [9], th. 6.5), a homomorphism $B_{\bar{K}} \to B_K$ is obtained. It can be shown that this homomorphism coincides with the one in th. 2.

Exercises

1. Keeping the hypotheses on K from th. 2, suppose further that the Brauer group of $\bar{K}$ is zero (cf. Chap. X, §7). Show that an element of B_K of order n is split by a cyclic unramified extension of degree n, and by only one.

2. Keeping the hypotheses on K from th. 2, let L/K be a finite extension, with ramification index e. Prove the commutativity of the diagram

$$\begin{array}{ccccccccc} 0 & \to & B_{\bar{K}} & \to & B_K & \to & X(\mathfrak{g}_K) & \to & 0 \\ & & \downarrow{\scriptstyle \mathrm{Res}} & & \downarrow{\scriptstyle \mathrm{Res}} & & \downarrow{\scriptstyle e.\,\mathrm{Res}} & & \\ 0 & \to & B_{\bar{L}} & \to & B_L & \to & X(\mathfrak{g}_L) & \to & 0. \end{array}$$

   Same question for the diagram:

$$\begin{array}{ccccccccc} 0 & \to & B_{\bar{L}} & \to & B_L & \to & X(\mathfrak{g}_L) & \to & 0 \\ & & \downarrow{\scriptstyle e.\,\mathrm{Cor}} & & \downarrow{\scriptstyle \mathrm{Cor}} & & \downarrow{\scriptstyle \mathrm{Cor}} & & \\ 0 & \to & B_{\bar{K}} & \to & B_K & \to & X(\mathfrak{g}_K) & \to & 0. \end{array}$$

3. Drop the hypothesis that $\bar{K}$ is perfect. Let B'_K be the subgroup of B_K of all elements split by K_{nr}.
   a) Show that the exact sequence of th. 2 applies to B'_K instead of B_K.
   b) Let $\bar{K}$ have characteristic p. Show that every element of B_K having order prime to p belongs to B'_K. (Represent such an element by a central division algebra D over K of rank n^2; use exer. 3 of Chap. X, §5, to show that $(n, p) = 1$; repeat the reasoning of §2 to deduce that D contains a maximal subfield which is unramified over K.)

CHAPTER XIII

Local Class Field Theory

Standard local class field theory is concerned with complete fields K whose residue field $\bar{K}$ is finite. As was shown by Moriya (see also Schilling [54] and Whaples [71]), the finiteness hypothesis on $\bar{K}$ can be weakened to quasi-finiteness (see §2), and it is in this framework that we will work.

The last section presents a result of Dwork [21], useful both for computing local symbols explicitly and for making a bridge with the "proalgebraic" viewpoint of [59].

§1. The Group $\hat{\mathbf{Z}}$ and Its Cohomology

We denote by $\hat{\mathbf{Z}}$ the completion of $\mathbf{Z}$ for the topology of subgroups of finite index; it is a compact totally disconnected group, equal to the *projective limit of the groups* $\mathbf{Z}/n\mathbf{Z}$. By decomposing these groups into their p-primary components, one sees that $\hat{\mathbf{Z}}$ is canonically isomorphic to the product $\prod \mathbf{Z}_p$ (p running through the prime numbers).

Set $\mathfrak{g} = \hat{\mathbf{Z}}$ and $\mathfrak{g}_n = n\hat{\mathbf{Z}}$; then $\mathfrak{g} = \lim \mathfrak{g}/\mathfrak{g}_n$ and every open subgroup of $\mathfrak{g}$ coincides with some $\mathfrak{g}_n$. Let A be topological $\mathfrak{g}$-module (in the sense of Chap. X, §3). The canonical generator $1 \in \mathbf{Z}$ defines an automorphism F of A; to say that A is a *topological* $\mathfrak{g}$-module then means that for each $a \in A$, there is a positive integer n such that $F^n a = a$. Thus the group A is the union of its subgroups $A^{\mathfrak{g}_n}$, the latter being $\mathfrak{g}/\mathfrak{g}_n$-modules. The cohomology groups of $\mathfrak{g}$ with values in A are defined by the formula

$$(*) \qquad H^q(\mathfrak{g}, A) = \varinjlim H^q(\mathfrak{g}/\mathfrak{g}_n, A^{\mathfrak{g}_n}).$$

Obviously $H^0(\mathfrak{g}, A) = A^{\mathfrak{g}}$. As for H^1, we have:

Proposition 1. *Let* A′ *be the subgroup of* A *consisting of those* $a \in A$ *for which there is a positive integer n satisfying* $(1 + F + \cdots + F^{n-1})a = 0$. *Then*

$$H^1(\mathfrak{g}, A) = A'/(F-1)A.$$

(The isomorphism is obtained by assigning to each 1-cocyle $\varphi: \mathfrak{g} \to A$ the coset of $\varphi(1)$ in $A'/(F-1)A$.)

The proposition follows from passage to the limit, using formula (*) and the determination of the H^1 of a cyclic group. □

Corollary. *The character group* $X(\mathfrak{g})$ *of* $\mathfrak{g}$ *can be identified with* $\mathbf{Q}/\mathbf{Z}$.

Indeed, this group is none other than $H^1(\mathfrak{g}, \mathbf{Q}/\mathbf{Z})$, with $\mathfrak{g}$ acting trivially on $\mathbf{Q}/\mathbf{Z}$.

Remark. The group A′ of prop. 1 *contains the torsion subgroup* A_f of A: indeed, if $a \in A_f$, then we have $na = 0$ for sufficiently large n, as well as $F^m a = a$ for sufficiently large m. It follows that

$$(1 + F + \cdots + F^{mn-1})a = nF^m a = 0$$

which shows that $a \in A'$.

Proposition 2. *If* A *is either a divisible group or a torsion group, then* $H^2(\mathfrak{g}, A) = 0$.

Suppose first A is finite. Then $H^2(\mathfrak{g}/\mathfrak{g}_n, A^{\mathfrak{g}_n}) = A^{\mathfrak{g}}/N_n A^{\mathfrak{g}^n}$, with

$$N_n = 1 + F + \cdots + F^{n-1}.$$

Let m be a positive integer. It is not difficult to check that the homomorphism

$$A^{\mathfrak{g}}/N_n A^{\mathfrak{g}^n} \to A^{\mathfrak{g}}/N_{nm} A^{\mathfrak{g}^{nm}}$$

which appears in the directed system (*) is induced by multiplication by m. If m is a multiple of the order of A, this homomorphism is zero, hence $\lim H^2(\mathfrak{g}/\mathfrak{g}_n, A^{\mathfrak{g}_n})$ is zero.

[Variant: let E be a compact totally disconnected group that is an extension of $\mathfrak{g}$ by A. Lifting to E the canonical generator of $\mathfrak{g}$, one defines a section homomorphism $\mathfrak{g} \to E$, which shows that E is the trivial extension. Hence $H^2(\mathfrak{g}, A) = 0$.]

If A is a torsion group, then $A = \varinjlim A_\alpha$, where the A_α are finite and stable under $\mathfrak{g}$, whence $H^2(\mathfrak{g}, A) = \varinjlim H^2(\mathfrak{g}, A_\alpha) = 0$.

Finally, suppose A is divisible. If $n \geq 1$, denote by ${}_nA$ the kernel of multiplication by n on A. The exact sequence

$$0 \to {}_nA \to A \xrightarrow{n} A \to 0$$

induces the exact cohomology sequence

$$H^2(\mathfrak{g}, {}_n A) \to H^2(\mathfrak{g}, A) \xrightarrow{n} H^2(\mathfrak{g}, A).$$

By the preceding argument, $H^2(\mathfrak{g}, {}_n A) = 0$. Hence multiplication by n is injective on $H^2(\mathfrak{g}, A)$. But this is a torsion group (e.g., by formula (*)), so it must be zero. □

EXERCISES

1. Using the computation of the cohomology of a cyclic group (Chap. VIII, §4), determine the homomorphisms

$$H^q(\mathfrak{g}/\mathfrak{g}_n, A^{\mathfrak{g}_n}) \to H^q(\mathfrak{g}/\mathfrak{g}_{nm}, A^{\mathfrak{g}_{nm}}).$$

Show that if $q = 2h$ or if $q = 2h + 1$, they are induced by multiplication by m^h. Deduce that for every topological $\mathfrak{g}$-module A and every $q \geq 3$, $H^q(\mathfrak{g}, A) = 0$.

2. Let L be a free group (non-abelian), and let $\mathfrak{g}$ be its completion for the topology of subgroups of finite index.
   a) Show that for any compact totally disconnected group $\mathfrak{h}$ and any continuous surjective homomorphism $f: \mathfrak{h} \to \mathfrak{g}$, there exists a continuous homomorphism $s: \mathfrak{g} \to \mathfrak{h}$ such that $f \circ s = 1$.
   b) Deduce that, for any topological $\mathfrak{g}$-module A that is either torsion or divisible, $H^2(\mathfrak{g}, A) = 0$.
   [For more details, see [20], [88], [93] or [113].

§2. Quasi-finite Fields

First let $\mathfrak{g}$ be any compact totally disconnected group, and let $s \in \mathfrak{g}$. Then there is a unique continuous homomorphism

$$f_s: \hat{\mathbf{Z}} \to \mathfrak{g}$$

such that $f_s(1) = s$. If the group $\mathfrak{g}$ is denoted multiplicatively, we will write s^ν instead of $f_s(\nu)$, $\nu \in \hat{\mathbf{Z}}$.

Let k_s be an algebraic closure of a field k, and let $F \in G(k_s/k)$. We say that F provides k with *a structure of quasi-finite field* if the following two conditions hold:

1) *k is perfect.*
2) *The map $\nu \mapsto F^\nu$ is an isomorphism of $\hat{\mathbf{Z}}$ onto the group $G(k_s/k)$.*

(Condition 2) also means that F is a *free generator* of $G(k_s/k)$.)

Thus a quasi-finite field is a perfect field k for which $G(k_s/k)$ is isomorphic to $\hat{\mathbf{Z}}$, with a specific isomorphism *given* as part of the structure.

Remark. The preceding definition seems to depend on the choice of an algebraic closure k_s. In fact, if k_s' is another algebraic closure of k, the group

$G(k'_s/k)$ is *canonically isomorphic* to $G(k_s/k)$ because it is *abelian*; hence the choice of a free generator of $G(k_s/k)$ automatically determines a free generator of $G(k'_s/k)$.

Galois theory shows that the only finite extensions of k within k_s are the cyclic extensions k_n consisting of the elements fixed by F^n. Conversely, if a perfect field k has the properties that for each n, there is a subextension k_n/k of k_s which is cyclic of degree n, that these extensions fit together and have union k_s, and that a generator F_n of $G(k_n/k)$ is given for each n, with F_n being the restriction of $F_{nm} \in G(k_{nm}/k)$, then k is a quasi-finite field.

EXAMPLES OF QUASI-FINITE FIELDS. a) *Finite fields.* If k has q elements, take F to be the "Frobenius substitution" $x \mapsto x^q$ (cf. Bourbaki, *Alg.*, Chap. V, §11).

b) Let C be an algebraically closed field of characteristic zero, and let $k = C((T))$, the power series field in one variable. For every integer $n \geq 1$, $k_n = C((T^{1/n}))$ is a cyclic extension of degree n of k, and the union k_s of the k_n is an algebraic closure of k (cf. Chap. IV, prop. 8). A generator F_n of $G(k_n/k)$ is defined by $T^{1/n} \mapsto a_n T^{1/n}$, where a_n is a primitive nth root of unity. For the F_n to fit together and define an F, the a_n must be chosen to satisfy $(a_{mn})^n = a_m$, which is possible (cf. Bourbaki, *loc. cit.*); when $C = \mathbf{C}$, one can take $a_n = \exp(2\pi i/n)$. Thus k is quasi-finite.

There exist other examples (cf. exer. 3), but the above two seem to be the only "non-pathological" ones.

Proposition 3. *Let k be a quasi-finite field, and let F be the given free generator of its Galois group. Let k'/k be a finite extension of degree n, and let $F' = F^n$. Then $F' \in G(k_s/k')$, and F' provides k' with a structure of quasi-finite field.*

Clear.

Whenever we have a finite extension of a quasi-finite field, it will be provided with the structure of quasi-finite field given by the preceding proposition.

Proposition 4. *Let k be a quasi-finite field, F the free generator of its Galois group $\mathfrak{g} = G(k_s/k)$.*

a) *If $w \in K_s^*$ is a root of unity, there exists $y \in k_s^*$ such that $w = y^{F-1}$* (i.e., $w = F(y)/y$).

b) *If k has characteristic $\neq 0$, the homomorphism $F - 1 : k_s \to k_s$ is surjective.*

We know $H^1(\mathfrak{g}, k_s^*) = 0$ (Chap. X, prop. 2); if we set $A = k_s^*$, then $A' = (F - 1)A$ by prop. 1; since w belongs to the torsion subgroup of A, we have $w \in A'$, whence *a*).

The same argument works for *b*), taking into account that $H^1(\mathfrak{g}, k_s) = 0$ (Chap. X, prop. 1) and now $A' = A$ by hypothesis. □

Proposition 5. *The Brauer group of a quasi-finite field is zero.*

Again put $\mathfrak{g} = G(k_s/k)$. The $\mathfrak{g}$-module k_s^* is divisible (since k_s is algebraically closed). By prop. 2, $H^2(\mathfrak{g}, k_s^*) = 0$. □

Corollary. *If k' is a finite extension of a quasi-finite field k, then the norm map $N: k'^* \to k^*$ is surjective.*

Indeed, since this extension is cyclic, $H^2(k'/k)$ is isomorphic to $k^*/N(k'^*)$ (Chap. VIII, §4, cor. to prop. 6). Apply the preceding result.

EXERCISES

1. Let k_s be an algebraic closure of a perfect field k. Suppose that for each integer $n \geq 1$, there is a unique subextension of k_s/k that has degree n over k. Show that k can be provided with a structure of quasi-finite field.
2. Let k be a quasi-finite field, F the given free generator of $G(k_s/k)$. For which values of $v \in \hat{\mathbf{Z}}$ is the element F^v a free generator?
3. a) Let L/K be a Galois extension with group isomorphic to $\hat{\mathbf{Z}}$, and let Ω be an algebraic closure of L. Show that there exists an extension E of K within Ω which is a quasi-finite field. (Lift a free generator of G(L/K) to $G(\Omega/K)$.)
   b) If K is a prime field, show that there is an extension L/K satisfying the hypothesis of a) (for $K = \mathbf{Q}$, consider the field L′ generated by all the roots of unity, and show that G(L′/K) admits $\hat{\mathbf{Z}}$ as a direct factor; deduce the construction of the desired field L).
   c) Let L and K be as in b), and let $\{T_\alpha\}$ be a family of indeterminates. Put $L' = L(T_\alpha)$, $K' = K(T_\alpha)$. Show that L′/K′ is Galois with group $\hat{\mathbf{Z}}$.
   d) Deduce that every algebraically closed field Ω admits a quasi-finite subfield E having Ω as its algebraic closure.

§3. The Brauer Group

From now until the end of this chapter, K denotes a field that is *complete under a discrete valuation v and having quasi-finite residue field* $\bar{K}$. If K_{nr} is the maximal unramified extension of K, put

$$\mathfrak{g} = G(K_{nr}/K) = G(\bar{K}_{nr}/\bar{K})$$

and denote by F the free generator of $\mathfrak{g}$ that defines the quasi-finite structure on $\bar{K}$.

By th. 2 of Chap. XII, there is an exact sequence

$$0 \to B_{\bar{K}} \to B_K \to X(\mathfrak{g}) \to 0$$

where $X(\mathfrak{g})$ is the character group of $\mathfrak{g}$. By the corollary to prop. 1, $X(\mathfrak{g})$ can be identified with $\mathbf{Q}/\mathbf{Z}$. This gives a homomorphism $B_K \to \mathbf{Q}/\mathbf{Z}$ which will be denoted inv_K.

Proposition 6. *The homomorphism* $\mathrm{inv}_K : B_K \to \mathbf{Q}/\mathbf{Z}$ *is an isomorphism.*

This follows from the exact sequence and the fact that $B_{\bar{K}} = 0$ (prop. 5). □

Let us make explicit the definition of inv_K, for that will be needed in the sequel. Consider the following isomorphisms:

$$B_K \xleftarrow{\alpha} H^2(\mathfrak{g}, K_{nr}^*) \xrightarrow{\beta} H^2(\mathfrak{g}, \mathbf{Z}) \xleftarrow{\delta} H^1(\mathfrak{g}, \mathbf{Q}/\mathbf{Z}) \xrightarrow{\gamma} \mathbf{Q}/\mathbf{Z},$$

where α is the canonical injection of $H^2(K_{nr}/K)$ into B_K, β is induced by $v: K_{nr}^* \to \mathbf{Z}$, δ is the coboundary induced by the exact sequence $0 \to \mathbf{Z} \to \mathbf{Q} \to \mathbf{Q}/\mathbf{Z} \to 0$, and, finally, γ assigns to each character χ of $\mathfrak{g}$ the element $\chi(F) \in \mathbf{Q}/\mathbf{Z}$. The construction of inv_K is then

$$\mathrm{inv}_K = \gamma \circ \delta^{-1} \circ \beta \circ \alpha^{-1}.$$

Proposition 7. *Let* L *be a finite extension of* K *of degree n, and let* $\mathrm{Res}_{K/L} : B_K \to B_L$ *be the canonical homomorphism of* B_K *into* B_L (cf. Chap. X, §4, as well as Chap. XI, §§1, 2). *Then*

$$\mathrm{inv}_L \circ \mathrm{Res}_{K/L} = n . \mathrm{inv}_K.$$

In other words, the diagram

$$\begin{array}{ccc} B_K & \to & \mathbf{Q}/\mathbf{Z} \\ \downarrow & & \downarrow n \\ B_L & \to & \mathbf{Q}/\mathbf{Z} \end{array}$$

is commutative.

PROOF. We begin with two special cases.

a) L/K *is unramified*

Then $K_{nr} = L_{nr}$; put $\mathfrak{g}_n = G(K_{nr}/L) = G(\bar{K}_{nr}/\bar{L})$: it is the unique subgroup of $\mathfrak{g}$ of index n.

Consider the diagram

$$\begin{array}{ccccccccc} B_K & \xleftarrow{\alpha} & H^2(\mathfrak{g}, K_{nr}^*) & \xrightarrow{\beta} & H^2(\mathfrak{g}, \mathbf{Z}) & \xleftarrow{\delta} & H^1(\mathfrak{g}, \mathbf{Q}/\mathbf{Z}) & \xrightarrow{\gamma} & \mathbf{Q}/\mathbf{Z} \\ \downarrow \mathrm{Res} & & \downarrow \mathrm{Res} & & \downarrow \mathrm{Res} & & \downarrow \mathrm{Res} & & \downarrow n \\ B_L & \xleftarrow{\alpha'} & H^2(\mathfrak{g}_n, K_{nr}^*) & \xrightarrow{\beta'} & H^2(\mathfrak{g}_n, \mathbf{Z}) & \xleftarrow{\delta'} & H^1(\mathfrak{g}_n, \mathbf{Q}/\mathbf{Z}) & \xrightarrow{\gamma'} & \mathbf{Q}/\mathbf{Z} \end{array}$$

where all the vertical homomorphisms are restrictions, except the one on the far right, which is multiplication by n. We must show the commutativity of this diagram. For the squares involving only the "Res", commutativity is obvious; there remains the last square

$$\begin{array}{ccc} H^1(\mathfrak{g}, \mathbf{Q}/\mathbf{Z}) & \xrightarrow{\gamma} & \mathbf{Q}/\mathbf{Z} \\ \downarrow \mathrm{Res} & & \downarrow n \\ H^1(\mathfrak{g}_n, \mathbf{Q}/\mathbf{Z}) & \xrightarrow{\gamma'} & \mathbf{Q}/\mathbf{Z}. \end{array}$$

If $\chi \in H^1(\mathfrak{g}, \mathbf{Q}/\mathbf{Z})$, then $\gamma(\chi) = \chi(F)$. On the other hand, $\mathrm{Res}(\chi)$ is simply the restriction of χ to the subgroup $\mathfrak{g}_n$ of $\mathfrak{g}$, and the distinguished generator of $\mathfrak{g}_n$ is F^n (cf. §2). Hence

$$\gamma'(\mathrm{Res}(\chi)) = \chi(F^n) = n\chi(F)$$

which gives the desired commutativity.

b) L/K *is totally ramified* (i.e., $\bar{L} = \bar{K}$)

Then the extension K_{nr}/K is linearly disjoint from L/K, and $L_{nr} = K_{nr}L$. The group $\mathfrak{g}$ is the same for L as for K. Consider the diagram

$$\begin{array}{ccccccccc}
B_K & \xleftarrow{\alpha} & H^2(\mathfrak{g}, K_{nr}^*) & \xrightarrow{\beta} & H^2(\mathfrak{g}, \mathbf{Z}) & \xleftarrow{\delta} & H^1(\mathfrak{g}, \mathbf{Q}/\mathbf{Z}) & \xrightarrow{\gamma} & \mathbf{Q}/\mathbf{Z} \\
{\scriptstyle \mathrm{Res}}\downarrow & & {\scriptstyle i}\downarrow & & {\scriptstyle n}\downarrow & & {\scriptstyle n}\downarrow & & {\scriptstyle n}\downarrow \\
B_L & \xleftarrow{\alpha'} & H^2(\mathfrak{g}, L_{nr}^*) & \xrightarrow{\beta'} & H^2(\mathfrak{g}, \mathbf{Z}) & \xleftarrow{\delta'} & H^1(\mathfrak{g}, \mathbf{Q}/\mathbf{Z}) & \xrightarrow{\gamma'} & \mathbf{Q}/\mathbf{Z},
\end{array}$$

where i is induced by the inclusion of K_{nr}^* into L_{nr}^*. Commutativity of the first square is obvious; that of the second follows from the fact that the valuation w of L_{nr} prolongs the valuation v of K_{nr} with ramification index $e = n$; that of the other squares is trivial.

The special cases being settled, the general case follows from the fact that L is a totally ramified extension of an intermediate field K′ that is unramified over K (cf. Chap. III, §5, cor. 3 to th. 3). □

Corollary 1. *Keeping the hypotheses and notation of prop. 7, a necessary and sufficient condition for an element* $a \in B_K$ *to be split by* L *is* $na = 0$.

Indeed, to say that a is split by L means that $\mathrm{Res}_{K/L}(a) = 0$, equivalently $0 = \mathrm{inv}_L \circ \mathrm{Res}_{K/L}(a) = \mathrm{inv}_K(na)$, i.e., $na = 0$.

Corollary 2. *Suppose* L/K *is Galois. Then the isomorphism* $\mathrm{inv}_K : B_K \to \mathbf{Q}/\mathbf{Z}$ *maps the subgroup* $H^2(L/K)$ *of* B_K *onto the subgroup* $(1/n)\mathbf{Z}/\mathbf{Z}$ *of* $\mathbf{Q}/\mathbf{Z}$.

Clear.

Corollary 3. *Let* D *be a central division algebra over* K *of rank* n^2, *and let* $a \in B_K$ *be the corresponding element of the Brauer group. Then* a *has order* n *in* B_K, *and every extension of* K *of degree* n *can be embedded in* D.

We know that every maximal subfield of D splits D; by cor. 1, $na = 0$ (this is a general fact—cf. Chap. X, §5, exer. 3). If d is the order of a, d divides n. Corollary 1 shows that D is split by an extension L/K of degree d, and the class of D contains a central simple algebra of rank d^2 (cf. Bourbaki, *Alg.*, Chap. VIII, §10, prop. 7). As D is a division ring, this implies $n \mid d$, whence $n = d$. Finally, if K′/K is any extension of degree n, cor. 1 shows that K′ is a splitting field for D, hence (cf. Bourbaki, *loc. cit.*) is isomorphic to a maximal subfield of D. □

Remark. The above corollary shows that $\mathrm{inv}_K(a) = m/n$, with $(m, n) = 1$. We conclude that, up to isomorphism, the number of central division algebras over K of rank n^2 is $\varphi(n)$.

EXERCISES

1. Let K satisfy the hypotheses of the beginning of this §, let D be as in cor. 3, and let $\mathrm{Nrd}: D^* \to K^*$ be the reduced norm.
   a) With the notation of exer. 3 of Chap. XII, §2, show that $e = n$, $f = 1$.
   b) If π is a uniformizer of D, deduce that $\mathrm{Nrd}(\pi)$ is a uniformizer of K.
   c) Show that every unit of K is the reduced norm of a unit of D. (Use the fact that D contains an unramified extension of degree n, and apply the corollary to prop. 3 of Chap. V.)
   d) Show that Nrd is surjective.

2. Suppose only that K is complete under a discrete valuation with perfect residue field $\bar{K}$, and that for each finite separable extension E of K, we are given an isomorphism

$$i_E : B_E \to \mathbf{Q}/\mathbf{Z}$$

   such that $i_{E'} \circ \mathrm{Res}_{E/E'} = [E' : E] i_E$ if $E' \supset E$. Show that there is a unique structure of quasi-finite field on $\bar{K}$ for which the corresponding isomorphism inv_E coincides with the given isomorphism i_E. (Use th. 2 and exer. 2 of Chap. XII, §3; show that $B_{\bar{E}} = 0$ because $\mathfrak{g}_E$ is abelian.)

   [In other words, the hypothesis that the residue field $\bar{K}$ is quasi-finite is *necessary and sufficient* for the validity of local class field theory.]

§4. Class Formation

We keep the hypotheses of §3 on K. We will associate to K a *class formation*, in the sense of Artin-Tate (cf. Chap. XI).

First choose a separable closure K_s of K, and let X be the set of finite subextensions of K_s. Let $G = G(K_s/K)$, and, for each $E \in X$, let $G_E = G(K_s/E)$ be the subgroup corresponding to E. The group G acts on K_s^*, and we obtain thereby a *formation* in the sense of Chap. XI, §1. Furthermore, if $E \in X$, the residue field $\bar{E}$ of E, being a finite extension of $\bar{K}$, is canonically provided with a structure of quasi-finite field (cf. §2). As in §3, we get an isomorphism

$$\mathrm{inv}_E : H^2(\ /E) \to \mathbf{Q}/\mathbf{Z}$$

since $H^2(\ /E) = B_E$, the Brauer group of E.

Theorem 1. *The preceding formation, together with the* inv_E, *is a class formation* (cf. Chap. XI, §2).

We need to check the following axioms:

I. *For every Galois extension* F/E, $H^1(F/E) = 0$.
   This is "Hilbert's theorem 90" (prop. 2 of Chap. X).

IIa). *The homomorphism* inv_E *is injective. If* F/E *is Galois,* inv_E *maps* $H^2(F/E)$ *onto the unique subgroup of* $\mathbf{Q}/\mathbf{Z}$ *of order* $[F:E]$.

This is cor. 2 to prop. 7.

IIb). *If* E′/E *is an arbitrary extension,*

$$\mathrm{inv}_{E'} \circ \mathrm{Res}_{E/E'} = [E':E]\,.\,\mathrm{inv}_E.$$

This is prop. 7. □

Having a class formation, we can apply to it the results of Chap. XI. We reproduce the most important ones:

First let F/E be a Galois extension (with $E, F \in X$ as always); denote by $G_{F/E}$ its Galois group. By IIa), there is a unique element

$$u_{F/E} \in H^2(F/E)$$

such that $\mathrm{inv}_E(u_{F/E}) = 1/n$, with $n = [F:E]$. It is called *the fundamental class* of the extension F/E.

Proposition 8. *For every integer n, the cup product with* $u_{F/E}$ *defines an isomorphism*

$$\hat{H}^n(G_{F/E}, \mathbf{Z}) \to \hat{H}^{n+2}(G_{F/E}, F^*).$$

This is Tate's theorem (th. 1 of Chap. XI). □

For $n = -2$, we have $\hat{H}^{-2}(G_{F/E},\mathbf{Z}) = G^a_{F/E}$ (i.e., the group $G_{F/E}$ made abelian). Therefore:

Corollary. *The cup product with* $u_{F/E}$ *defines an isomorphism*

$$\theta_{F/E}: G^a_{F/E} \to E^*/NF^* \quad (\text{N } \textit{denoting the norm in } F/E).$$

In particular, $(E^*:NF^*) = [F:E]$ *if* F/E *is an abelian extension.*

Conversely (cf. Chap. XI, cor. to prop. 3):

Proposition 9. *If* F/E *is a finite separable extension, then* NF^* *has finite index in* E^*; *this index divides* $[F:E]$, *and is equal to it if and only if* F/E *is abelian.*

Returning to the isomorphism

$$\theta_{F/E}: G^a_{F/E} \to E^*/NF^*$$

of the corollary above, we call the inverse isomorphism $\omega = \theta^{-1}_{F/E}$ *the reciprocity isomorphism.* If $x \in E^*$ has image $\bar{x}$ in E^*/NF^*, we put

$$(x, F/E) = \omega(\bar{x}),$$

which is an element of $G^a_{F/E}$. In particular, if F/E is abelian (case to which one can always reduce), (x, F/E) belongs to $G_{F/E}$. Then

$$(xx', F/E) = (x, F/E).(x', F/E)$$
$$(x, F/E) = 1 \quad \text{if and only if } x \in NF^*.$$

(Because of the latter property, (x, F/E) is often called the *norm residue symbol* of x in F/E.)

Every $s \in G^a_{F/E}$ is of the form (x, F/E), with $x \in E^*$.

The functorial properties of the reciprocity isomorphism are given by the three following propositions, proved in Chap. XI, §3:

Proposition 10. *Consider an extension* $F \supset E' \supset E$, *with* F/E *Galois. The group* $G_{F/E'}$ *is a subgroup of* $G_{F/E}$.

a) *If* $x \in E'^*$ *and* $y = N_{E'/E}(x)$, *then the image of* $(x, F/E') \in G^a_{F/E'}$ *in* $G^a_{F/E}$ *is equal to* (y, F/E).
b) *If* $x \in E^*$, *the image of* $(x, F/E) \in G^a_{F/E}$ *in* $G^a_{F/E'}$ *under the transfer is equal to* (x, F/E').

Proposition 11. *If* F/E *is a Galois extension and* $s \in G_{F/E}$, *then*

$$(sx, sF/sE) = s \circ (x, F/E) \circ s^{-1} \qquad \textit{for all } x \in E^*.$$

Proposition 12. *Consider an extension* $F' \supset F \supset E$, *with* F/E *and* F'/E *Galois. If* $x \in E^*$, *then the image of* (x, F'/E) *in* $G^a_{F/E}$ *is equal to* (x, F/E).

This last proposition allows us to define (x, F/E) for a Galois extension F/E that is not finite; it also shows that we can confine ourselves to *abelian* extensions. When F is the *maximal abelian extension* E^a of E, we write (x, */E) instead of $(x, E^a/E)$; it is an element of the Galois group $\mathfrak{A}_E = G(E^a/E)$. The reciprocity map allows the *identification of* $\mathfrak{A}_E$ *with the Hausdorff completion of* E^* *for the topology defined by the groups of norms* (cf. Chap. XI, §4). We will determine this topology in the next chapter, for the special case where $\bar{K}$ is finite.

The explicit computation of (x, F/E) is easy in the unramified case:

Proposition 13. *Let* L/K *be an unramified extension. If we identify the groups* $G_{L/K}$ *and* $G_{\bar{L}/\bar{K}}$, *then*

$$(x, L/K) = F_K^{v(x)},$$

where F_K *denotes the canonical generator of* $G_{\bar{L}/\bar{K}}$, *and where* v *is the valuation of* K.

Set $g = G_{L/K}$ and $F = F_K$ for brevity. Let χ be a character of g. We must check that

$$\chi((x, L/K)) = \chi(F^{v(x)}).$$

By prop. 2 of Chap. XI, the left side is equal to $\mathrm{inv}_K(x.\delta\chi)$. Recall that the isomorphism inv_K was defined in §3 as the composite

$$H^2(\mathfrak{g}, L^*) \xrightarrow{\beta} H^2(\mathfrak{g}, \mathbf{Z}) \xrightarrow{\delta^{-1}} H^1(\mathfrak{g}, \mathbf{Q}/\mathbf{Z}) \xrightarrow{\gamma} \mathbf{Q}/\mathbf{Z}.$$

Clearly the image of $x.\delta\chi$ in $H^2(\mathfrak{g}, \mathbf{Z})$ is equal to $v(x).\delta\chi$; its image in $H^1(\mathfrak{g}, \mathbf{Q}/\mathbf{Z})$ is therefore $v(x).\chi$, and its image in $\mathbf{Q}/\mathbf{Z}$ is then $v(x)\chi(F) = \chi(F^{v(x)})$. □

Corollary. *Let* F/E *be an abelian extension with Galois group* G. *Let* $U_E \subset E^*$ *be the group of units of* E. *Then the reciprocity map* $E^* \to G$ *sends* U_E *onto the inertia subgroup of* G.

Let T be the inertia subgroup of G, and let E′ be the corresponding subextension of F. The extension E′/E is unramified; applying the preceding proposition to it, we see that U_E is mapped trivially into G/T, hence the image of U_E is contained in T. Conversely, let $t \in T$, and let $a \in E^*$ be such that $(a, F/E) = t$. Put $f = [E':E]$. As $(a, F/E)$ is trivial on E′, the above proposition shows that f divides $v(a)$. Hence there exists $b \in F^*$ such that $N(b)$ and a have the same valuation. If we put $u = a.N(b)^{-1}$, then $u \in U_E$, and $(u, F/E) = (a, F/E) = t$. Thus U_E maps onto T. □

Remark. One has an analogous result for the higher ramification: the reciprocity map $E^* \to G$ sends the U_E^n onto the ramification groups G^n (with upper numbering); cf. Chap. XV, §2.

Prop. 13 can be stated more nicely by "passing to the limit". Let $\mathfrak{A}_K$ be the Galois group of the maximal abelian extension K^a of K; if K_{nr} is the maximal unramified extension of K, then $K_{nr} \subset K^a$, whence the exact sequence

$$0 \to \mathfrak{T}_K \to \mathfrak{A}_K \to \hat{\mathbf{Z}} \to 0$$

where $\mathfrak{T}_K$ is the *inertia group of* K^a/K (defined by passage to the limit from the finite case).

On the other hand, we have the exact sequence

$$0 \to U_K \to K^* \to \mathbf{Z} \to 0.$$

Prop. 13 can then be expressed by the commutativity of the diagram

$$\begin{array}{ccccccccc} 0 & \to & U_K & \to & K^* & \to & \mathbf{Z} & \to & 0 \\ & & \downarrow{\scriptstyle \omega_T} & & \downarrow{\scriptstyle \omega} & & \downarrow{\scriptstyle i} & & \\ 0 & \to & \mathfrak{T}_K & \to & \mathfrak{A}_K & \to & \hat{\mathbf{Z}} & \to & 0 \end{array}$$

where ω is the reciprocity map $x \to (x, */K)$, and i is the canonical injection of $\mathbf{Z}$ into $\hat{\mathbf{Z}}$. As for ω_T, it allows us to identify $\mathfrak{T}_K$ with the completion of U_K for the topology induced by the norm groups; we will see in Chap. XIV that it is an isomorphism in the classical case (i.e., when $\bar{K}$ is finite).

EXERCISES

Let E'/E be a finite extension (not necessarily separable), let F/E be a Galois extension, and let $F' = E'F$. Identify the Galois group $G_{F'/E'}$ with a subgroup of the group $G_{F/E}$.

a) Let $y \in E'^*$ and let $x = N_{E'/E}(y)$. Show that $(x, F/E) \in G^a_{F/E}$ is the canonical image of $(y, F'/E') \in G^a_{F'/E'}$.
b) Put $d = [F':F] = [E':F \cap E']$. Let $x \in E^*$. Show that $(x, F'/E')$ is equal to the dth power of the transfer of $(x, F/E)$ in $G^a_{F'/E'}$. (Handle separately the cases where E'/E is separable or purely inseparable; argue directly in the latter case, whereas in the former, apply props. 10 and 12 to a Galois extension of E containing F'.)
c) Suppose that F/E is abelian. Show that, in order for an element $y \in E'^*$ to belong to $N_{F'/E'}(F'^*)$, it is necessary and sufficient that $N_{F'/E}(y)$ belong to $N_{F/E}(F^*)$.

§5. Dwork's Theorem

Throughout this section, L/K denotes a totally ramified Galois extension, with Galois group G. If K_{nr} is the maximal unramified extension of K, and if we put $L_{nr} = LK_{nr}$, we know that L_{nr}/K_{nr} is Galois with group G; the same holds for the completed extension $\hat{L}_{nr}/\hat{K}_{nr}$. Let v (resp. w) be the discrete valuation of $\hat{K}_{nr}$ (resp. $\hat{L}_{nr}$). We use the exponential notation x^s to denote $s(x)$, for $s \in G$, $x \in \hat{L}_{nr}$.

We begin with a general result (which is valid without assuming $\bar{K}$ to be quasi-finite).

Proposition 14. *For every $q \in \mathbf{Z}$, $\hat{H}^q(G, \hat{L}^*_{nr}) = 0$.*

This follows from prop. 11 of Chap. X combined with prop. 7 of Chap. V.

Corollary 1. *The exact sequence $0 \to \hat{U}_{nr} \to \hat{L}^*_{nr} \to \mathbf{Z} \to 0$ defines an isomorphism of $G^a = \hat{H}^{-2}(G, \mathbf{Z})$ onto $\hat{H}^{-1}(G, U_{nr})$.*

Obvious.

(Notice that this isomorphism assigns to an element $s \in G$ the class of π^{s-1} in $\hat{H}^{-1}(G, \hat{U}_{nr})$, π being a uniformizer of $\hat{L}_{nr}$.)

Corollary 2. *Let $s_i \in G$ and $z_i \in \hat{L}^*_{nr}$ be elements such that*

$$\prod z_i^{s_i - 1} = 1.$$

Then $\prod s_i^{w(z_i)} = 1$ in G^a.

Let π uniformize $\hat{L}_{nr}$. Then $z_i = u_i \pi^{n_i}$, with $n_i = w(z_i)$. We have

$$\pi^{\Sigma n_i(s_i - 1)} \prod u_i^{s_i - 1} = 1.$$

Let I be the augmentation ideal of $\mathbf{Z}[G]$. If we set $s = \prod s_i^{n_i}$, then

$$s - 1 \equiv \sum n_i(s_i - 1) \text{ mod. } I^2, \quad \text{and} \quad \pi^{I^2} \subset \hat{U}_{nr}^I.$$

The above formula thus means that the image of π^{s-1} in $\hat{H}^{-1}(G, \hat{U}_{nr})$ is zero, and by the preceding corollary, this implies that the image of s in G^a is trivial. □

Back to the case where $\bar{K}$ is quasi-finite. Put

$$\mathfrak{g} = G(K_{nr}/K) = G(L_{nr}/L)$$

and let F be the canonical generator of $\mathfrak{g}$. The elements of $\mathfrak{g}$—in particular F—extend by continuity to automorphisms of $\hat{L}_{nr}$ and $\hat{K}_{nr}$ which commute with the elements of G.

Lemma 1. *In order that an element x of $\hat{L}_{nr}$ (resp. of $\hat{K}_{nr}$) belong to* L (resp. *to* K), *it is necessary and sufficient that* $Fx = x$.

Necessity is obvious. Let us prove sufficiency for L (that for K is similar). Suppose $x \in \hat{L}_{nr}$ is such that $Fx = x$. Let π be a uniformizer of L. Write $x = \pi^n u$, with $w(u) = 0$, so that $Fu = u$; thus we are reduced to the case where x is a unit. If $\bar{x} \in \bar{L}_{nr}$ is the residue class of x, then $F\bar{x} = \bar{x}$, whence $\bar{x} \in \bar{L}$, and there exists $a_0 \in U_L$ such that $\bar{a}_0 = \bar{x}$. Thus we can write $x = a_0 + \pi x_1$, with $w(x_1) \geq 0$, and $Fx_1 = x_1$. Applying the same reasoning to x_1, and iterating this procedure, we get series expansion of x:

$$x = a_0 + \pi a_1 + \cdots + \pi^n a_n + \cdots, \qquad a_i \in L,\ w(a_i) \geq 0,$$

and since L is complete, this shows that x belongs to L. □

In the theorem below, N denotes the norm for the extension $\hat{L}_{nr}/\hat{K}_{nr}$; it extends $N_{L/K}$.

Theorem 2 (Dwork). *Let* L/K *be a totally ramified Galois extension, having abelian Galois group* G. *Suppose that the residue field* $\bar{K} = \bar{L}$ *is quasi-finite. Let* $x \in K^*$; *let* $y \in \hat{L}_{nr}^*$ *be such that* $Ny = x$; *let* $z_i \in \hat{L}_{nr}^*$, $s_i \in G$ *be such that*

$$y^{F-1} = \prod z_i^{s_i - 1}.$$

Put $s = \prod s_i^{w(z_i)}$. *Then* $(x, L/K) = s^{-1}$.

[Given x, there always exists y such that $Ny = x$ as above, because $N(\hat{L}_{nr}^*) = \hat{K}_{nr}^*$ (one could even choose y in L_{nr}^* if desired). Then $N(y^{F-1}) = x^{F-1}$; by prop. 14, there exist z_i and s_i such that $y^{F-1} = \prod z_i^{s_i-1}$, and by cor. 2 to prop. 14, the element $s = \prod s_i^{w(z_i)}$ does not depend on the choice of the z_i and s_i. Thus th. 2 provides *a method for computing* $(x, L/K)$.]

Put $y = y_0 . y'$, with $y_0 \in L^*$, and y' unit of $\hat{L}_{nr}$. Then $x = x_0 . x'$, with $x_0 = Ny_0$, $x' = Ny'$, whence

$$(x, L/K) = (x', L/K).$$

On the other hand, $y'^{F-1} = y^{F-1} = \prod z_i^{s_i - 1}$. Thus it suffices to prove the theorem for x' and y', i.e., *we may assume that x and y are units.*

The Galois group of $L_{nr} = L \otimes_K K_{nr}$ over K can be identified with the direct product $G \times \mathfrak{g}$; the elements of this group act on $\hat{L}_{nr}$ (extended by continuity). Put

$$t = (x, L/K)$$

and consider the subfield L′ of L_{nr} fixed by $t.F = t \otimes F$. As the closed subgroup $\mathfrak{g}'$ of $G \times \mathfrak{g}$ generated by $t.F$ is a complement of G, Galois theory implies that the extension L′/K is linearly disjoint from K_{nr}/K, and that $L'K_{nr} = L_{nr}$. In particular, $N_{L'/K}$ is just the restriction of N to L′.

Let π be a uniformizer of L, and let $\pi_K = N\pi$; it is a uniformizer of K. Then

$$t.F = (x\pi_K, L_{nr}/K).$$

Indeed, $(x, L_{nr}/K)$ is equal to t on L and to the identity on K_{nr}, since x is a unit; on the other hand, $(\pi_K, L_{nr}/K)$ is the identity on L since π_K is a norm, and it is F on K_{nr} by prop. 13.

Since $t.F$ is trivial on L′, there exists $z \in L'^*$ such that $Nz = x\pi_K$. Hence

$$Nz = N(y\pi), \quad \text{i.e., } N(z^{-1}y\pi) = 1.$$

Applying prop. 14 with $q = -1$, we see that there exist $b_j \in \hat{L}_{nr}^*$ and $s_j \in G$ such that

$$z^{-1}y\pi = \prod b_j^{s_j - 1}.$$

We have $z^{tF-1} = 1$, $\pi^{tF-1} = \pi^{t-1}$, and

$$y^{tF-1} = y^{-1}(y.\prod z_i^{s_i - 1})^t = y^{t-1}\prod z_i^{t(s_i - 1)}.$$

We deduce that

$$y^{t-1}\pi^{t-1}\prod z_i^{t(s_i-1)} = \prod b_j^{(tF-1)(s_j-1)}.$$

Applying the second corollary to prop. 14 to this identity, and taking into account that

$$w(y) = 0,\ w(\pi) = 1,\ w(z_i^t) = w(z_i),\ w(b_j^{tF-1}) = 0,$$

we get $t.s = 1$. □

Corollary. *Let $x \in K^*$, $y \in \hat{L}_{nr}^*$, $s \in G$ be such that*

$$x = Ny, \quad y^{F-1} = \pi^{s-1}, \qquad \pi \text{ being a uniformizer of } \hat{L}_{nr}.$$

Then $(x, L/K) = s^{-1}$.

Clear.

When $\bar{K}$ has characteristic $p > 0$, which is by far the most interesting case, the above corollary suffices to determine the reciprocity isomorphism.

Indeed, we first have:

Proposition 15. *Suppose* $\bar{K}$ *has positive characteristic. Let* V *be the group of units of* $\hat{L}_{nr}$. *For every integer* $m \geq 1$, *the homomorphism* $x \mapsto x^{F-1}$ *maps* $V^{(m)}$ *onto itself. If* x *is an element of* V *whose residue class* $\bar{x} \in \bar{K}^*_{nr}$ *is a root of unity, then there exists* $y \in V$ *such that* $x = y^{F-1}$.

(Here $V^{(m)}$ denotes the set of $x \in V$ such that $w(x-1) \geq m$.)

The group $V^{(m)}$ is complete Hausdorff for the filtration by the $V^{(m+k)}$, and the quotients $V^{(m+k)}/V^{(m+k+1)}$ can be identified with the additive group $\bar{K}_{nr}$. As $F - 1: \bar{K}_{nr} \to \bar{K}_{nr}$ is surjective (prop. 4), lemma 2 of Chap. V shows that the restriction of $F - 1$ to $V^{(m)}$ is also surjective. Furthermore, if $x \in V$ is such that $\bar{x}$ is a root of unity, prop. 4, a) shows the existence of $z \in V$ such that $\bar{z}^{F-1} = \bar{x}$, whence $x = z^{F-1}x'$, with $x' \in V^{(1)}$; by the above, $x' = y'^{F-1}$. □

Corollary. *If* $\bar{K}$ *is a finite field, then* $V = V^{F-1}$.

(If the proalgebraic structure of V were used, this would also follow from a general theorem of Lang [39] on algebraic groups.)

We go back to the situation of theorem 1, still assuming that $\bar{K}$ has positive characteristic. Let $s \in G$ and let π be a uniformizer of $\hat{L}_{nr}$. Consider π^{s-1}; it is an element of V, and its image in $\bar{K}^*_{nr}$ is a root of unity (cf. Chap. IV, §2). By prop. 15, there exists $y \in V$ such that $y^{F-1} = \pi^{s-1}$. Take the norm of both sides, and put $x = Ny$; we find $x^{F-1} = 1$, whence $x \in K^*$ (lemma 1). Therefore we may apply the above corollary, and we see that $(x, L/K) = s^{-1}$. Thus *the corollary does suffice to determine the reciprocity map.*

Exercises

1. Let V be the group of units of $\hat{L}_{nr}$; show that V/V^{F-1} is a torsion-free divisible group; deduce that it is a cohomologically trivial G-module.

2. Let L/K be a finite Galois extension, of group G, not necessarily totally ramified. Define L_{nr} to be the tensor product $K_{nr} \otimes_K L$, and similarly for $\hat{L}_{nr}$. These are products of fields. Make G and F act on L_{nr} and $\hat{L}_{nr}$ in the obvious way.
   a) Let $\hat{L}^*_{nr}$ be the multiplicative group of invertible elements of $\hat{L}_{nr}$, let $w: L^*_{nr} \to \mathbf{Z}$ be the sum of the discrete valuations of the different components of $\hat{L}_{nr}$, and let $V = \mathrm{Ker}(w)$. Let V′ be the set of all y^{F-1}, $y \in \hat{L}^*_{nr}$; then $V' \subset V$. Put $H = V/V'$. Show that H and L^*_{nr} are cohomologically trivial G-modules. (Use exercise from Chap. VII, §5, to reduce to the totally ramified case.)
   b) Using the coboundaries of the exact sequences

$$0 \to L^* \to \hat{L}^*_{nr} \xrightarrow{F-1} V' \to 0$$

$$0 \to V \to \hat{L}^*_{nr} \xrightarrow{\omega} Z \to 0$$

   as well as the isomorphism $\hat{H}^q(G, V') \to \hat{H}^q(G, V)$, define isomorphisms

$$\gamma: \hat{H}^q(G, \mathbf{Z}) \to \hat{H}^{q+2}(G, L^*), \qquad q \in \mathbf{Z}.$$

Show that γ is given by the cup product with the class $v_{L/K} = \gamma(1)$ of $H^2(G, L^*)$.

c) Show that the $v_{L/K}$ are fundamental classes for a class formation (cf. Chap. XI, §3).

d) Let $u_{L/K}$ be the fundamental class defined in §4. Show that $u_{L/K} = -v_{L/K}$. (Reduce to the unramified case by means of c), and make a direct computation in this case.)

e) Make γ explicit for $q = -2$, and deduce another proof of Dwork's theorem, thanks to d).

CHAPTER XIV

Local Symbols and Existence Theorem

§1. General Definition of Local Symbols

Let K be a field, K_s a separable closure of K, and let $G = G(K_s/K)$. If χ is a *character* of G—i.e., an element of $H^1(G, \mathbf{Q}/\mathbf{Z})$—then $\delta\chi$ is an element of $H^2(G, \mathbf{Z})$. If $b \in K^*$, the cup product $b\,.\,\delta\chi$ is an element of the Brauer group $H^2(G, K_s^*) = B_K$. We denote this element by the symbol (χ, b).

Proposition 1

(i) $(\chi + \chi', b) = (\chi, b) + (\chi', b)$.
(ii) $(\chi, bb') = (\chi, b) + (\chi, b')$.

Obvious.

We work with a character χ of G. Let H_χ be its kernel, and let L_χ be the extension of K corresponding to H_χ. It is a cyclic extension. More precisely, if n is equal to the order of χ, L_χ/K is cyclic of degree n, and we choose as generator of its Galois group an element s such that $\chi(s) = 1/n$. This choice of s allows us to identify K^*/NL_χ^* with the subgroup $H^2(L_\chi/K)$ of B_K.

Proposition 2. *Let* $b \in K^*$. *The element of* $H^2(L_\chi/K)$ *which corresponds to* b *is none other than* (χ, b).

This follows from what has been said in Chap. VIII, §4.

Corollary 1. *For* (χ, b) *to be zero, it is necessary and sufficient that* b *be a norm in the extension* L_χ/K.

Corollary 2. *For an element of* B_K *to have the form* (χ, b), *it is necessary and sufficient that it be split by* L_χ.

Let us now pass to the case of local class fields, i.e., suppose that K is *complete* for a discrete valuation v with *quasi-finite* residue field $\bar{K}$. We have defined in the previous chapter an isomorphism

$$\mathrm{inv}_K : B_K \to \mathbf{Q}/\mathbf{Z}.$$

We set

$$(\chi, b)_v = \mathrm{inv}_K(\chi, b)$$

which is an element of $\mathbf{Q}/\mathbf{Z}$. The next proposition gives a more direct definition.

Proposition 3. *Let* $s_b = (b, */K)$ *be the element of* G^a *defined by the reciprocity map. Then*

$$(\chi, b)_v = \chi(s_b).$$

This follows from prop. 2 of Chap. XI, §3.

Corollary. *If* χ *is a character such that* $(\chi, b)_v = 0$ *for all* $b \in K^*$, *then* $\chi = 0$.

Indeed, we then have $\chi(s_b) = 0$ for all b, and as the elements of G^a of the form s_b are dense in G^a, it follows that χ is trivial.

Remark. Proposition 3 shows that *the knowledge of the* $(\chi, b)_v$ *for all* χ *is equivalent to that of the* $(b, */K)$. This is why these symbols are interesting.

In order to make the above more explicit, it remains to construct some characters χ. There are essentially two methods for that:

i) Kummer theory (cf. Chap. X, §3), which leads to the symbols (a, b),
ii) Artin-Schreier theory (*loc. cit.*), which leads to the symbols $[a, b)$.

This will be the subject of the next two sections. Afterwards, we will see how the results obtained yield the existence theorem for local fields with finite residue field.

Exercise

Let L be a finite separable extension of K, and let $H = G(K_s/L)$. Consider the homomorphisms

$$\mathrm{Res} : B_K \to B_L, \qquad \mathrm{Res} : X(G) \to X(H)$$
$$\mathrm{Cor} : B_L \to B_K, \qquad \mathrm{Cor} : X(H) \to X(G).$$

Show that

$$\mathrm{Res}(\chi, b) = (\mathrm{Res}\,\chi, b) \qquad \chi \in X(G),\ b \in K^*,$$
$$\mathrm{Cor}(\psi, b) = (\mathrm{Cor}\,\psi, b) \qquad \psi \in X(H),\ b \in K^*,$$
$$\mathrm{Cor}(\chi, c) = (\chi, Nc) \qquad \chi \in X(G),\ c \in L^*.$$

How must these formulas be modified when the extension L/K is no longer assumed to be separable?

§2. The Symbol (a, b)

Let n be a positive integer. Throughout this section we will assume that *n is prime to the characteristic of* K, and that K *contains the group* μ_n *of* nth *roots of unity*.

We have seen in Chap. X, §3 that each element $a \in K^*$ defines a homomorphism $\varphi_a : G \to \mathbf{Z}/n\mathbf{Z}$ as follows: we choose in K_s a root α of the equation $\alpha^n = a$, and set $\varphi_a(s) = s(\alpha)/\alpha$. Next we choose a *primitive* nth *root of unity* w; that has the effect of identifying μ_n with the group $\mathbf{Z}/n\mathbf{Z}$; if we then set

$$\chi_a(s) = \frac{1}{n}\,\varphi_a(s)$$

we obtain a homomorphism of G into the subgroup $(1/n)\mathbf{Z}/\mathbf{Z}$ of $\mathbf{Q}/\mathbf{Z}$, i.e., a *character* of G. We know (*loc. cit.*) that the map $a \to \chi_a$ defines an isomorphism of K^*/K^{*n} onto the group of those characters of G having order dividing n.

Now let $b \in K^*$. Put

$$(a, b) = (\chi_a, b).$$

This is an element of B_K of order dividing n.

Proposition 4.

i) $(aa', b) = (a, b) + (a', b)$.
ii) $(a, bb') = (a, b) + (a, b')$.
iii) *In order that* $(a, b) = 0$, *it is necessary and sufficient that* b *be a norm in the extension* $K(a^{1/n})/K$.
iv) *If* $a \in K^*$ *and* $x \in K$ *are such that* $x^n - a \neq 0$, *then* $(a, x^n - a) = 0$. *In particular*, $(a, -a) = 0$ *and* $(a, 1 - a) = 0$.
v) $(a, b) + (b, a) = 0$.

Properties i) and ii) result from prop. 1; iii) results from cor. 1 to prop. 2. As for iv), we rely on the identity

$$x^n - a = \prod_{i=0}^{n-1} (x - w^i\alpha), \qquad \alpha^n = a.$$

Let d be the largest divisor of n for which the equation $a = y^d$ has a solution $y \in K$, and put $n = dm$. The extension $K(\alpha)/K$ is cyclic of degree m, and the conjugates of $x - w^i\alpha$ are none other than the elements $x - w^j\alpha$, with $j \equiv i \bmod d$. Hence

$$x^n - a = \prod_{i=0}^{d-1} N(x - w^i\alpha)$$

which shows that $x^n - a$ is indeed a norm in the extension $K(\alpha)/K$. Setting

$x = 0$ gives $(a, -a) = 0$, and setting $x = 1$ gives $(a, 1 - a) = 0$. Finally,

$$\begin{aligned} 0 = (ab, -ab) &= (a, -ab) + (b, -ab) \\ &= (a, -a) + (a, b) + (b, a) + (b, -b) \\ &= (a, b) + (b, a) \end{aligned}$$

which proves v). □

Remarks. 1) The symbol (a, b) *depends on the choice of* w; if we wished to avoid this dependence, it would be necessary to redefine it to be an element of $\mu_n \otimes \mathrm{B}_{\mathrm{K},n}$ where $\mathrm{B}_{\mathrm{K},n}$ denotes the group of all $\xi \in \mathrm{B}_{\mathrm{K}}$ such that $n\xi = 0$.

2) Properties i) and ii) show that (a, b) depends only on the cosets of a and b mod. K^{*n}.

3) For $n = 2$, we have $\mu_n = \{\pm 1\}$ and the symbol (a, b) is defined under the single hypothesis that K have characteristic different from 2. The corresponding Severi-Brauer variety is the conic having homogeneous equation $x^2 - ay^2 - bz^2 = 0$. In particular, $(a, b) = 0$ if and only if this conic has a rational point in K.

The symbol (a, b) can be interpreted as the cup product of two cohomology classes of degree 1. To be precise, consider the exact sequence

$$0 \to \mathbf{Z}/n\mathbf{Z} \to \mathrm{K}_s^* \xrightarrow{v} \mathrm{K}_s^* \xrightarrow{v} 0,$$

where $\mathbf{Z}/n\mathbf{Z}$ has been identified with μ_n, and $v(x) = x^n$. As $\mathrm{H}^1(\mathrm{G}, \mathrm{K}_s^*) = 0$, we get the exact sequence

$$0 \to \mathrm{H}^2(\mathrm{G}, \mathbf{Z}/n\mathbf{Z}) \to \mathrm{B}_{\mathrm{K}} \xrightarrow{n} \mathrm{B}_{\mathrm{K}}.$$

We denote by i the injection thus defined of $\mathrm{H}^2(\mathrm{G}, \mathbf{Z}/n\mathbf{Z})$ into the Brauer group B_{K}.

Now let a and b be two elements of K^*. By what was said at the beginning of this section, we can associate to them the elements φ_a, φ_b of the group $\mathrm{H}^1(\mathrm{G}, \mathbf{Z}/n\mathbf{Z})$. The cup product then yields an element $\varphi_a . \varphi_b$ of $\mathrm{H}^2(\mathrm{G}, \mathbf{Z}/n\mathbf{Z})$.

Proposition 5. $(a, b) = i(\varphi_a . \varphi_b)$.

To facilitate the computation, we will write all the G-modules additively—in particular, K_s^*. Moreover, if φ denotes a function on G with values in $\mathbf{Z}/n\mathbf{Z}$, we will denote by $\bar{\varphi}$ a lifting of this function to $\mathbf{Z}$.

Let $\beta \in \mathrm{K}_s^*$ be such that $n\beta = b$. By definition, (a, b) is equal to $b . \delta\chi_a$, and can therefore be represented by the cocycle

$$(s, t) \mapsto b . \left[\frac{1}{n}(\bar{\varphi}_a(s) + \bar{\varphi}_a(t) - \bar{\varphi}_a(st))\right] = \beta . (\bar{\varphi}_a(s) + \bar{\varphi}_a(t) - \bar{\varphi}_a(st)).$$

On the other hand, the anticommutativity of the cup product shows that $\varphi_a . \varphi_b = -\varphi_b . \varphi_a$. In view of the formula for the cup product of two cocycles (cf. Cartan-Eilenberg [13], p. 221), the class $-i(\varphi_b . \varphi_a)$ can be represented by the cocycle

$$(s, t) \mapsto -i(\bar{\varphi}_b(s), \bar{\varphi}_a(t)) = -(s\beta - \beta)\,\bar{\varphi}_a(t).$$

The difference of these two cocycles is

$$(s, t) \mapsto s\beta \,.\, \bar{\varphi}_a(t) - \beta \,.\, \bar{\varphi}_a(st) + \beta \,.\, \bar{\varphi}_a(s)$$

and this is the coboundary of the cochain $s \mapsto \beta \,.\, \bar{\varphi}_a(s)$. □

Remark. The preceding proposition shows that *the anticommutativity of the symbol (a, b) derives from that of the cup product.* Similarly, a cohomological proof of the triviality of the symbol $(a, -a)$ can be given—cf. exer. 1.

Let us now return to local class field theory, i.e., assume that K is complete under a discrete valuation v with quasi-finite residue field $\bar{K}$. If $a, b \in K^*$, we can apply inv_K to (a, b), and we obtain an element $\mathrm{inv}(a, b)$ of the subgroup $(1/n)\mathbf{Z}/\mathbf{Z}$ of $\mathbf{Q}/\mathbf{Z}$. We could take this element to be the "local symbol"; actually, it is more convenient (and more traditional) to replace it with:

$$(a, b)_v = w^{n\,.\,\mathrm{inv}(a, b)}$$

which makes sense because $n\,.\,\mathrm{inv}(a, b)$ is an element of $\mathbf{Z}/n\mathbf{Z}$. Thus *$(a, b)_v$ is an nth root of unity.*

Proposition 6. *If $\alpha \in K_s^*$ is an nth root of a, and if $s_b = (b, */K)$, then*

$$(a, b)_v = s_b(\alpha)/\alpha.$$

Indeed

$$\begin{aligned}(a, b)_v &= w^{n\,.\,\mathrm{inv}(a, b)} = w^{n\,.\,\mathrm{inv}(\chi_a, b)}\\ &= w^{n\chi_a(s_b)}. \quad \text{by prop. 3,}\\ &= w^{\varphi_a(s_b)} = s_b(\alpha)/\alpha, \quad \text{by definition of } \varphi_a.\end{aligned}$$

Corollary. *The symbol $(a, b)_v$ does not depend on the choice of the root of unity w.*

In fact, the formula in prop. 6 characterizes $(a, b)_v$, and w does not occur in this formula.

Proposition 7

i) $(aa', b)_v = (a, b)_v \,.\, (a', b)_v$.
ii) $(a, bb')_v = (a, b)_v \,.\, (a, b')_v$.
iii) *In order that $(a, b)_v = 1$, it is necessary and sufficient that b be a norm in the extension $K(a^{1/n})/K$.*
iv) $(a, -a)_v = 1 = (a, 1 - a)_v$.
v) $(a, b)_v \,.\, (b, a)_v = 1$.
vi) *If $(a, b)_v = 1$ for all $b \in K^*$, then $a \in K^{*n}$.*

Properties i) to v) follow from the corresponding properties of the symbol

(a, b)—cf. prop. 4. With the hypothesis of vi), we have $(\chi_a, b)_v = 0$ for all $b \in K^*$, whence $\chi_a = 0$ (cor. to prop. 3), i.e., $a \in K^{*n}$.

Corollary. *If an element $b \in K^*$ is a norm in every cyclic extension of K of degree dividing n, then $b \in K^{*n}$.*

By iii), we then have $(a, b)_v = 1$ for all a, whence $(b, a)_v = 1$ by v), and vi) implies $b \in K^{*n}$.

Remark. The symbol $(a, b)_v$ defined here is the inverse of the one in Artin-Tate ([8], Chap. XII), but ooincides with the one in Hasse's Bericht ([31], Teil II, §11), thanks to two compensating changes of sign. Properties i), ii) and iv) of prop. 7 show that $(a, b)_v$ is a "symbol" in the sense of algebraic K-theory; see Moore [100] and Tate [120, 121] for more on this.

EXERCISES

1. Assume that n is even, and let $m = n/2$.
   a) Let G be any group and let $\varphi \in H^1(G, \mathbf{Z}/n\mathbf{Z})$. Show that the square $\varphi . \varphi \in H^2(G, \mathbf{Z}/n\mathbf{Z})$ is equal to $m . \pi(\delta\varphi)$, where δ is the coboundary in the exact sequence $0 \to \mathbf{Z} \xrightarrow{n} \mathbf{Z} \to \mathbf{Z}/n\mathbf{Z} \to 0$, and π is the canonical homomorphism of $H^2(G, \mathbf{Z})$ into $H^2(G, \mathbf{Z}/n\mathbf{Z})$. (Reduce to the case where G is cyclic of order n, and make an explicit computation of cocycles.)
   b) Let $a \in K^*$. Applying the preceding result to $\varphi = \varphi_a$, recover the formula $(a, a) = (a, -1)$.

2. Let $a, b \in K^*$, with $a + b \neq 0$. Show that

$$(a, b) = (a, a + b) + (a + b, b) + (-1, a + b).$$

3. Assume that n is odd. Let $a, b, c \in K^*$ be three elements such that $a + b + c = 0$. Prove the formula

$$(a, b) + (b, c) + (c, a) = 0.$$

4. Let L/K be a finite separable extension. Denote by $(a, b)_K$ (resp. $(a, b)_L$) the symbol (a, b) computed in K (resp. L). Show that

$$\operatorname{Res}(a, b)_K = (a, b)_L \quad \text{if } a \in K^*, b \in K^*,$$
$$\operatorname{Cor}(a, c)_L = (a, Nc)_K \quad \text{if } a \in K^*, c \in L^*.$$

   Deduce analogous formulas for the symbols $(a, b)_v$.

5. With the hypotheses of local class field theory, show that given $a \in K^*$, a necessary and sufficient condition that $(a, b)_v = 1$ for every unit $b \in K^*$ is that $K(a^{1/n})/K$ be unramified.

§3. Computation of the Symbol $(a, b)_v$ in the Tamely Ramified Case

We remain in the set-up of local class fields, i.e., we suppose that K is a field that is complete under a discrete valuation v with quasi-finite residue field $\bar{K}$. We further assume that *n is an integer prime to the characteristic of* $\bar{K}$; the extensions $K(a^{1/n})/K$ then do not involve higher ramification: they are "tamely ramified".

Lemma 1. K *contains the* nth *roots of unity if and only if* $\bar{K}$ *does; in that case the canonical homomorphism* $U_K \to \bar{K}^*$ *induces an isomorphism of the group* μ_n *of* nth *roots of unity of* K *onto the group* $\bar{\mu}_n$ *of* nth *roots of unity of* $\bar{K}$.

If $\bar{K}$ has characteristic zero, then K is isomorphic to $\bar{K}((T))$, and the lemma is obvious. So suppose $\bar{K}$ has characteristic $p \neq 0$. If $\bar{K}$ contains the group $\bar{\mu}_n$ of nth roots of unity, the multiplicative representative (in the sense of Chap. II, §4) of these roots form the subgroup μ_n of U_K that is mapped isomorphically onto $\bar{\mu}_n$; thus K contains the nth roots of unity. Conversely, if K contains μ_n, the elements of μ_n must be multiplicative representatives (by assertion ii) of prop. 8 of Chap. II), and the image $\bar{\mu}_n$ of μ_n in $\bar{K}^*$ is isomorphic to μ_n, so that $\bar{K}$ contains the nth roots of unity. □

For the rest of this section we will assume that *the conditions of lemma* 1 *hold*, and we will agree to identify the groups μ_n and $\bar{\mu}_n$. The symbol $(a, b)_v$ takes its values in μ_n; we propose to compute it explicitly.

We will need the following auxiliary result.

Lemma 2. *Let k be a quasi-finite field containing the group μ_n of* nth *roots of unity (n being prime to the characteristic of k). Given $x \in k^*$, let $y \in k_s^*$ be a solution to the equation $y^n = x$, and let $z = Fy/y$. Then z belongs to μ_n, and does not depend on the choice of solution y; if we set $P_n(x) = z$, then the map P_n defines by passage to the quotient an isomorphism of k^*/k^{*n} onto μ_n.*

This follows—*via* Kummer theory—from the fact that k has one and only one cyclic extension of degree n. □

Example. If k is a *finite field* with q elements, the hypothesis that k contains μ_n is equivalent to saying that n *divides* $q - 1$. The map P_n is then given by the following formula:

$$P_n(x) = x^{\frac{q-1}{n}}.$$

Indeed, if $y^n = x$, then $P_n(x) = Fy/y = y^{q-1} = x^{(q-1)/n}$.

For $n = 2$, we recover the *Legendre symbol*.

We return to the computation of $(a, b)_v$.

Proposition 8. *Let a and b be two elements of K^*, and let α and β be their valuations. Set*

$$c = (-1)^{\alpha\beta} \frac{a^\beta}{b^\alpha}.$$

Then c is a root of unity in K, *and if* $\bar{c}$ *denotes its image in* $\bar{K}^*$, *then*

$$(a, b)_v = P_n(\bar{c}).$$

Corollary. *If* $\bar{K}$ *is a finite field with q elements, then* $(a, b)_v = \bar{c}^{(q-1)/n}$

PROOF OF PROP. 8. It is clear that c is a unit, and that $P_n(\bar{c})$ depends bilinearly (in the multiplicative sense) on a and b; as this is also the case for the symbol $(a, b)_v$, we are reduced to proving the formula when a and b are uniformizers. So put $a = \pi$, $b = -\pi u$, where u is a unit. Both sides equal 1 for $a = \pi$, $b = -\pi$; this reduces us to the case $a = \pi$, $b = u$, then by symmetry to the case $a = u$, $b = \pi$. Set $x = \bar{u}$, and let $\bar{K}'$ be a finite extension of $\bar{K}$ containing an nth root y of x; let K′ be the unramified extension of K corresponding to $\bar{K}'$. Since the equation $T^n = x$ admits the simple root $T = y$ in $\bar{K}'$, the equation $T^n = u$ admits a root v projecting onto y (cf. Chap. II, §4, prop. 7). By prop. 13 of Chap. XIII, $(\pi, K'/K)$ is equal to F, the canonical generator of the Galois group G(K′/K). If $w = (u, \pi)_v$, prop. 6 shows that $w = Fv/v$, whence $\bar{w} = Fy/y = P_n(x) = P_n(\bar{u})$. □

EXERCISE

Let K be a field complete under a discrete valuation with perfect residue field $\bar{K}$. Suppose that K contains the nth roots of unity, n being prime to the characteristic of $\bar{K}$. Determine explicitly the symbol $(a, b) \in B_K$ by using the decomposition $B_K = B_{\bar{K}} \times X(g)$ obtained from the choice of a uniformiser of K (cf. Chap. XII, §3, theorem 2).

§4. Computation of the Symbol $(a, b)_v$ for the Field $\mathbf{Q}_p$ $(n = 2)$

(We confine ourselves to this particularly simple example. The reader will find others, less trivial, in Artin-Tate [8], Chap. XII, in Hasse [31], Teil II, §§15–21 and in Iwasawa [85]; see also Chap. XV, §3.)

We take $n = 2$ and $K = \mathbf{Q}_p$, the usual p-adic field. Write $(a, b)_p$ instead of $(a, b)_v$. We distinguish two cases:

Case (i). $p \neq 2$.

Then we are in the tame case and can apply prop. 8. If $x \in \mathbf{F}_p^*$, then

$$P_2(x) = x^{(p-1)/2} = \left(\frac{x}{p}\right),$$

the Legendre symbol. If we then write a and b in the form

$$a = p^\alpha a', \qquad b = p^\beta b',$$

where a' and b' are units, we have

$$c = (-1)^{\alpha\beta} \frac{a'^\beta}{b'^\alpha}$$

whence

$$(a,b)_p = (-1)^{\frac{p-1}{2}\alpha\beta}\left(\frac{b'}{p}\right)^{\alpha}\left(\frac{a'}{p}\right)^{\beta}$$

In particular, $(p,p)_p = (-1)^{(p-1)/2}$, and

$$(p,b)_p = \left(\frac{b}{p}\right) \quad \text{if } b \text{ is a unit.}$$

Case (ii). $p = 2$.

Then we are no longer in the tame case and must argue directly. As $(a,b)_2$ is a bilinear form on K^*/K^{*2}, the first thing to do is to determine this group. We have:

Lemma 3. *Every element x of $\mathbf{Z}_2$ which is congruent to* 1 *mod.* 8 *is a square.*

Let us accept this lemma for the time being. Let U be the group of units of $\mathbf{Q}_2$, and let U′ be the subgroup of U of all x congruent to 1 mod. 8. A system of representatives of U/U′ is $\{1, -1.5, -5\}$; the square of each of these elements belongs to U′. Hence $U' = U^2$, and the group $\mathbf{Q}_2^*/\mathbf{Q}_2^{*2}$ is a group of type (2,2,2) generated by $\{2, 5, -1\}$. Thus it suffices to determine $(a,b)_2$ for these particular values.

We have $(-1,x)_2 = 1$ if and only if x is a norm from the extension $\mathbf{Q}_2(\sqrt{-1})/\mathbf{Q}_2$, i.e., if x can be written in the form $y^2 + z^2$, with $y, z \in \mathbf{Q}_2$. As $5 = 4 + 1$ and $2 = 1 + 1$, we have $(-1,2)_2 = (-1,5)_2 = 1$. If we had $(-1,-1)_2 = 1$, we would then have $(-1,x)_2 = 1$ for all x, and -1 would be a square in $\mathbf{Q}_2^*$, which is not the case. Thus $(-1,-1)_2 = -1$.

We have $(2,2)_2 = (2,-1)_2 = 1$ and similarly $(5,5)_2 = (5,-1)_2 = 1$. It remains to determine $(2,5)_2$. If it were equal to 1, we would have $(2,x)_2 = 1$ for all x and 2 would be a square, which it isn't. Hence $(2,5)_2 = -1$, which achieves the computation of $(a,b)_2$ in $\mathbf{Q}_2$. The result may be stated as follows:

If u is a unit of $\mathbf{Q}_2$, let $\omega(u)$ be the coset mod. 2 of $(u^2 - 1)/8$, and let $\varepsilon(u)$ be the coset mod. 2 of $(u-1)/2$. It can easily be verified that ω and ε are homomorphisms of U into $\mathbf{Z}/2\mathbf{Z}$. With these notations, we have:

$$(2,u)_2 = (-1)^{\omega(u)} \quad \text{if } u \text{ is a unit,}$$
$$(u,v)_2 = (-1)^{\varepsilon(u)\varepsilon(v)} \quad \text{if } u \text{ and } v \text{ are units.}$$

It remains to prove lemma 3. More generally:

Proposition 9. *Let K be a field complete under a discrete valuation v; suppose that K has characteristic zero and its residue field $\bar{K}$ has characteristic $p \neq 0$. Let $e = v(p)$ be the absolute ramification index of* K (cf. Chap. II, §5). *For every positive integer m, denote by $U^{(m)}$ the multiplicative group of those x in* K *such that $v(x-1) \geq m$* (cf. Chap. IV, §2). *Then, for $m > e/(p-1)$, the map $x \mapsto x^p$ is an isomorphism of $U^{(m)}$ onto $U^{(m+e)}$.*

(When $K = \mathbf{Q}_2$, we have $p = 2$, $e = 1$, and taking $m = 2$, we see that $x \mapsto x^2$ is an isomorphism of $U^{(2)}$ onto $U^{(3)}$, which proves lemma 3.)

Let π be a uniformizer of K, and let $y \in U^{(m+e)}$. We can write y in the form

$$y = 1 + a\pi^{m+e}, \quad \text{with } v(a) \geq 0.$$

We have $p = b\pi^e$, with $v(b) = 0$. We seek a root $x \in U^{(m)}$ of the equation $x^p = y$. Putting $x = 1 + z\pi^m$, with $v(z) \geq 0$, we are led to the equation

$$1 + a\pi^{m+e} = 1 + pz\pi^m + \cdots + z^p\pi^{mp}.$$

In view of the inequality $m > e/(p-1)$, all the terms on the right, except the first two, have valuation $> m + e$. Hence the preceding equation may be written in the form

$$a = bz + g(z)$$

where $g(z)$ is a polynomial whose coefficients are divisible by π. Reducing mod. π yields the equation $\bar{a} = \bar{b}.\bar{z}$, and since $\bar{b} \neq 0$, this equation has a unique root. By prop. 7 of Chap. II, the same holds for the equation $x^p = y$. □

The preceding proposition is the starting point of the study of the structure of the group U: cf. Hasse [34], §15 or [59], §1.8. We will give one simple application.

The group $U^{(1)}$ is the projective limit of the groups $U^{(1)}/U^{(m)}$, each of which is annihilated by a power of p; by passing to the limit, we conclude that $U^{(1)}$ *is a module over the ring* $\mathbf{Z}_p$ *of p-adic integers.*

Proposition 10. *With the hypotheses and notation of prop. 9, assume further that* K *is a finite extension of degree n of the field* $\mathbf{Q}_p$. *If* $m > e/(p-1)$, *then the group* $U^{(m)}$ *is a free* $\mathbf{Z}_p$*-module of rank n, and the group* $U^{(1)}$ *is the product of a free* $\mathbf{Z}_p$*-module of rank n and a cyclic group of order a power of p.*

Let $q = p^f$ be the cardinality of $\bar{K}$; we have $ef = n$. The group $U^{(m)}/U^{(m+e)}$ is an abelian group of type $(p, \ldots, p)$ and of order $q^e = p^n$. Let $x_1, \ldots, x_n$ be elements of $U^{(m)}$ generating $U^{(m)}$ mod. $U^{(m+e)}$. They define a homomorphism

$$\theta: \mathbf{Z}_p^n \to U^{(m)}.$$

Suppose $m > e/(p-1)$. Filter $\mathbf{Z}_p^n$ by the $(p^k\mathbf{Z}_p)^n$ and $U^{(m)}$ by the $U^{(m+ke)}$. Prop. 9 shows that θ is compatible with these filtrations and defines an isomorphism of the associated graded modules. By lemma 2 of Chap. V, §1, θ is an isomorphism, which proves the first part of the proposition. [One could also replace θ with an exponential.] The second follows from the remark that $U^{(1)}/U^{(m)}$ is a finite group, hence $U^{(1)}$ and $U^{(m)}$ have the same rank, and that the torsion submodule of $U^{(1)}$ must be cyclic. □

EXAMPLES. i) If $K = \mathbf{Q}_p$, with $p \neq 2$, we can take $m = 1$, and we see that $U^{(1)}$ is isomorphic to $\mathbf{Z}_p$; an element x of $U^{(1)}$ is a generator if and only if x does not belong to $U^{(2)}$; $x = 1 + p$ is an example.

ii) If $K = \mathbf{Q}_2$, we can take $m = 2$. The group $U = U^{(1)}$ can be identified with the direct product $\{\pm 1\} \times U^{(2)}$. The group $U^{(2)}$ is isomorphic to $\mathbf{Z}_2$; an element x of $U^{(2)}$ is a generator if and only if it does not belong to $U^{(3)}$; $x = 1 + 2^2 = 5$ is an example.

(We thereby recover well-known results; cf. Bourbaki, *Alg.*, Chap. VII, §2, no. 4.)

EXERCISES

1. Interpret the homomorphisms ω and ε of the group U of units of $\mathbf{Q}_2$ into $\mathbf{Z}/2\mathbf{Z}$ by means of the direct product decomposition $U = \{\pm I\} \times U^{(2)}$.

2. Let K be a field complete under a discrete valuation with finite residue field. If K has characteristic zero, show that every subgroup of finite index in K^* is closed. If K has characteristic p 0, show that there exist subgroups of index p in K^* that are not closed; what is the cardinality of this set of subgroups?

3. With the hypotheses of exer. 2, let m be an integer prime to the characteristic of K. Let μ_m be the group of mth roots of unity contained in K^*, and let

$$\varphi(K^*) = (K^* : K^{*m})\mathrm{Card}(\mu_m);$$

this is the Herbrand quotient of K^* for the group $\mathbf{Z}/m\mathbf{Z}$ acting trivially. Show that

$$\varphi(K^*) = m/\|m\|_K,$$

where $\|x\|_K$ denotes the normalised absolute value of K (cf. Chap. II, §1). (Use prop. 10 to show that $\varphi(U^{(1)}) = I/\|m\|_K$.)

§5. The Symbol $[a, b)$

Throughout this paragraph, K denotes a field *of characteristic* $p \neq 0$. Put $\wp(x) = x^p - x$; it is an endomorphism of the additive group of K.

Let K_s be a separable closure of K, and let $G = G(K_s/K)$. We have seen in Chap. X, §3 that every element $a \in K$ defines a homomorphism $\varphi_a : G \to \mathbf{Z}/p\mathbf{Z}$ as follows: we choose a root α in K_s of the equation $\wp(\alpha) = a$, and we set $\varphi_a(s) = s\alpha - \alpha$. If $\psi_a(s) = (1/p)\varphi_a(s)$, the map ψ_a is a homomorphism of G into the subgroup $(1/p)\mathbf{Z}/\mathbf{Z}$ of $\mathbf{Q}/\mathbf{Z}$, i.e., a *character of* G. We know (*loc. cit.*) that the map $a \to \psi_a$ defines an isomorphism of $K/\wp(K)$ onto the group of those characters ψ of G such that $p\psi = 0$.

Now let $b \in K^*$. According to the general definitions of §1, we set

$$[a, b) = (\psi_a, b).$$

This is an element of B_K; we have $p[a, b) = 0$.

Proposition 11.

i) $[a + a', b) = [a, b) + [a', b)$.
ii) $[a, bb') = [a, b) + [a, b')$.
iii) *In order that* $[a, b) = 0$, *it is necessary and sufficient that* b *be a norm in the extension* $K(\alpha)/K$, *where* $\wp(\alpha) = a$.
iv) $[a, a) = 0$ *for all* $a \in K^*$.

Properties i) and ii) result from prop. 1; iii) is a consequence of cor. 1 to prop. 2. Assertion iv) is obvious if a has the form $\wp(\alpha)$, with $\alpha \in K$. Otherwise the extension $K(\alpha)/K$ has degree p, and the conjugates of α are the elements $\alpha + i$, $i \in \mathbf{Z}/p\mathbf{Z}$. Since

$$\alpha(\alpha - 1) \cdots (\alpha - p + 1) = \alpha^p - \alpha = a$$

we see that a is equal to the norm of α in $K(\alpha)/K$, whence $[a, a) = 0$ by iii). □

EXAMPLE. For $p = 2$, the Severi-Brauer variety corresponding to $[a, b)$ is the conic having homogeneous equation $x^2 + xy + ay^2 + bz^2 = 0$. In particular, $[a, b) = 0$ if and only if this conic has a rational point in K.

We now pass to the case where K is the field $k((t))$ of formal series over a field k of characteristic p. If $\omega = f\,dt$ is a differential form of K, the coefficient of t^{-1} in f is called the *residue* of ω, and is denoted $\operatorname{Res}(\omega)$. It can be shown (cf. for example [56], Chap. II, prop. 5′) that it does not depend on the choice of uniformiser t.

Proposition 12. *Let* $K = k((t))$, *where* k *is a perfect field of characteristic* $p \neq 0$. *If* $a \in K$ *and* $b \in K^*$, *let* $c = \operatorname{Res}(a\,db/b) \in k$. *Then*

$$[a, b) = [c, t) \quad \text{in } B_K.$$

As both sides are bilinear, it suffices to consider the case where b is a uniformizer, i.e., the case $b = t$ (changing the variable if needed). If $a = \sum a_n t^n$, we decompose a into $a_0, \sum_{n<0} a_n t^n, \sum_{n>0} a_n t^n$, and we consider separately each of these cases:

Case i). *a is constant.* Then $\operatorname{Res}(a\,dt/t) = a$, and the formula to be proved becomes $[a, t) = [a, t)$.

Case ii). $a = ut^{-n}$, *with* $u \in k$ *and* $n > 0$. Then $\operatorname{Res}(a\,dt/t) = 0$, and we must check that $[a, t) = 0$. We argue by induction on n:

If n is prime to p, we have the relation

$$-n[ut^{-n}, t) = [ut^{-n}, t^{-n}) = [ut^{-n}, ut^{-n}) - [ut^{-n}, u).$$

Now $[ut^{-n}, ut^{-n}) = 0$ by prop. 11, and $[ut^{-n}, u) = 0$ since u is a pth power. As n is prime to p, we deduce $[ut^{-n}, t) = 0$.

If $n = mp$, we put $u = v^p$, whence

$$ut^{-n} = \wp(vt^{-m}) + vt^{-m}, \quad \text{and} \quad [ut^{-n}, t) = [vt^{-m}, t) = 0$$

by inductive hypothesis.

Case iii). $a = \sum_{n>0} a_n t^n$. Then $\mathrm{Res}(a\,dt/t) = 0$, and we must check that $[a, t) = 0$. That follows from the formula $a = -\wp(a')$, where

$$a' = a + a^p + a^{p^2} + \cdots. \quad \square$$

Remark. The above proposition reduces the computation of $[a, b)$ to that of $[c, t)$, where c is constant. The latter computation can be done in the general case (cf. exer. 1); we will still restrict ourselves to the case of local class field theory.

Suppose therefore that k is a *quasi-finite* field. Then we can transform the element $[a, b)$ by means of the isomorphism $\mathrm{inv}_K : \mathrm{B}_K \to \mathbf{Q}/\mathbf{Z}$. We get an element $\mathrm{inv}\,[a, b)$ of $(1/p)\mathbf{Z}/\mathbf{Z}$. Put

$$[a, b)_v = p \,.\, \mathrm{inv}[a, b).$$

This is an element of $\mathbf{Z}/p\mathbf{Z}$.

Proposition 13. *If* $\alpha \in \mathrm{K}_s$ *is a root of the equation* $\wp(\alpha) = a$, *and if* $s_b = (b, */\mathrm{K})$, *then*

$$[a, b)_v = s_b(\alpha) - \alpha.$$

This follows from prop. 3. $\square$

Proposition 14.

i) $[a + a', b)_v = [a, b)_v + [a', b)_v$.
ii) $[a, bb')_v + [a, b')_v$.
iii) *In order that* $[a, b)_v = 0$, *it is necessary and sufficient that* b *be a norm in the extension* $\mathrm{K}(\alpha)/\mathrm{K}$, *where* $\wp(\alpha) = a$.
iv) $[a, a)_v = 0$ *for all* $a \in \mathrm{K}^*$.
v) $[a, b)_v = 0$ *for all* $b \in \mathrm{K}^*$, *then* $a \in \wp(\mathrm{K})$.

Properties i) to iv) follow from the corresponding properties of the symbol $[a, b)$—cf. prop. 11. Property v) follows from the corollary to prop. 3. $\square$

We now compute $[a, b)_v$ explicitly. We need the following result which is similar to lemma 2:

Lemma 4. *Let* k *be a quasi-finite field of characteristic* $p \neq 0$. *Given* $x \in k$, *let* $y \in k_s$ *be a solution to the equation* $\wp(y) = x$, *and let* $z = \mathrm{F}y - y$. *Then* $z \in \mathbf{F}_p = \mathbf{Z}/p\mathbf{Z}$, *and* z *does not depend on the choice of solution* y; *if we put* $\mathrm{S}(x) = z$, *then the map* S *defines by passage to the quotient an isomorphism of* $k/\wp(k)$ *onto* $\mathbf{Z}/p\mathbf{Z}$.

That is a consequence of Artin-Schreier theory and of the fact that k has one and only one cyclic extension of degree p. $\square$

EXAMPLE. If k is a *finite field* with $q = p^f$ elements, the map S is given by the formula

$$S(x) = x^{p^{f-1}} + x^{p^{f-2}} + \cdots + x^p + x = \mathrm{Tr}_{k/\mathbf{F}_p}(x).$$

Indeed:

$$\begin{aligned} S(x) &= Fy - y = y^{p^f} - y \\ &= (y^{p^f} - y^{p^{f-1}}) + \cdots + (y^p - y) \\ &= x^{p^{f-1}} + \cdots + x^p + x = \mathrm{Tr}_{k/\mathbf{F}_p}(x). \end{aligned}$$

We go back to the computation of $[a, b)_v$.

Proposition 15. *Let* $K = k((t))$, *where* k *is a quasi-finite field of characteristic* p. *If* $a \in K$ *and* $b \in K^*$, *then*

$$[a, b)_v = S\left(\mathrm{Res}\left(a\frac{db}{b}\right)\right).$$

Corollary. *If* k *is a finite field, then* $[a, b)_v = \mathrm{Tr}_{k/\mathbf{F}_p}(\mathrm{Res}(a\, db/b))$.

In view of prop. 12, it suffices to consider the case a constant (i.e., an element of k) and $b = t$. We must show that $[a, t)_v = S(a)$.

Let k' be a finite extension of k containing a root α of the equation $\wp(\alpha) = a$, and let $K' = k'((t))$. The extension K'/K is unramified, and prop. 13 of Chap. XIII shows that $(t, K'/K) = F$, the canonical generator of the Galois group $G(K'/K) = G(k'/k)$. By prop. 13,

$$[a, t)_v = F\alpha - \alpha = S(a). \quad \square$$

Proposition 16. *Let* $K = k((t))$, *where* k *is a quasi-finite field of characteristic* p. *If an element* $b \in K^*$ *is a norm in every cyclic extension of* K *of degree* p, *then* $b \in K^{*p}$.

The hypothesis amounts to saying that $[a, b)_v = 0$ for all $a \in K$. If b were not a pth power, the differential form db/b would not be identically zero; for every constant c, one could then choose a so that $a\, db/b = c\, dt/t$; by prop. 15 we would have $S(c) = 0$, which is absurd since $S: k/\wp(k) \to \mathbf{Z}/p\mathbf{Z}$ is surjective. $\square$

Remarks. 1) Prop. 15 is due to H. Schmid [55]; a proof from the "global" viewpoint can be found in [56], Chap. VI, no. 30.

2) All the results of this section extend to the characters of order a power of p given by Witt vectors; thus one is led to symbols $[a, b)$ with $a \in W_n(k)$, $b \in K^*$; cf. Witt [73].

EXERCISES

1. Let k be a perfect field of characteristic p, let $\mathfrak{g} = G(k_s/k)$, and let $X(\mathfrak{g})$ be the character group of $\mathfrak{g}$. Let $K = k((t))$. The p-primary component of the Brauer group of k is

zero (cf. Chap. X, §4, exer. 1). Theorem 2 of Chap. XII then shows that the p-primary component of B_K may be identified with that of $X(\mathfrak{g})$. Let $c \in k$. Show that the image of $[c, t)$ under this isomorphism is the element ψ_c of $X(\mathfrak{g})$ defined as in the beginning of this section.

2. Let k be a field of characteristic p, let $K = k((t))$, and let $c \in k$, $c \notin k^p$. Show that $[t^{-1}, c)$ is an element of B_K of order p which is not split by any unramified extension of K.

§6. The Existence Theorem

We now place ourselves in the *usual* framework of local class field theory: K denotes a field complete under a discrete valuation v with *finite* residue field (K is then *locally compact*—cf. Chap. II, §1).

Theorem 1. *Every closed subgroup of finite index in* K^* *is a norm group* (in the sense of Chap. XI, §4).

(In other words, if H is such a subgroup, there exists a finite abelian extension L/K such that $N_{L/K}(L^*) = H$; moreover, if such an extension exists, we know that it is unique.)

Consider the class formation of Chap. XIII, §4. We show that it satisfies the conditions of th. 2 of Chap. XI, §5. There are three axioms to check.

III.1. *For every finite extension* F/E (E *being a finite extension of* K), *the map* $N_{F/E}: F^* \to E^*$ *is proper.*

Indeed, every compact subset of E is contained in a finite number of translates of the group of units U_E, and $N_{F/E}^{-1}(U_E) = U_F$, which is compact.

III.2. *For every prime number p, there exists a field* E_p *with the property that if* E *contains* E_p, *then the map* $x \to x^p$ *of* E^* *into itself has a compact kernel and an image that contains the group of universal norms.*

If p is distinct from the characteristic of K, take E_p to be the field obtained from K by adjoining the pth roots of unity. If E contains E_p, the kernel of $x \mapsto x^p$ is cyclic of order p, hence compact. If $x \in E^*$ is a universal norm, the corollary to prop. 7 shows that $x \in E^{*p}$.

If p equals the characteristic of K, take $E_p = K$. The kernel of $x \mapsto x^p$ is trivial. If $x \in E^*$ is a universal norm, prop. 16 shows that $x \in E^{*p}$.

III.3. *There exists a compact subgroup* U_E *of* E^* *such that every closed subgroup of finite index of* E^* *that contains* U_E *is a norm group.*

Take U_E to be the *group of units* of E. The subgroups of finite index of E^* that contain U_E are the inverse images under the discrete valuation v_E of the subgroups $n\mathbf{Z}$ of $\mathbf{Z}$. Prop. 13 of Chap. XIII (or prop. 3 of Chap. V) shows that these groups are norm groups for unramified extensions of K. □

Corollary 1. *In order that a subgroup of finite index of* K^* *be a norm group, it is necessary and sufficient that it contain* $U_K^{(n)}$ *for* n *sufficiently large.*

Indeed, to say that a subgroup H of K^* contains $U_K^{(n)}$ is equivalent to saying that H is *open*; if H has finite index in K^*, the latter is equivalent to H being closed.

Remark. Given a norm group H, there is obviously a *least* integer n such that H contains $U_K^{(n)}$; in the next chapter we will see how to determine this integer using the ramification groups of the extension corresponding to H.

Corollary 2

i) *The intersection of the norm groups of* K^* *is* $\{1\}$.
ii) *If* $\mathfrak{T}_K$ *denotes the inertia group of the maximal abelian extension* K^a/K, *then the reciprocity map* $U_K \to \mathfrak{T}_K$ *is an isomorphism of topological groups* (cf. Chap. XIII, §3).
iii) *The Galois group* $\mathfrak{A}_K = G(K^a/K)$ *of the maximal abelian extension is an extension of* $\hat{\mathbf{Z}}$ *by* U_K.

Let π be a uniformizer of K; if m and n are positive integers, let $V_{m,n}$ be the subgroup of K^* generated by π^m and $U_K^{(n)}$. These subgroups are closed, of finite index, and their intersection is $\{1\}$; since, by theorem 1, these are norm groups, i) follows. As for ii), it suffices to remark that the topology of U_K is defined by its closed subgroups of finite index, and that U_K is *compact* in this topology. Finally, iii) follows from ii), since $\mathfrak{A}_K/\mathfrak{T}_K = \hat{\mathbf{Z}}$. □

(Loosely speaking, one could say that $\mathfrak{A}_K$ is obtained from K^* by replacing the direct factor $\mathbf{Z}$ with its completion $\hat{\mathbf{Z}}$.)

Remarks. 1) When K has characteristic zero, every subgroup of finite index is closed (cf. §4, exer. 2); this is no longer the case when K has positive characteristic (*idem*).

2) Theorem 1 can be presented in a more striking manner for function fields, proceeding as in Weil [67], [123]:

Let K^a be the maximal abelian extension of K, which contains K_{nr}. Let $\mathfrak{A}_K^0$ be the subgroup of $\mathfrak{A}_K = G(K^a/K)$ consisting of those automorphisms that induce on K_{nr} an *integer* power of the Frobenius substitution F. We have $\mathfrak{T}_K \subset \mathfrak{A}_K^0$ and $\mathfrak{A}_K^0/\mathfrak{T}_K = \mathbf{Z}$; one topologises $\mathfrak{A}_K^0$ by considering $\mathfrak{T}_K$ as an *open* subgroup of $\mathfrak{A}_K^0$. Then the reciprocity map $x \mapsto (x, */K)$ defines an isomorphism of the topological group K^* onto the "modified Galois group" $\mathfrak{A}_K^0$.

3) Let K be a field complete under a discrete valuation with *quasi-finite* residue field $\bar{K}$. Then theorem 1 is still valid for the subgroups of K^* of finite index prime to the characteristic of $\bar{K}$ (cf. exer. 2), but it no longer

holds for those of index a power of this characteristic; cf. Whaples [71], where one can find various characterisations of the norm groups, based essentially on the structure of "proalgebraic group" on the group of units of K; see also Chap. XV, §1, exer. 2.

EXERCISES

1. Let K′ be a finite extension of K; show that the natural map $\mathfrak{A}_{K'} \to \mathfrak{A}_K$ and the transfer $\mathfrak{A}_K \to \mathfrak{A}_{K'}$ map the subgroups $\mathfrak{A}_K^0$ and $\mathfrak{A}_{K'}^0$ into each other. Make props. 10–13 of Chap. XIII explicit by means of the isomorphism $K^* \to \mathfrak{A}_K^0$.
2. Let K be a field complete under a discrete valuation with quasi-finite residue field. Let H be a subgroup of finite index of K^*, and let $n = (U_K : H \cap U_K)$. Assume that n is prime to the characteristic of $\bar{K}$.
   a) Show that H contains $U_K^{(1)}$. The group $H \cap U_K$ is therefore the inverse image of a subgroup $\bar{H}$ of index n of $\bar{K}^*$.
   b) Using exer. 1 of §1 of Chap. XV, show that $\bar{H} = \bar{K}^{*n}$, that K and $\bar{K}$ contain the nth roots of unity, and that H is a norm group of an abelian extension of K.

§7. Example: The Maximal Abelian Extension of $\mathbf{Q}_p$

Theorem 2. *Let p be prime, and let* E *be the field obtained by adjoining all the roots of unity to* $\mathbf{Q}_p$. *Then* E *is the maximal abelian extension of* $\mathbf{Q}_p$.

Put $K = \mathbf{Q}_p$ and $K^a =$ the maximal abelian extension of K. Obviously $E \subset K^a$. We have seen in Chap. IV, §4 that E is the composite of the linearly disjoint extensions K_{nr} and K_{p^∞}, whose Galois groups have been determined. We have

$$K^a \supset E \supset K_{nr}$$

whence a surjective homomorphism $\pi : G(K^a/K_{nr}) \to G(E/K_{nr})$. By cor. 2 to th. 1, $G(K^a/K_{nr})$ is isomorphic to the group U_p of units of $\mathbf{Q}_p$; the same holds for $G(E/K_{nr}) = G(K_{p^\infty}/K)$ by remark 1) of Chap. IV, §4. The homomorphism π can therefore be considered as a *surjective endomorphism* of U_p. But U_p is a $\hat{\mathbf{Z}}$-module isomorphic to the product of a finite group with $\mathbf{Z}_p$ (cf. §4); in particular, it is a Noetherian $\hat{\mathbf{Z}}$-module. Now any surjective endomorphism of a Noetherian module is bijective (Bourbaki, *Alg.*, Chap. VIII, §2, lemma 3). Thus π is bijective, which implies $E = K^a$. □

Remark. In fact one has $\pi(x) = x^{-1}$. This is immediate from the global viewpoint by using Artin's reciprocity law (cf. the appendix), and Dwork [21] has found a "local" proof (cf. also Lubin-Tate [98]).

EXERCISE

Show that p is a norm in each of the extensions K_{p^n}/K. Deduce that $(p, K_{p^\infty}/K) = 1$.

APPENDIX

Global Case (Statement of Results)

Let K be an *algebraic number field*, i.e., a finite extension of **Q**. Each absolute value v of K determines a completion K_v of K for the topology defined by v. These fields are of two types: for ultrametric v, K_v is a field of the sort considered in the preceding section, whereas for non-ultrametric v, K_v is isomorphic to **R** or to **C**. Note that local class field theory applies (trivially!) to **R** and to **C**; in particular, there is a "reciprocity isomorphism"

$$f: \mathbf{R}^*/N(\mathbf{C}^*) \to G(\mathbf{C}/\mathbf{R})$$

defined in an obvious way.

An *idèle* of K is, by definition, a family $\{x_v\}$, with $x_v \in K_v^*$ for all v, but x_v being a unit for almost all v. The idèles form a group I_K under multiplication; the multiplicative group K^* embeds diagonally into I_K.

Now let L/K be a finite abelian extension, with Galois group G. For each absolute value v of K, choose an absolute value w of L prolonging v. Then the decomposition group D_v of w does not depend on the choice of w; it is the Galois group of the local extension L_w/K_v. Hence for every v ultrametric or not) we have a reciprocity isomorphism

$$f_v: K_v^*/NL_w^* \to D_v.$$

If $x = \{x_v\}$ is an idèle of K, the $f_v(x_v)$ are almost all equal to 1: indeed, almost all the local extensions are unramified, and almost all the x_v are units; our assertion results from prop. 13 of Chap. XIII. Hence we can multiply the $f_v(x_v)$, defining thereby a global "reciprocity map"

$$f: I_K \to G.$$

Artin's reciprocity law can be expressed as follows:

Theorem. *The map f is surjective, and its kernel is generated by* K^* *and* $N_{L/K}(I_L)$.

For the proof, see Artin-Tate [8], Chap. VII, §3. (Other references are Cassels-Fröhlich [75], Lang [94] and Weil [123].) One of the essential points is the fact that K^* is contained in the kernel of f; that would no longer be true if one changed the selection of Frobenius substitutions in the local fields K_v.

The above theorem contains as a special case the one we called "the reciprocity law" in Chap. I, §8. More precisely, let v_i be the discrete valuations of K which ramify in L. In the local extension L_{w_i}/K_{v_i}, the norm group N_i is an open subgroup of $K^*_{v_i}$, and therefore an integer n_i exists such that $v_i(x-1) \geq n_i$ implies $x \in N_i$—cf. cor. 1 to th. 1. Consider then an element $x \in K^*$ that satisfies the two following conditions:

i) $v_i(x-1) \geq n_i$ for all i.
ii) x is positive in every real embedding of K that is not induced by a real embedding of L.

Let $\mathfrak{a} = (x)$ be the principal ideal generated by x. *Then the Artin symbol* $(\mathfrak{a}, L/K)$ *is equal to* 1; this sharpens the statement in Chap. I (by telling how to choose the n_i).

The proof is immediate: in the equation $1 = f(x) = \prod f_v(x)$, the terms corresponding to the v_i and to the non-ultrametric valuations are trivial, because of conditions i) and ii); the others are computed by means of prop. 13 of Chap. XIII, and give $(\mathfrak{a}, L/K)$ as their product. □

Another application: Suppose K contains the nth roots of unity, and let $a, b \in K^*$. Let $L = K(a^{1/n})$. Applying to $b \in K^*$ the formula $f(b) = 1$, we get the relation

$$(*) \qquad \prod_v (a, b)_v = 1.$$

[When $K_v = \mathbf{R}$, which can only happen for $n = 2$, the corresponding symbol $(a, b)_v$ is equal to -1 if a and b are negative, and to $+1$ otherwise.]

EXAMPLE. Take $K = \mathbf{Q}$, $n = 2$, and choose a and b to be distinct odd primes. By §5, we have

$$(a, b)_a = \left(\frac{b}{a}\right), \qquad (a, b)_b = \left(\frac{a}{b}\right), \qquad (a, b)_2 = (-1)^{\frac{a-1}{2}\cdot\frac{b-1}{2}},$$

and the other $(a, b)_v$ are equal to 1. Formula (*) then gives another proof of *the quadratic reciprocity law*:

$$\left(\frac{a}{b}\right)\cdot\left(\frac{b}{a}\right) = (-1)^{\frac{a-1}{2}\cdot\frac{b-1}{2}}.$$

For other examples, see Artin-Tate [8], Chap. XII, §4, as well as Hasse [31].

CHAPTER XV

Ramification

Let K be a field complete under a discrete valuation with quasi-finite residue field $\bar{K}$. Let L/K be a finite abelian extension, with Galois group G. In Chapter XIII, §4, we have defined the reciprocity isomorphism

$$\omega : K^*/NL^* \to G.$$

The goal of the present chapter is to determine the relations existing between ω and the filtration of G given by the ramification groups; we will see, in particular, that the image of U_K^n under ω is equal to G^n (cf. §2).

§1. Kernel and Cokernel of an Additive (resp. Multiplicative) Polynomial

Throughout this section, K denotes a *quasi-finite* field.

Let P be a non-constant *additive polynomial* with coefficients in k (cf. Chap. V, §5). If k_s denotes the algebraic closure of k, then the homomorphism $P : k_s \to k_s$ is surjective.

Proposition 1. *Let* N *be the kernel of* P *in* k_s. *If* $x \in k$, *let* $y \in k_s$ *be a solution to the equation* $P(y) = x$, *and let* $z = Fy - y$. *Then* $z \in N$, *and the image* $\bar{z}$ *of* z *in* $N/(F-1)N$ *does not depend on the choice of* y; *if we set* $\delta_P(x) = \bar{z}$, *then the map* δ_P *defines by passage to the quotient an isomorphism of* $k/P(k)$ *onto* $N/(F-1)N$.

Changing y amounts to replacing it with $y + t$, with $t \in N$, and z is then replaced with $z + (Ft - t)$; this shows that $\bar{z}$ does not depend on the choice of y. It is immediate that $\delta_P(x) = 0$ if and only if $x = P(y)$ for some $y \in k$. It remains to show that

$$\delta_P : k/P(k) \to N/(F - 1)N$$

is surjective. This is trivial if k has characteristic zero: P has the form aX, $a \in k^*$, and $N = 0$. If k has characteristic p, let $z \in N$; by prop. 4 of Chap. XIII, there exists $y \in k_s$ such that $Fy - y = z$. If we put $x = P(y)$, then

$$Fx - x = P(Fy - y) = P(z) = 0$$

whence $x \in k$, and it is clear that $\delta_P(x) = \bar{z}$.

[*Variant.* Let $\mathfrak{g} = G(k_s/k)$. Apply the exact cohomology sequence to the exact sequence of $\mathfrak{g}$-modules

$$0 \to N \to k_s \xrightarrow{P} k_s \to 0.$$

obtaining the exact sequence

$$k \xrightarrow{P} k \to H^1(\mathfrak{g}, N) \to 0$$

and since $H^1(\mathfrak{g}, N) = N/(F - 1)N$, we get prop. 1.] □

Corollary 1. *If* N *is contained in* k, *the map* δ_P *is an isomorphism of* $k/P(k)$ *onto* N.

Indeed, we then have $Fz = z$ for all $z \in N$.

Corollary 2. *The order of* $k/P(k)$ *is finite; it is a divisor of the separable degree* $d_s(P)$ *of* P, *and is equal to the latter if and only if* N *is contained in* k.

Indeed, we know that the order of N is equal to $d_s(P)$—cf. Chap. V, §5.

Corollary 3. *The kernel and cokernel of* $P : k \to k$ *are finite groups of the same order.*

Let N_k be the kernel of $P : k \to k$, i.e., the intersection of N with k. There is an exact sequence

$$0 \to N_k \to N \xrightarrow{F-1} N \to N/(F - 1)N \to 0.$$

As N is finite, it follows that N_k and $N/(F - 1)N$ are equipotent.

EXAMPLES. 1. Assume the characteristic is $p \neq 0$, and take P to be the polynomial $\wp(X) = X^p - X$. Then $N = \mathbf{Z}/p\mathbf{Z}$, and the isomorphism $\delta_\wp : k/\wp(k) \to \mathbf{Z}/p\mathbf{Z}$ is the isomorphism S of Chap. XIV, §5.

2. More generally, take $P = aX^p + bX$, $a, b \in k^*$. If P has no non-trivial zero in k, then $N/(F - 1)N = 0$, whence $k = P(k)$. If P has a non-trivial zero $c \in k$, the equation $P(y) = x$ can be written

$$ac^p(y^p/c^p) + bc(y/c) = x,$$

or

$$(y/c)^p - y/c = a^{-1}c^{-p}x.$$

Therefore,

$$\mathrm{P}(x) = \mathrm{F}y - y = c.(\mathrm{F}(y/c) - y/c) = c.\mathrm{S}(a^{-1}c^{-p}x) = -c.\mathrm{S}(b^{-1}c^{-1}x)$$

which reduces the computation of δ_P to that of S.

Let us now pass to the multiplicative case.

Proposition 2. *Let* $\mathrm{P} = \mathrm{X}^n$, *with* $n \geq 1$, *and let* N *be the kernel of* $\mathrm{P}: k_s^* \to k_s^*$. *If* $x \in k$, *let* $y \in k_s^*$ *be a solution to the equation* $\mathrm{P}(y) = x$, *and let* $z = y^{\mathrm{F}-1}$. *Then* $z \in \mathrm{N}$, *and the image* $\bar{z}$ *of* z *in* $\mathrm{N}/\mathrm{N}^{\mathrm{F}-1}$ *does not depend on the choice of* y; *if we put* $\delta_\mathrm{P}(x) = \bar{z}$, *then the map* δ_P *defines by passage to the quotient an isomorphism of* $k^*/\mathrm{P}(k^*)$ *onto* $\mathrm{N}/\mathrm{N}^{\mathrm{F}-1}$.

(To keep the analogy with the additive case, we have used the exponential notation: $y^\mathrm{F} = \mathrm{F}(y)$.)

The proof is identical to that of prop. 1.

Corollary 1. *If* N *is contained in* k^*, *then the map* δ_P *is an isomorphism of* $k^*/\mathrm{P}(k^*)$ *onto* N.

Corollary 2. *The order of* $k^*/\mathrm{P}(k^*)$ *is finite; it is a divisor of the separable degree* $d_s(\mathrm{P})$ *of* P, *and is equal to the latter if and only if* N *is contained in* k^*.

Corollary 3. *The kernel and cokernel of* $\mathrm{P}: k^* \to k^*$ *are finite groups of the same order.*

The proofs are the same as for the corresponding corollaries in the additive case.

EXAMPLE. Suppose n is prime to the characteristic of k, and assume that k^* contains the group μ_n of all nth roots of unity. Then $\mathrm{N} = \mu_n$, and the isomorphism $\delta_\mathrm{P}: k^*/k^{*n} \to \mu_n$ is the isomorphism P_n of Chap. XIV, §3.

EXERCISES

Let k be a quasi-finite field of characteristic p.

1. a) Let m be a positive integer prime to p, and let μ_m be the group of all mth roots of unity in k_s^*. Show that there is a unique divisor n of m such that $\mu_m \cap k^* = \mu_n$. Show that $k^{*m} = k^{*n}$, and deduce that k^*/k^{*m} is cyclic of order n.
   b) Let H be a subgroup of k^* of finite index m. Show that m is prime to p. If n is the integer defined in a), show that $m = n$ and $\mathrm{H} = k^{*n}$ (note that H contains k^{*m}).
   c) Deduce from b) that the subgroups of finite index of k^* are the k^{*n}, where n runs through the set of those integers prime to p such that $\mu_n \subset k^*$.

2. Let A be a commutative linear absolutely irreducible algebraic k-group (without nilpotent elements); thus it is the product of a group of multiplicative type with a unipotent group. Denote by A_k (resp. A_{k_s}) the group of rational points of A in k (resp. k_s).
   a) Let $\mathfrak{g} = G(k_s/k)$. Show that $H^q(\mathfrak{g}, A_{k_s}) = 0$ for all $q \geq 1$.
   b) Let A′ be an algebraic group satisfying the same hypotheses as A and of the same dimension. Let $P: A' \to A$ be a homomorphism with finite kernel N. Define an isomorphism
   $$\delta_P: A_k/P(A'_k) \to N/(F-1)N$$
   by the method of propositions 1 and 2. Generalise corollaries 1, 2, 3.
   c) A subgroup of A_k is called *normic* if it has the form $P(A'_k)$ for some homomorphism $P: A' \to A$ as in b). Show that the intersection of two normic subgroups is normic, and that every subgroup containing a normic subgroup is normic.
   d) When k is a *finite* field, show that the hypothesis "A is linear" is superfluous, and that every subgroup of A_k is normic (cf. Lang [39]). Give an example to show that this is not true in the general case.

   The normic subgroups play a role in the *existence theorem* of Whaples [71]: if K is complete under a discrete valuation with quasi-finite residue field $\bar{K}$, then a subgroup H of finite index of K^* is a norm group if and only if it contains U_K^n for sufficiently large n, and if the image of $H \cap U_K$ in U_K/U_K^n is a normic subgroup of U_K/U_K^n (the latter being considered as the set of rational points in $\bar{K}$ of the algebraic group $U_{K_{nr}}/U_{K_{nr}}^n$; cf. Greenberg [25]).

§2. The Norm Groups

From now on we assume that K is complete under a discrete valuation with quasi-finite residue field $\bar{K}$.

Let L/K be a *totally ramified* Galois extension, with Galois group G. It is immediately seen that U_K/NU_L can be identified with K^*/NL^*. We are going to study the filtration of U_K/NU_L defined by the images of the U_K^n.

By prop. 9 of Chap. VI, there is an exact sequence

$$0 \to G_{\psi(n)}/G_{\psi(n)+1} \to U_L^{\psi(n)}/U_L^{\psi(n)+1} \xrightarrow{N_n} U_K^n/U_K^{n+1}$$

where N_n is defined by an additive (resp. multiplicative) polynomial if $n \geq 1$ (resp. if $n = 0$). The separable degree of this polynomial is equal to $(G_{\psi(n)}: G_{\psi(n)+1})$, which allows us to apply cor. 1 to prop. 1 (resp. to prop. 2) to it. We get:

Proposition 3. *The group* $U_K^n/U_K^{n+1}NU_L^{\psi(n)}$ *is isomorphic to* $G_{\psi(n)}/G_{\psi(n)+1}$.

Put $h_n = (G_{\psi(n)}: G_{\psi(n)+1})$.

Corollary 1. $(U_K^n : U_K^{n+1}NU_L^{\psi(n)}) = h_n$.

Clear.

Corollary 2. *The group* $NU_L^{\psi(n)}$ *is a subgroup of finite index of* U_K^n. *If* v_n *denotes this index, then* $v_n = 1$ *for sufficiently large* n; *moreover,* v_n *divides* $v_{n+1}h_n$, *with equality taking place if and only if the canonical homomorphism*

$$(*_n) \qquad U_K^{n+1}/NU_L^{\psi(n+1)} \to U_K^n/NU_L^{\psi(n)}$$

is injective.

We know that $U_K^n = NU_L^{\psi(n)}$ if n is sufficiently large (cf. Chap. V. §6, cor. 3 to prop. 9). On the other hand, there is an exact sequence

$$U_K^{n+1}/NU_L^{\psi(n+1)} \to U_K^n/NU_L^{\psi(n)} \to U_K^n/U_K^{n+1}NU_L^{\psi(n)} \to 0.$$

This exact sequence shows that if v_{n+1} is finite, so is v_n, and that v_n divides $v_{n+1}h_n$; the quotient $v_{n+1}h_n/v_n$ is equal to the order of the kernel of homomorphism $(*_n)$, whence the corollary follows.

Corollary 3. *The integer* $v_0 = (U_K : NU_L) = (K^* : NL^*)$ *divides the product of the* h_n.

Indeed, v_0 divides v_1h_0, which divides $v_2h_1h_0, \ldots$, which divides $v_nh_{n-1} \cdots h_0$. Taking n large enough so that $v_n = 1$, we get the result we seek.

Remark. As the product of the h_n divides $[L:K]$, we recover the fact that $(K^* : NL^*)$ divides $[L:K]$—cf. Chap. XIII, prop. 9.

Theorem 1. *Suppose* G *is abelian. Then*:

a) $G_m = G_{m+1}$ *if* $\varphi(m)$ *is not an integer.*
b) $v_n = v_{n+1}h_n$ *for all* n.
c) *The canonical map of* $U_K^n/NU_L^{\psi(n)}$ *into* K^*/NL^* *is injective.*

By prop. 9 of Chap. XIII, $v_0 = [L:K]$, or, what comes to the same,

$$v_0 = \prod_{m=0}^{\infty} (G_m : G_{m+1}).$$

On the other hand, cor. 3 to prop. 3 shows that v_0 divides the product $\prod(G_{\psi(n)} : G_{\psi(n)+1})$. It follows that $(G_m : G_{m+1}) = 1$ if m does not have the form $\psi(n)$, i.e., if $\varphi(m)$ is not an integer, whence a) follows. Similarly, if v_n were a proper divisor of $v_{n+1}h_n$ for one integer n then v_0 would be a proper divisor of the product of the h_n, which is impossible; thus b) holds. By cor. 2 to prop. 3, the homomorphisms

$$(*_n) : U_K^{n+1}/U_L^{\psi(n+1)} \to U_K^n/NU_L^{\psi(n)}$$

are all injective. By composing them, we see that

$$U_K^{n+1}/NU_L^{\psi(n+1)} \to U_K/NU_L$$

is injective, whence c) follows, since $U_K/NU_L = K^*/NL^*$. □

Remark. Assertion a) is none other than the Hasse-Arf theorem, of which we thus obtain a second proof (valid only when the residue field is quasi-finite).

Corollary 1. *The groups* $U_K^n/NU_L^{\psi(n)}$ *form a decreasing filtration of* K^*/NL^*. *We have* $U_K^n/NU_L^{\psi(n)} = 0$ *if and only if* $G^n = \{1\}$.

The first assertion is just a reformulation of c). As for the second, note that $v_n = 1$ is equivalent to $h_n = h_{n+1} = \cdots = 1$, i.e., $G^n = G^{n+1} = \cdots = \{1\}$.

Corollary 2. *Let c be the largest integer for which* $G_c \neq \{1\}$, *and let* $f = \varphi(c) + 1$. *Then* $U_K^f \subset NL^*$, *and f is the least integer enjoying this property.*

[The ideal $\mathfrak{p}_K^f$ is called the *conductor* of the extension L/K; when L/K is the cyclic extension defined by a character χ of degree 1 of $G(K_s/K)$, the corollary to prop. 6 of Chap. VI shows that $\mathfrak{p}_K^f$ coincides with the *Artin conductor* $\mathfrak{f}(\chi)$ of χ.]

We know that U_K^f is contained in NL^* (cf. Chap. V, §6, cor. 3 to prop. 9). On the other hand, corollary 1 shows that U_K^{f-1} is not contained in NL^*, because $G^{f-1} = G_c$ is non-trivial.

Corollary 3. *The reciprocity map* $\omega: K^*/NL^* \to G$ *transforms the filtration by the* $U_K^n/NU_L^{\psi(n)}$ *into the filtration by the* G^n.

It suffices to show that for every subgroup H of G, the relations

$$\omega(U_K^n) \subset H \quad \text{and} \quad G^n \subset H$$

are equivalent. Passing to the quotient by H, this comes down to saying that $\omega(U_K^n) = \{1\}$ is equivalent to $G^n = \{1\}$. Now according to cor. 2, the first relation is equivalent to $n \geq f$; the second relation signifies that $G_{\psi(n)} = \{1\}$, i.e., $\psi(n) > c$, or

$$n \geq \varphi(c) + 1 = f. \quad \square$$

Theorem 2. *Let* L/K *be an abelian extension, with Galois group* G. *Then the image of* U_K^n *under the reciprocity map* $\omega: K^* \to G$ *is dense in* G^n.

In view of the definition of G^n, that comes down to saying that $\omega(U_K^n) = G^n$ when L/K is finite. Let T then be the inertia group of G, and K'/K the corresponding abelian extension. Denote by ω' the reciprocity map for the extension L/K'. By cor. 3 to theorem 1, we have

$$\omega'(U_{K'}^n) = T^n = G^n.$$

On the other hand, $\omega \circ N_{K'/K} = \omega'$ (Chap. XIII, prop. 10), and $N_{K'/K}(U_{K'}^n) = U_K^n$ (Chap. V, prop. 3). It follows that $\omega(U_K^n) = G^n$. $\square$

Remark. In the "usual" case where $\bar{K}$ is *finite*, we have $\omega(U_K^n) = G^n$ because U_K^n is compact.

§3. Explicit Computations

We return to the situation of th. 1. Thus L/K is a finite, totally ramified abelian extension, with Galois group G. We have seen (cor. 3 to th. 1) that the reciprocity isomorphism

$$\omega : K^*/NL^* \to G$$

transforms the subgroups $U_K^n/NU_L^{\psi(n)}$ of K^*/NL^* into the ramification groups G^n of G. Passing to the quotient, we get isomorphisms

$$\omega_n : U_K^n/U_K^{n+1}NU_L^{\psi(n)} \to G^n/G^{n+1}.$$

On the other hand, we have seen that the homomorphism

$$N_n : U_L^{\psi(n)}/U_L^{\psi(n)+1} \to U_K^n/U_K^{n+1}$$

is induced by an additive (resp. multiplicative) polynomial if $n \geq 1$ (resp. if $n = 0$); moreover, this polynomial is canonically determined (i.e., is invariant under residue extensions—cf. Chap. V, §6, remark 2). The method of §1 associated to this polynomial an isomorphism

$$\delta_n : U_K^n/U_K^{n+1}NU_L^{\psi(n)} \to G_{\psi(n)}/G_{\psi(n)+1}$$

cf. prop. 3. We have $G_{\psi(n)} = G^n$, and the Hasse-Arf theorem shows that $G_{\psi(n)+1} = G^{n+1}$. Thus *ω_n and δ_n both map $U_K^n/U_K^{n+1}NU_L^{\psi(n)}$ into G^n/G^{n+1}* Comparing them is indispensable:

Proposition 4. $\omega_n(\alpha) = \delta_n(\alpha^{-1})$ *for all* $\alpha \in U_K^n/U_K^{n+1}NU_L^{\psi(n)}$.

Put $L_0 = \hat{L}_{nr}$, $K_0 = \hat{K}_{nr}$ (cf. Chap. XIII, §5). We begin with the computation of $\delta_n(\alpha)$. In view of the definitions of §1, we must choose an element $\beta \in U_{L_0}^{\psi(n)}/U_{L_0}^{\psi(n)+1}$ such that $N_n(\beta) = \alpha$; form $\zeta = \beta^{F-1}$, which belongs to the kernel of N_n, and therefore is of the form π^{s-1} mod. $U_{L_0}^{\psi(n)+1}$, with $s \in G^n$, π being a uniformizer of L; then $\delta_n(\alpha) \equiv s$ mod. G^{n+1}. To be more explicit, choose a representative $x \in U_K^n$ of α, and $y \in U_{L_0}^{\psi(n)}$ such that $Ny = x$ (which is possible by Chap. V, §6, cor. 3 to prop. 9). Put $z = y^{F-1}$; then $z = \pi^{s-1}z'$, with $s \in G^n$ and $z' \in U_{L_0}^{\psi(n)+1}$ and

$$\delta_n(\alpha) \equiv s \quad \text{mod. } G^{n+1}.$$

We now distinguish two cases:

Case a). $\bar{K}$ has positive characteristic

By prop. 15 of Chap. XIII, there exists $y' \in U_{L_0}^{\psi(n)+1}$ such that $z' = y'^{F-1}$. We have $(yy'^{-1})^{F-1} = \pi^{s-1}$. If $x' = Ny'$, then $(xx'^{-1})^{F-1} = 1$, whence $xx'^{-1} \in K^*$ and $x' \in K^*$. Furthermore, Dwork's theorem (Chap. XIII, §5, cor. to th. 2) shows that $\omega(xx'^{-1}) = s^{-1}$. As x' belongs to $U_{K_0}^{n+1} \cap K^* = U_K^{n+1}$, we have $\omega(x') \in G^{n+1}$ and $\omega(x) \equiv s^{-1}$ mod. G^{n+1}, which proves the proposition in this case.

Case b). $\bar{K}$ has characteristic zero

Then $G^n = \{1\}$ if $n \geq 1$, so we can assume $n = 0$. Moreover, G is cyclic, and if r is its order, the field $\bar{K}$ contains the group of all rth roots of unity (cf. Chap. IV, §2). We can choose x and y to be *multiplicative representatives* (cf. Chap. II, §4); the element $z = y^{F-1}$ is then an rth root of unity, and it is easy to show that it can be put into the form π^{s-1}, where π is a uniformizer of L and s is a suitably chosen element of G. We then apply Dwork's theorem as in case *a*). □

The preceding proposition reduces the computation of ω_n to that of δ_n, i.e., ultimately to that of N_n. We give an example:

Proposition 5. *Let* L/K *be a cyclic totally ramified extension of prime degree p equal to the characteristic of* $\bar{K}$; *let* G *be its Galois group, and let s be a generator of* G; *let t be the largest integer for which* $G_t \neq \{1\}$. *Let π uniformize* L, *and set* $M = s(\pi)/\pi - 1$. *Let* $x \in U_K^t$, *and let* $c(x) = (x - 1)/\mathrm{Tr}(M)$. *Then* $c(x)$ *belongs to the valuation ring of* K, *and if* $\bar{c}(x)$ *denotes its image in* $\bar{K}$, *we have*

$$(x, L/K) = s^{S(\bar{c}(x))}.$$

[For the definition of $S : \bar{K} \to \mathbf{Z}/p\mathbf{Z}$, see Chap. XIV, §4.]

Let π' be an element of K such that $v_K(\pi') = t$. Then $\mathrm{Tr}(M) = b\pi'$ and $N(M) = a\pi'$, with $a, b \in U_K$—cf. Chap. V, §3. If we put $y = 1 + \eta M$, with $\eta \in A_L$, then

$$N(y) \equiv 1 + (a\eta^p + b\eta)\pi' \quad \mathrm{mod.}\ U_K^{t+1},$$

cf. Chap. V, §3. The map N_t is therefore represented by the additive polynomial $P_t(\eta) = \bar{a}\eta^p + \bar{b}\eta$, and since $N(\pi^{s-1}) = 1$, $\eta = 1$ is an element of the kernel of P_t. According to example 2 of §1, $\delta_{P_t}(\bar{\xi})$ is equal to—$S(\bar{b}^{-1}\bar{\xi})$. If $x = 1 + \xi\pi'$, then $\delta_t(x) = s^{-m}$, with $m = -S(\bar{b}^{-1}\bar{\xi}) = -\bar{c}(x)$, whence the result follows, taking prop. 4 into account. □

Prop. 5 can itself be used to compute some local symbols $(a, b)_v$. We give only one example:

Proposition 6. *Suppose that* K (resp. $\bar{K}$) *has characteristic zero* (resp. p), *and that* K *contains the group* μ_p *of all* p*th roots of unity. Let* w *be a generator of* μ_p. *Let* $e = v_K(p)$ *be the absolute ramification index of* K, *and let* $t = ep/(p-1)$, *which is an integer. If* $a \in K^*$ *and* $b \in U_K^t$, *then*

$$(a, b)_v = w^{v_K(a) \cdot m(b)}, \quad \textit{with } m(b) = S\left(\frac{b-1}{p(w-1)}\right).$$

[We write $S(c)$ instead of $S(\bar{c})$ if $c \in A_K$.]

We know from Chap. IV, prop. 17 that $v_K(w - 1) = e/(p - 1)$, which shows that t is indeed an integer. On the other hand, the bilinearity of the symbol

$(a, b)_v$ allows us to restrict to the case where a is a uniformizer of K. Let π be a pth root of a, and let $\mathrm{L} = \mathrm{K}(\pi)$; we can apply prop. 5 to the extension L/K. Choose the generator s of G(L/K) so that $\pi^{s-1} = w$; we then see that

$$t = ep/(p-1), \quad \mathrm{M} = w - 1, \quad \mathrm{Tr}(\mathrm{M}) = p(w-1).$$

Hence

$$(b, \mathrm{L}/\mathrm{K}) = s^{m(b)}, \quad \text{with } m(b) = \mathrm{S}\left(\frac{b-1}{p(w-1)}\right).$$

Therefore $(a, b)_v = \pi^{(b, \mathrm{L}/\mathrm{K}) - 1} = w^{m(b)}$. □

The reader will find more examples of computations of local symbols in Fosenko-Vostokov [129], Iwasawa [85], Lubin-Tate [98] and Wiles [124].

Exercises

1. With the notation of prop. 5, show that

$$c(x) \equiv -\frac{x-1}{\mathrm{N}(\mathrm{M})} \equiv -\frac{x-1}{\mathrm{M}^p} \quad \mathrm{mod.}\ \mathfrak{p}_{\mathrm{L}}.$$

2. With the notation of prop. 6, show that the extension $\mathrm{K}(b^{1/p})/\mathrm{K}$ is unramified, and deduce from that another proof of prop. 6.

3. With the notation of prop. 6, let i and j be integers ≥ 0 such that $i + j = t$. Let $a \in \mathrm{U}_{\mathrm{K}}^i$ and $b \in \mathrm{U}_{\mathrm{K}}^j$. Prove the formula

$$(a,b)_v = w^{i \cdot \mathrm{S}[(a-1)(b-1)/p(w-1)]}$$

Bibliography

1. S. AMITSUR. *Generic splitting fields of central simple algebras*. Ann. of. Maths., **62**, 1955, p. 8–43.
2. C. ARF. *Untersuchungen über reinverzweigte Erweiterungen diskret bewerteter perfekter Körper*. J. reine ang. Math., **181**, 1940, p. 1–44.
3. E. ARTIN. *Beweis des allgemeinen Reziprozitätsgesetzes*. Hamb. Abh., **5**, 1927, p. 353–363.
4. E. ARTIN. *Idealklassen in Oberkörpern und allgemeines Reziprozitätsgesetz*. Hamb. Abh., **7**, 1929, p. 46–51.
5. E. ARTIN. *Zur Theorie der L-Reihen mit allgemeinen Gruppencharakteren*. Hamb. Abh., **8**, 1930, p. 292–306.
6. E. ARTIN. *Die gruppentheoretische Struktur des Diskriminanten algebraischer Zahlkörper*. J. reine ang. Math., **164**, 1931, p. 1–11.
7. E. ARTIN, C. NESBITT et R. THRALL. *Rings with minimum condition*. University of Michigan Press, Ann Arbor, 1948.
8. E. ARTIN et J. TATE. *Class field theory*. Harvard, 1961.
9. M. AUSLANDER et O. GOLDMAN. *The Brauer group of a commutative ring*. Trans. Amer. Math. Soc., **97**, 1960, p. 367–409.
10. G. AZUMAYA. *On maximally central algebras*. Nagoya Math. J., **2**, 1951, p. 119–150.
11. R. BRAUER et J. TATE. *On the characters of finite groups*. Ann. of Maths., **62**, 1955, p. 1–7.
12. H. CARTAN et C. CHEVALLEY. *Géométrie algébrique*. Séminaire E.N.S., 1955–1956.
13. H. CARTAN et S. EILENBERG. *Homological algebra*. Princeton Math. Ser., n^0 **19**, Princeton, 1956.
14. J. CASSELS. *Arithmetic on curves of genus* 1. *I. On a conjecture of Selmer*. J. reine ang. Math., **202**, 1959, p. 52–99. *II. A general result*. *Ibid.*, **203**, 1960, p. 174–208. See also *III, IV, V, VI, VII, VIII*.
15. F. CHATELET. *Variations sur un thème de H. Poincaré*. Annales E.N.S., **61**, 1944, p. 249–300.
16. F. CHATELET. *Géométrie diophantienne et théorie des algèbres*. Séminaire DUBREIL, 1954–1955, exp. 17.
17. C. CHEVALLEY. *Class field theory*. Nagoya, 1954.
18. I. COHEN. *On the structure and ideal theory of complete local rings*. Trans. Amer. Math. Soc., **59**, 1946, p. 54–106.

19. M. Deuring. *Algebren*. Ergebn. der Math., IV-1, 1935.
20. A. Douady. *Cohomologie des groupes compacts totalement discontinus*. Séminaire Bourbaki, 1959–1960, exp. 189.
21. B. Dwork. *Norm residue symbol in local number fields*. Hamb. Abh., **22**, 1958, p. 180–190.
22. B. Eckmann. *Cohomology of groups and transfer*. Ann. of Maths., **58**, 1953, p. 481–493.
23. B. Eckmann. *Homotopie et dualité*. Colloque de Topologie algébrique, Louvain, 1956, p. 41–53.
24. J. Frenkel. *Cohomologie non abélienne et espaces fibrés*. Bull. Soc. Math. France, **85**, 1957, p. 135–220.
25. M. Greenberg. *Schemata over local rings*. Ann. of Maths., **73**, 1961, p. 624–648.
26. A. Grothendieck. *Sur quelques points d'algébre homologique*. Tohoku Math. J., **9**, 1957, p. 119–221.
27. A. Grothendieck. *A general theory of fibre spaces with structure sheaf*. Univ. of Kansas, Report n^0 **4**, 1955.
28. A. Grothendieck. *Technique de descente et théorèmes d'existence en géométrie algébrique. I. Généralités. Descente par morphismes fidèlement plats*. Séminaire Bourbaki, 1959–1960, exp. 190.
29. A. Grothendieck. *Séminaire de géométrie algébrique*. Inst. Htes. Et. Scient., 1960–1961-. . .
30. M. Hall. *The theory of groups*. The Macmillan Cy., New York, 1959.
31. H. Hasse. *Bericht über neuere Untersuchungen und Probleme aus der Theorie der algebraischen Zahlkörper*. Jahr. der D. Math. Ver., **35**, 1926, p. 1–55; *ibid.*, **36**, 1927, p. 255–311; *ibid.*, **39**, 1930, p. 1-204.
32. H. Hasse. *Führer, Diskriminante und Verzweigungskörper relativ Abelscher Zahlkörper*. J. reine ang. Math., **162**, 1930, p. 169–184.
33. H. Hasse. *Normenresttheorie galoisscher Zahlkörper mit Anwendungen auf Führer und Diskriminante abelscher Zahlkörper*. J. Fac. Sci. Tokyo, **2**, 1934, p. 477–498.
34. H. Hasse. *Zahlentheorie*. Berlin, Akademie-Verlag, 1949.
35. G. Hochschild. *Relative homological algebra*. Trans. Amer. Math. Soc., **82**, 1956, p. 246–269.
36. G. Hochschild et T. Nakayama. *Cohomology in class field theory*. Ann. of. Maths., **55**, 1952, p. 348–366.
37. G. Hochschild et J.-P. Serre. *Cohomology of group extensions*. Trans. Amer. Math. Soc., **74**, 1953, p. 110–134.
38. S. Lang. *On quasi-algebraic closure*. Ann. of Maths., **55**, 1952, p. 373–390.
39. S. Lang. *Algebraic groups over finite fields*. Amer. J. of Maths., **78**, 1956, p. 555–563.
40. S. Lang. *Abelian varieties*. Interscience Tracts n^0 **7**, New York, 1959.
41. S. Lang et J. Tate. *Principal homogeneous spaces over abelian varieties*. Amer. J. of. Maths., **80**, 1958, p. 659–684.
42. M. Lazard. *Détermination des anneaux p-adiques et π-adiques dont les anneaux de restes sont parfaits*. Séminaire Krasner, 1953–1954, exp. 9.
43. M. Lazard. *Bemerkungen zur Theorie der bewerteten Körper und Ringe*. Math. Nach., **12**, 1954, p. 67–73.
44. M. Lazard. *Sur les groupes nilpotents et les anneaux de Lie*. Annales E.N.S., **71**, 1954, p. 101–190.
45. R. Mackenzie et G. Whaples. *Artin-Schreier equations in characteristic zero*. Amer. J. of Maths., **78**, 1956, p. 473–485.
46. Y. Nakai. *On the theory of differentials in commutative rings*. J. Math. Soc. Jap., **13**, 1961, p. 63–84.
47. T. Nakayama. *Cohomology of class field theory and tensor product modules. I*. Ann. of Maths., **65**, 1957, p. 255–267.

48. T. NAKAYAMA. *On modules of trivial cohomology over a finite group. I.* Illinois J. Math., **1**, 1957, p. 36–43; *II*, Nagoya Math. J., **12**, 1957, p. 171–176.
49. O. ORE. *Abriss einer arithmetischen Theorie der Galoisschen Körper. I.* Math. Ann., **100**, 1928, p. 650–673; *II, ibid.*, **102**, 1930, p. 283–304.
50. O. ORE. *On a special class of polynomials.* Trans. Amer. Math. Soc., **35**, 1933, p. 559–584 (*Errata, ibid.*, **36**, 1934, p. 275).
51. D. RIM. *Modules over finite groups.* Ann. of Maths., **69**, 1959, p. 700–712.
52. P. ROQUETTE. *Abspaltung des Radikals in vollständigen lokalen Ringen.* Hamb. Abh., **23**, 1959, p. 75–113.
53. P. SAMUEL et O. ZARISKI. *Commutative Algebra.* Van Nostrand.
54. O. SCHILLING. *The theory of valuations.* Math. Surveys IV, New-York, 1950.
55. H. SCHMID. *Ueber das Reziprozitätsgesetz in relativ-zyklischen algebraischen Funktionenkörpern mit endlichem Konstantenkörper.* Math. Zeit., **40**, 1936, p. 94–109.
56. J.-P. SERRE. *Groupes algébriques et corps de classes.* Paris, Hermann, 1959.
57. J.-P. SERRE. *Sur la rationalité des représentations d'Artin.* Ann. of Maths., **72**, 1960, p. 406–420.
58. J.-P. SERRE. *Groupes finis á cohomologie périodique* (*d' après* R. SWAN). Séminaire BOURBAKI, 1960–1961, exp. 209.
59. J.-P. SERRE. *Sur les corps locaux á corps résiduel algébriquement clos.* Bull. Soc. Math. France, **89**, 1961, p. 105–154.
60. A. SPEISER. *Zahlentheoretische Sätze aus der Gruppentheorie.* Math. Zeit., **5**, 1919, p. 1–6.
61. A. SPEISER. *Die Zerlegungsgruppe.* J. reine ang. Math., **149**, 1920, p. 174–188.
62. R. SWAN. *Induced representations and projective modules.* Ann. of Maths., **71**, 1960, p. 552–578.
63. J. TATE. *The higher dimensional cohomology groups of class field theory.* Ann. of Maths., **56**, 1952, p. 294–297.
64. J. TATE. *WC-groups over p-adic fields.* Séminaire BOURBAKI, 1957–1958, exp. 156.
65. B. VAN DER WAERDEN. *Moderne Algebra.* I. 5te Auflage, Springer, 1960.
66. A. WEIL. *Sur les courbes algébriques et les variétés qui s'en déduisent.* Hermann, Paris, 1948.
67. A. WEIL. *Sur la théorie du corps de classes.* J. Math. Soc. Jap., **3**, 1951, p. 1–35.
68. H. WEYL. *Algebraic theory of numbers.* Ann. of Math. St., n^0 1, Princeton, 1940.
69. G. WHAPLES. *Additive polynomials.* Duke Math. J., **21**, 1954, p. 55–66.
70. G. WHAPLES. *Galois cohomology of additive polynomial and n-th power mappings of fields.* Duke Math. J., **24**, 1957, p. 143–150.
71. G. WHAPLES. *Generalized local class field theory. I.* Duke Math. J., **19**, 1952, p. 505–517; *II, ibid.*, 21, 1954, p. 247–256; *III, ibid.*, p. 575–581; *IV, ibid.*, p. 583–586.
72. E. WITT. *Schiefkörper über diskret bewerteten Körpern.* J. reine ang. Math., **176**, 1936, p. 153–156.
73. E. WITT. *Zyklische Körper und Algebren der Charakteristik p vom Grade p^n.* J. reine ang. Math., **176**, 1936, p. 126–140.

Supplementary Bibliography for the English Edition

Topics:

Textbooks: CASSELS [126]; CASSELS-FRÖHLICH [75]; FOSENKO-VOSTOKOV [129]; IWASAWA [131]; LANG [94]; NEUKIRCH [132]; SERRE [114], [115]; WEIL [123].
Group cohomology: KOCH [88]; LANG [93]; POITOU [102]; SERRE [110]; TATE [117], [118], [121].
Brauer groups: GROTHENDIECK [83]; SERRE [110]; WEIL [123].
Non abelian cohomology: GIRAUD [80]; SERRE [113].
Structure theorems for Galois groups of local fields: DEMUŠKIN [77]; IWASAWA [84]; KOCH [88]; LABUTE [91], [92]; ŠAFAREVIČ [104], [105]; WEIL [123].
Number of extensions of a local field: KRASNER [89]; SERRE [116].
Ramification subgroups: FONTAINE [78]; MAUS [99]; SEN [106], [107]; SEN-TATE [109]; WYMAN [125].
Artin conductors and applications: FONTAINE [78]; MARTINET [79]; OGG [101]; RAYNAUD [103]; SERRE [112].
Local symbols, explicit reciprocity laws: COATES-WILES [76]; FOSENKO-VOSTOKOV [129]; IWASAWA [85]; LUBIN-TATE [98]; TATE [120]; WILES [124].
Formal groups, p-divisible groups and Hodge-Tate modules: CASSELS-FRÖHLICH [75]; FONTAINE [128]; LUBIN [97]; LUBIN-TATE [98]; SEN [108]; SERRE [111]; TATE [119]; WILES [124].
Equations with coefficients in a local field: AX-KOCHEN [74]; GREENBERG [81], [82]; TERJANIAN [122].
p-adic Lie groups: DIXON-DU SAUTOY-MANN-SEGAL [127]; LAZARD [95]; MOORE [100].
p-adic zeta functions: COATES [79]; IWASAWA [86]; KOBLITZ [87]; KUBOTA-LEOPOLDT [90].
Reviews: LEVEQUE [96]; GUY [130].

74. J. AX and S. KOCHEN, *Diophantine problems over local fields, I.* Amer. J. of Math. **87** (1965), 605–630; *II*, *ibid.* 631–648; *III*, Ann. of Math. **83** (1966), 437–456.

75. J. CASSELS and A. FRÖHLICH, *Algebraic number theory*, Acad. Press, New York, 1967.
76. J. COATES and A. WILES, *Explicit reciprocity laws*, Astérisque **41–42** (1977), 7–17.
77. S. DEMUŠKIN, *The group of a maximal p-extension of a number field* (in Russian), Izv. Akad. Nauk SSR Ser. Mat. **25** (1961), 329–346.
78. J-M. FONTAINE, *Groupes de ramification et représentations d'Artin*, Ann. Sci. E.N.S. (4), **4** (1971), 337–392.
79. A. FRÖHLICH (edit.), *Algebraic number fields* (L-*functions and Galois properties*), Acad. Press, London, 1977.
80. J. GIRAUD, *Cohomologie non abélienne*, Grund. math. Wiss. 179, Springer-Verlag, 1971.
81. M. GREENBERG, *Rational points in Henselian discrete valuation rings*, Publ. Math. I.H.E.S. **31** (1966), 59–64.
82. M. GREENBERG, *Lectures on forms in many variables*, Benjamin, New York, 1969.
83. A. GROTHENDIECK, *Le groupe de Brauer I, II, III*, Dix exposés sur la cohomologie des schémas (J. GIRAUD *et al.*), 46–188, Masson, North-Holland, 1968.
84. K. IWASAWA, *On Galois groups of local fields*, Trans. A.M.S. **80** (1955), 448–449.
85. K. IWASAWA, *On explicit formulas for the norm residue symbol*, J. Math. Soc. Japan **20** (1968), 151–165.
86. K. IWASAWA, *Lectures on p-adic* L-*Functions*, Ann. Math. Studies **74**, Princeton Univ. Press, Princeton, 1972.
87. N. KOBLITZ, *p-adic numbers, p-adic analysis, and zeta-functions*, G.T.M. **58**, Springer-Verlag, 1977.
88. H. KOCH, *Galoissche Theorie der p-Erweiterungen*, Math. Mon. **10**, V.E.B. Deutscher Verlag der Wiss., Berlin, 1970.
89. M. KRASNER, *Nombre des extensions d'un degré donné d'un corps p-adique*, Colloque C.N.R.S. **143**, Clermont-Ferrand (1966), 143–169.
90. T. KUBOTA and H. LEOPOLDT, *Eine p-adische Theorie der Zetawerte I*, Crelle J. **214/215** (1964), 328–339.
91. J. LABUTE, *Classification of Demuškin groups*, Canad. J. of Math. **19** (1967), 106–132.
92. J. LABUTE, *Demuškin groups of rank* $\aleph_0$, Bull. Soc. Math. France **94** (1966), 211–244.
93. S. LANG, *Rapport sur la cohomologie des groupes*, Benjamin, New York, 1966.
94. S. LANG, *Algebraic number theory*, Addison-Wesley, 1970.
95. M. LAZARD, *Groupes analytiques p-adiques*, Publ. Math. I.H.E.S. **26** (1965), 1–219.
96. W. J. LEVEQUE (edit.), *Reviews in number theory*, vol. 5, chap. S, 304–354, Amer. M.S., 1974.
97. J. LUBIN, *One-parameter formal Lie groups over p-adic integer rings*, Ann. of Math. **80** (1964), 464–484 (Correction: ibid. **84** (1966), 372).
98. J. LUBIN and J. TATE, *Formal complex multiplication in local fields*, Ann. of Math. **81** (1965), 380–387.
99. E. MAUS, *Die gruppentheoretische Struktur der Verzweigungsgruppenreihen* Crelle J. **230** (1968), 1–28.
100. C. C. MOORE, *Group extensions of p-adic and adelic linear groups*, Publ. Math. I.H.E.S. **35** (1968), 157–221.
101. A. OGG, *Elliptic curves and wild ramification*, Amer. J. of Math. **89** (1967), 1–21.
102. G. POITOU, *Cohomologie galoisienne des modules finis*, Dunod, Paris, 1967.
103. M. RAYNAUD, *Caractéristique d'Euler-Poincaré d'un faisceau et cohomologie des variétés abéliennes*, Sém. Bourbaki, 1964–65, exp. 286.
104. I. ŠAVAREVIČ, *On Galois groups of p-adic fields* (in Russian), Dokl. Akad. Nauk SSSR **53** (1946), 15–16.
105. I. ŠAVAREVIČ, *On p-extensions* (in Russian), Math. Sbornik **20** (1947), 351–363.
106. S. SEN, *On automorphisms of local fields*, Ann. of Math. **90** (1969), 33–46.

107. S. Sen, *Ramification in p-adic Lie extensions*, Inv. Math. **17** (1972), 44–50.
108. S. Sen, *Lie algebras of Galois groups arising from Hodge-Tate modules*, Ann. of Math. **97** (1973), 160–170.
109. S. Sen and J. Tate, *Ramification groups of local fields*, J. Ind. Math. Soc. **27** (1963), 197–202.
110. J.-P. Serre, *Applications algébriques de la cohomologie des groupes*. Sém. H. Cartan, E.N.S. 1950/51, exp. 5, 6, 7, Benjamin, New York, 1967.
111. J.-P. Serre, *Sur les groupes de Galois attachés aux groupes p-divisibles*, Proc. Conf. Local Fields, 118–131, Springer-Verlag, 1967.
112. J.-P. Serre, *Conducteurs d' Artin des caractères réels*, Inv. Math. **14** (1971), 173–183.
113. J.-P. Serre, *Cohomologie Galoisienne*, Lect. Notes in Math. **5**, 5th revised edit., Springer-Verlag, 1994.
114. J.-P. Serre, *Linear representations of finite groups*, GTM **42**, Springer-Verlag, 1977.
115. J.-P. Serre, *A course in arithmetic*, GTM **7**, 2nd edit., Springer-Verlag, 1978.
116. J.-P. Serre, *Une "formule de masse" pour les extensions totalement ramifiées de degré donné d'un corps local*, C.R. Acad. Sci. Paris, Série A, **286** (1978), 1031–1036.
117. J. Tate, *Duality theorems in Galois cohomology over number fields*, Proc. Int. Congress, Stockholm 1962, 288–295.
118. J. Tate, *Cohomology groups of tori in finite Galois extensions of Algebraic Number Fields*, Nagoya Math. J. **27** (1966), 709–719.
119. J. Tate, *p-divisible groups*, Proc. Conf. Local Fields, 158–183, Springer-Verlag, 1967.
120. J. Tate, *Symbols in arithmetic*, Actes Congrès Int. Nice 1970, t. I, 201–211.
121. J. Tate, *Relations between* K_2 *and Galois cohomology*, Inv. Math. **36** (1976), 257–274.
122. G. Terjanian, *Un contre-exemple à une conjecture d'Artin*, C. R. Acad. Sci. Paris **262** (1966), 612.
123. A. Weil, *Basic number theory*, Grund. math. Wiss., 3rd edit., Springer-Verlag, 1974.
124. A. Wiles, *Higher explicit reciprocity laws*, Ann. of Math. **107** (1978), 235–254.
125. B. Wyman, *Wildly ramified gamma extensions*, Amer. J. of Math. **91** (1969), 135–152.
126. J. Cassels, *Local Fields*, Cambridge Univ. Press, London, 1986.
127. J. Dixon, M. du Sautoy, A. Mann and D. Segal, *Analytic pro-*p*-groups*, LMS Lect. Notes 157, Cambridge, 1991.
128. J.-M. Fontaine, *Groupes* p*-divisibles sur les corps locaux*, SMF, Astérisque n^0 **47–48**, 1977.
129. I. B. Fosenko and S. V. Vostokov, *Local Fields and Their Extensions. A Constructive Approach*, AMS Transl. Monographs **121**, 1993.
130. R. K. Guy (edit.), *Reviews in Number Theory* (1973–1983), vol. 5, chap. S, AMS, 1984.
131. K. Iwasawa, *Local Class Field Theory*, Oxford Univ. Press, 1986.
132. J. Neukirch, *Class Field Theory*, Springer-Verlag, 1986.

Index

F

G

H

I

J

K

L

M

N

O

P

Q

R

S

T

U

V

W

Graduate Texts in Mathematics

(continued from page ii)

120 ZIEMER. Weakly Differentiable Functions: Sobolev Spaces and Functions of Bounded Variation.
121 LANG. Cyclotomic Fields I and II. Combined 2nd ed.
122 REMMERT. Theory of Complex Functions. *Readings in Mathematics*
123 EBBINGHAUS/HERMES et al. Numbers. *Readings in Mathematics*
124 DUBROVIN/FOMENKO/NOVIKOV. Modern Geometry—Methods and Applications. Part III.
125 BERENSTEIN/GAY. Complex Variables: An Introduction.
126 BOREL. Linear Algebraic Groups. 2nd ed.
127 MASSEY. A Basic Course in Algebraic Topology.
128 RAUCH. Partial Differential Equations.
129 FULTON/HARRIS. Representation Theory: A First Course. *Readings in Mathematics*
130 DODSON/POSTON. Tensor Geometry.
131 LAM. A First Course in Noncommutative Rings.
132 BEARDON. Iteration of Rational Functions.
133 HARRIS. Algebraic Geometry: A First Course.
134 ROMAN. Coding and Information Theory.
135 ROMAN. Advanced Linear Algebra.
136 ADKINS/WEINTRAUB. Algebra: An Approach via Module Theory.
137 AXLER/BOURDON/RAMEY. Harmonic Function Theory.
138 COHEN. A Course in Computational Algebraic Number Theory.
139 BREDON. Topology and Geometry.
140 AUBIN. Optima and Equilibria. An Introduction to Nonlinear Analysis.
141 BECKER/WEISPFENNING/KREDEL. Gröbner Bases. A Computational Approach to Commutative Algebra.
142 LANG. Real and Functional Analysis. 3rd ed.
143 DOOB. Measure Theory.
144 DENNIS/FARB. Noncommutative Algebra.
145 VICK. Homology Theory. An Introduction to Algebraic Topology. 2nd ed.
146 BRIDGES. Computability: A Mathematical Sketchbook.
147 ROSENBERG. Algebraic *K*-Theory and Its Applications.
148 ROTMAN. An Introduction to the Theory of Groups. 4th ed.
149 RATCLIFFE. Foundations of Hyperbolic Manifolds.
150 EISENBUD. Commutative Algebra with a View Toward Algebraic Geometry.
151 SILVERMAN. Advanced Topics in the Arithmetic of Elliptic Curves.
152 ZIEGLER. Lectures on Polytopes.
153 FULTON. Algebraic Topology: A First Course.
154 BROWN/PEARCY. An Introduction to Analysis.
155 KASSEL. Quantum Groups.
156 KECHRIS. Classical Descriptive Set Theory.
157 MALLIAVIN. Integration and Probability.
158 ROMAN. Field Theory.
159 CONWAY. Functions of One Complex Variable II.
160 LANG. Differential and Riemannian Manifolds.
161 BORWEIN/ERDÉLYI. Polynomials and Polynomial Inequalities.
162 ALPERIN/BELL. Groups and Representations.
163 DIXON/MORTIMER. Permutation Groups.
164 NATHANSON. Additive Number Theory: The Classical Bases.
165 NATHANSON. Additive Number Theory: Inverse Problems and the Geometry of Sumsets.
166 SHARPE. Differential Geometry: Cartan's Generalization of Klein's Erlangen Program.
167 MORANDI. Field and Galois Theory.
168 EWALD. Combinatorial Convexity and Algebraic Geometry.
169 BHATIA. Matrix Analysis.
170 BREDON. Sheaf Theory. 2nd ed.
171 PETERSEN. Riemannian Geometry.
172 REMMERT. Classical Topics in Complex Function Theory.
173 DIESTEL. Graph Theory. 2nd ed.
174 BRIDGES. Foundations of Real and Abstract Analysis.
175 LICKORISH. An Introduction to Knot Theory.
176 LEE. Riemannian Manifolds.
177 NEWMAN. Analytic Number Theory.
178 CLARKE/LEDYAEV/STERN/WOLENSKI. Nonsmooth Analysis and Control Theory.
179 DOUGLAS. Banach Algebra Techniques in Operator Theory. 2nd ed.

180 SRIVASTAVA. A Course on Borel Sets.
181 KRESS. Numerical Analysis.
182 WALTER. Ordinary Differential Equations.
183 MEGGINSON. An Introduction to Banach Space Theory.
184 BOLLOBAS. Modern Graph Theory.
185 COX/LITTLE/O'SHEA. Using Algebraic Geometry.
186 RAMAKRISHNAN/VALENZA. Fourier Analysis on Number Fields.
187 HARRIS/MORRISON. Moduli of Curves.
188 GOLDBLATT. Lectures on the Hyperreals: An Introduction to Nonstandard Analysis.
189 LAM. Lectures on Modules and Rings.
190 ESMONDE/MURTY. Problems in Algebraic Number Theory.
191 LANG. Fundamentals of Differential Geometry.
192 HIRSCH/LACOMBE. Elements of Functional Analysis.
193 COHEN. Advanced Topics in Computational Number Theory.
194 ENGEL/NAGEL. One-Parameter Semigroups for Linear Evolution Equations.
195 NATHANSON. Elementary Methods in Number Theory.
196 OSBORNE. Basic Homological Algebra.
197 EISENBUD/ HARRIS. The Geometry of Schemes.
198 ROBERT. A Course in p-adic Analysis.
199 HEDENMALM/KORENBLUM/ZHU. Theory of Bergman Spaces.
200 BAO/CHERN/SHEN. An Introduction to Riemann–Finsler Geometry.
201 HINDRY/SILVERMAN. Diophantine Geometry.

Printed in the United States
By Bookmasters